# Microsatellites

# Microsatellites

Evolution and Applications

Edited by

## David B. Goldstein

Department of Zoology,
University of Oxford

and

## Christian Schlötterer

Institut fur Tierzucht und Genetik
Veterinämedizinsche Universität
Wien

# OXFORD

UNIVERSITY PRESS

Great Clarendon Street, Oxford OX2 6DP

Oxford University Press is a department of the University of Oxford
and furthers the University's aim of excellence in research, scholarship,
and education by publishing worldwide in

Oxford New York

Athens  Auckland  Bangkok  Bogotá  Buenos Aires  Calcutta
Cape Town  Chennai  Dar es Salaam  Delhi  Florence  Hong Kong  Istanbul
Karachi  Kuala Lumpur  Madrid  Melbourne  Mexico City  Mumbai
Nairobi  Paris  São Paulo  Singapore  Taipei  Tokyo  Toronto  Warsaw
and associated companies in  Berlin  Ibadan

Oxford is a registered trade mark of Oxford University Press

Published in the United States
by Oxford University Press, Inc., New York

British Library Cataloguing in Publication Data

Data available

Library of Congress Cataloging in Publication Data

1. Microsatellites (Genetics)   2. Genetic markers.   I. Goldstein,
David B.   II. Schlötterer, Christian.
QH452.2.M53 1999        572.8'6–dc21        98-49876
ISBN 0 19 850408 X (Hbk)
ISBN 0 19 850407 1 (Pbk)

1 3 5 7 9 10 8 6 4 2

Typeset by Newgen Imaging Systems (P) Ltd., Chennai, India
Printed in Great Britain
on acid-free paper by
St. Edmundsbury Press, Suffolk

# Preface

In a seminal report in 1980, Botstein and colleagues speculated that a new kind of genetic marker, termed restriction fragment length polymorphism (RFLP), could be used to develop a dense map of polymorphic markers for linkage analysis in humans. With the prospect of elucidating the genetic and physiological bases of inherited disorders clearly in sight, however, genetic discoveries and a critical technological breakthrough provided the ingredients for a new and dramatically superior class of genetic marker. It had been known for some time that repetitive DNA has peculiar mutational properties, but with the advent of PCR, the superiority of microsatellites over earlier markers was quickly demonstrated. Here, finally, was a highly polymorphic codominant marker, densely distributed throughout all eukaryotic genomes. The PCR format meant not only convenience and accessibility to laboratories with only modest facilities, but was a breakthrough for conservation biology and behavioural ecology. Analyses of relatedness, and population surveys, could now be carried out non-invasively, using hair, roots, or even faeces. The simple genetic basis of microsatellite variation, moreover, meant that inferences could be made using explicit population genetic models based on the observation of allele frequencies at identified loci.

Although we expect that microsatellites will eventually go the way of earlier markers, microsatellite analyses are in a phase of dramatic growth, and it is now clear that a huge collection of microsatellite data sets will be amassed before the next generation of genetic marker replaces microsatellites. We therefore decided to produce a reference volume that would serve the interests of the widest possible range of biologists using microsatellite markers. One aim of the book is to provide a feeling both for what microsatellites are currently being used for, as well as what they might be used for in the future. Thus, popular and well-known uses such as gene mapping, forensics, and behavioural ecology are included, along with more innovative and speculative approaches with a view toward inspiring novelty in future projects. Beyond representing the applications of microsatellites, however, we wanted to provide a sound and comprehensive review of how microsatellite data are, and should be, analysed. In many cases, the biological details of the marker prove critical in the development of inferential methods, and we have therefore included a substantial amount of detail concerning the biology of microsatellites, and more generally, repetitive DNA. Finally, beyond the specifics of microsatellite markers, we have tried to convey a sense of the remarkable range of biological questions that can be profitably explored using genetic markers. This more general aspect of the book can be extracted fairly easily from the introductions of the chapters, even by those with relatively little background in the use of genetic markers. We hope that this aspect of the book will serve as a resource not only for graduate students, but also for working scientists considering the introduction of genetic markers in their areas of research.

Finally, we would like to express our gratitude to the European Science Foundation (ESF) for funding a workshop on microsatellites in the Department of Zoology at the University of Oxford in 1997. The ESF workshop not only inspired us to put together this volume, but also provided us with many interesting contributions. We also want to thank the staff of OUP, and G. Partridge for her help in managing the collection of the chapters.

*Oxford and Vienna*　　　　　　　　　　　　　　　　　　　　　D.B.G. and C.S.
October 1998

# Contents

# Contributors

**Santos Alonso Alegre**  Department of Genetics, University of Nottingham, Queen's Medical Centre, Nottingham NG7 2UH, UK

**Chester Alper**  The Center for Blood Research and Harvard Medical School, Boston, Massachusetts 02118, USA

**William Amos**  Department of Zoology, University of Cambridge, Downing Street, Cambridge CB2 3EJ, UK, Tel/Fax: +44 (0) 1223 336616, email: w.amos@zoo.cam.ac.uk

**John A.L. Armour**  Department of Genetics, University of Nottingham, Queen's Medical Centre, Nottingham NG7 2UH, UK, email: John.Armour@Nottingham.ac.uk

**Zuheir Awdeh**  The Center for Blood Research and Harvard Medical School, Boston, Massachusetts 02118, USA

**Richard M. Badge**  Department of Genetics, University of Nottingham, Queen's Medical Centre, Nottingham NG7 2UH, UK

**David Balding**  Department of Applied Statistics, University of Reading, PO Box 240, Reading RG6 6FN, UK, email: d.j.balding@reading.ac.uk

**Lisa Barcellos**  Department of Integrative Biology, University of California, Berkeley, California 94720-3140, USA

**Mark A. Beaumont**  Institute of Zoology, Regent's Park, London NW1 4RY, UK, email: m.beaumont@ucl.ac.uk

**Michael W. Bruford**  Institute of Zoology, Regent's Park, London NW1 4RY, UK, email: m.bruford@ucl.ac.uk

**Mary Carrington**  IRSP, SAIC Frederick, National Cancer Institute-Frederick Cancer Research and Development Center, Frederick, Maryland 21702, USA, Tel: 301-846-1390, Fax: 301-846-1909, e-mail: carringt@mail.ncifcrf.gov

**Ranajit Chakraborty**  Human Genetics Center, University of Texas Health Science Center, P.O. Box 20334, Houston, TX 77225, USA, email: rc@hgc9.sph.uth.tmc.edu

**John Chen**   Department of Integrative Biology, University of California, Berkeley, California 94720-3140, USA

**D.W. Coltman**   Institute of Cell, Animal and Population Biology, University of Edinburgh, Ashworth Laboratories, West Mains Road, Edinburgh EH9 3JT, UK

**Jean-Marie Cornuet**   Laboratoire de Modélisation et Biologie Evolutive, URLB-INRA, 488 Rue Croix de Lavit, 34090 Montpellier, France

**T.N. Coulson**   Institute of Zoology, Zoological Society of London, Regent's Park, London NW1 4RY, UK

**Peter Donnelly**   Department of Statistics, University of Oxford, 1 South Parks Road, Oxford, OX1 3TG, UK, email: donnelly@stats.ox.ac.uk

**Jonathan A. Eisen**   The Institute for Genomic Research, 9712 Medical Center Drive, Rockville, MD 20850, USA, email: jeisen@tigr.org

**Arnaud Estoup**   Laboratoire de Génétique des Poissons, INRA, 78352 Jouy-en-Josas, France, Tel: 33 1 34 65 27 57, Fax: 33 1 34 65 23 90, email: estoup@jouy.inra.fr

**Marcus W. Feldman**   Center for Computational Genetics and Biological Modeling, Department of Biological Sciences, Stanford University, Stanford, CA 94305, USA, email: marc@charles.stanford.edu

**David B. Goldstein**   Galton laboratory, Department of Biology, University College London, Wolfson House, 4 Stephenson Way, London NW1 2HE, email: d.goldstein@ucl.ac.uk

**Michael Hammer**   Laboratory of Molecular Systematics and Evolution, Department EEB, Biosciences West, University of Arizona, Tucson, AZ 85721, USA, Tel: (520) 621-9828, Fax: (520) 626-8050, email: mhammer@u.arizona.edu

**John M. Hancock**   Comparative Sequence Analysis Group, MRC Clinical Sciences Centre, Imperial College School of Medicine, Hammersmith Hospital, Du Cane Road, London W12 0NN, UK, email: jhancock@rpms.ac.uk

**Gavin Huttley**   John Curtin School of Medical Research, The Australian National University, Canberra, ACT 0200, Australia

**T. Karafet**   Laboratory of Molecular Systematics and Evolution, University of Arizona, Tucson, Arizona, USA, and Laboratory of Human Molecular and Evolutionary Genetics, Institute of Cytology and Genetics, Novosibirsk, Russia, email: tkarafet@u.arizona.edu

**Yechezkel Kashi**   Department of Food Engineering and Biotechnology, The Technion, Technion City, 32000 Haifa, Israel, email: Kashi@TX.Technion.ac.il

**Marek Kimmel**  Department of Statistics, Rice University, P.O. Box 1892, Houston, TX 77251, USA, email: kimmel@rice.edu

**William Klitz**  School of Public Health, University of California, Berkeley, California 94720-3140, USA

**Jochen Kumm**  Center for Computational Genetics and Biological Modeling, Department of Biological Sciences, Stanford University, Stanford, CA 94305, USA

**Andrés Ruiz Linares**  Departamento de Bioquímica, Facultad de Medicina, Universidad de Antioquia, Medellín, Colombia, *Present address*: Imperial College School of Medicine, Department of Medical and Community Genetics, Northwick Park and St. Mark's NHS Trust Level 8V, Harrow HA1 3UJ, UK, Tel/Fax: +44 (181) 869 3537/869 3167, e-mail: andres.ruiz@ic.ac.uk

**Darlene Marti**  IRSP, SAIC Frederick, National Cancer Institute-Frederick Cancer Research and Development Center, Frederick, Maryland 21702, USA

**Sue Miles**  Department of Genetics, University of Nottingham, Queen's Medical Centre, Nottingham NG7 2UH, UK

**Stephen J. O'Brien**  Laboratory of Genomic Diversity, National Cancer Institute, Frederick, MD 21702-1201, USA, email: obrien@ncifcrf.gov

**L.P. Osipova**  Laboratory of Human Molecular and Evolutionary Genetics, Institute of Cytology and Genetics, Novosibirsk, Russia, email: osipova@bionet.nsc.ru

**J.M. Pemberton**  Institute of Cell, Animal and Population Biology, University of Edinburgh, Ashworth Laboratories, West Mains Road, Edinburgh EH9 3JT, UK, email: j.pemberton@ed.ac.uk

**O.L. Posukh**  Laboratory of Human Molecular and Evolutionary Genetics, Institute of Cytology and Genetics, Novosibirsk, Russia, email: posukh@bionet.nsc.ru

**Jonathan Pritchard**  Center for Computational Genetics and Biological Modeling, Department of Biological Sciences, Stanford University, Stanford, CA 94305, USA

**David E. Reich**  Department of Zoology, University of Oxford, South Parks Road, Oxford OX1 3UJ, UK

**David C. Rubinsztein**  Department of Medical Genetics, Cambridge Institute for Medical Research, Wellcome/MRC Building, Addenbrooke's Hospital, Hills Road, Cambridge, CB2 2XY, email: dcr1000@cus.cam.ac.uk

**Christian Schlötterer**   Institut für Tierzucht und Genetik, Veterinärmedizinische Universität Wien, Josef Baumann Gasse 1, 1210 Wien, Austria, email: schlotc@i122mc03.vu-wien.ac.at

**Darryl Shibata**   Associate Professor of Pathology, University of Southern California, School of Medicine, Los Angeles, CA 90033, USA, email: dshibata@hsc.usc.edu

**Hyoung Doo Shin**   Intramural Research Support Program, SAIC-Frederick, National Cancer Institute, Frederick, MD 21702-1201, USA, email: shinh@mail.ncifcrf.gov

**J. Slate**   Institute of Cell, Animal and Population Biology, University of Edinburgh, Ashworth Laboratories, West Mains Road, Edinburgh EH9 3JT, UK

**Michael W. Smith**   Intramural Research Support Program, SAIC-Frederick, National Cancer Institute, Frederick, MD 21702-1201, USA, email: smithm@mail.ncifcrf.gov

**Morris Soller**   Department of Genetics, The Silberman Life Sciences Institute, The Hebrew University of Jerusalem, 91904 Jerusalem, Israel, email: soller@vms.huji.ac.il

**J. Claiborne Stephens**   Laboratory of Genomic Diversity, National Cancer Institute, Frederick, MD 21702-1201, USA, email: joels@fcrtv1.ncifcrf.gov

**Gunnar Sturfelt**   Department of Medical Microbiology, Clinical Immunology Section and Department of Rheumatology, Lund University, Lund, Sweden

**Glenys Thomson**   Department of Integrative Biology, University of California, Berkeley, California 94720-3140, USA

**Lennart Truedsson**   Department of Medical Microbiology, Clinical Immunology Section and Department of Rheumatology, Lund University, Lund, Sweden

**Judy Wade**   TTH Regional Histocompatibility Laboratory, University of Toronto, Ontario, Canada

**V. Wiebe**   Laboratory of Human Molecular and Evolutionary Genetics, Institute of Cytology and Genetics, Novosibirsk, Russia, email: wiebe@zi.biologie.uni-muenchen.de

**Thomas Wiehe**   Institut für Molekulare Biotechnologie, Abteilung Genomanalyse, Postfach 100813, D 07708 Jena, Germany

**Louise J. Williams**   Department of Genetics, University of Nottingham, Queen's Medical Centre, Nottingham NG7 2UH, UK

# 1 Microsatellites and other simple sequences: genomic context and mutational mechanisms

John M. Hancock

## Chapter contents

## Abstract

'Microsatellite' is now the commonest term used to describe tandem repeats of short sequence motifs (no more than six bases long). Microsatellites have been found in every organism investigated so far. They may be highly polymorphic, especially if long and uninterrupted, and they are therefore useful genetic markers. Microsatellite-like sequences made up of concentrations of specific motifs that are not tandemly repeated are known as cryptically simple, and such sequences are also widespread, although generally more common in organisms with large genomes. The functional significance of such sequences is still unknown although they are sometimes found within exons and have been associated with diseases. Microsatellites appear to be more or less uniformly distributed across eukaryotic genomes, but are under-represented in coding regions, and perhaps telomeres. Repeats of poly(A)/poly(T) are the most common microsatellites in all genomes, but beyond that different genomes show subtly different frequency distributions. Genetic and other studies suggest that slipped-strand mispairing during replication is the predominant mechanism of mutation of microsatellites, but recently an increasing amount of circumstantial evidence has also implicated recombination.

## 1.1 Introduction

Microsatellites are sequences made up of a single sequence motif, no more than six bases long, that is tandemly repeated, that is, arranged head-to-tail without interruption by any other base or motif. Historically, the term microsatellite has often been applied solely to repeats of the dinucleotide motif CA (GT) (Litt and Luty 1989; Weber and May 1989) and a variety of other names have been used to describe tandemly repeated sequences, including simple sequences (Tautz 1989) and STRs (short tandem repeats; Edwards *et al.* 1991). This leads to potential confusion, especially in literature searches. The term microsatellite has now become by far the most common designation for these kinds of sequence, and it is probably appropriate that this be adopted as general usage. It may be appropriate to use the term simple sequences to describe all sequences based on repeats of short motifs, including minisatellites and cryptically simple regions (see below for a definition of cryptic simplicity).

Microsatellites have been detected within the genomes of every organism so far analysed, and are often found at frequencies much higher than would be predicted purely on the basis of base composition. This has been shown formally for TA repeats in yeast (Valle 1993) and polypurine and polypyrimidine tracts in a variety of genomes (Behe 1987, 1995). Microsatellites show high levels of polymorphism (Litt and Luty 1989; Weber and May 1989; Tautz 1989). The most polymorphic, and therefore the most useful for many purposes, are uninterrupted arrays (Weber 1990), but many microsatellite loci amplified by PCR in fact contain interruptions (and, in many cases, the interruption status of the locus is unknown). Microsatellites may also be compound, that is, made up of contiguous or adjacent tandem arrays of different motifs. Indeed, many microsatellite loci are poorly characterized at the sequence level, as they may have been isolated from genomic libraries using oligonucleotide probes and then tested for polymorphism. This can complicate the interpretation of their patterns of length variation at the population level (see for example Freimer and Slatkin 1996).

## 1.2 Cryptic simplicity → *how to define ?* *> 2 repeat* *< 3 repeat?*

A microsatellite that acquired numerous point mutations would eventually degrade to a non-repetitive sequence, but an intermediate state is also possible in which a sequence is made up of a few, intermixed motifs which show little sign of a tandem arrangement. This pattern, known as cryptic simplicity (Tautz *et al.* 1986), has been shown to be common in many genomes, particularly large eukaryotic ones (Tautz *et al.* 1986; Sarkar *et al.* 1991; Hancock 1995*a*, 1996*a*). Cryptically simple regions are more common in genomes than expected by chance (Tautz *et al.* 1986; Hancock 1995*a*, 1996*c*) even when the base doublet composition of the genomes is taken into account (Hancock 1995*a*). Like microsatellites, cryptically simple regions of sequence are found preferentially in non-coding regions of genomes, but they can also be found in coding regions

(Tautz *et al.* 1986; Hancock 1995*a*). Using the program SIMPLE34 (Hancock and Armstrong 1994) the motifs making up simple sequences can be summarized in terms of the most common trinucleotide motifs within a cryptically simple region. This kind of analysis shows that cryptically simple regions are rich in the same motifs as those making up the most common microsatellites (Hancock 1995*a*). This is consistent with an origin as corrupt microsatellites, but it is not clear that this is the only way that cryptically simple sequences can arise. A recent analysis of a cryptically simple region within the small subunit ribosomal RNA gene of a group of tiger beetle species (Vogler *et al.* 1997; J.M. Hancock and A.P. Vogler, manuscript in preparation) showed that these regions evolve by insertion and deletion of single bases in a manner suggestive of slippage. Thus cryptically simple regions may undergo slippage in a manner similar to microsatellites, although other analyses of variable region evolution suggest that this probably occurs at lower rates (Hancock 1995*b*).

Genomes show different contents of cryptically simple sequences, roughly in proportion to the logarithm of their size (Hancock 1995*a*, 1996*a*). In species whose genomes have high contents of cryptically simple sequences, such as human, mouse and *Drosophila melanogaster*, significant numbers of coding regions also contain cryptically simple regions (Hancock 1995*a*; J.M. Hancock, unpublished observations). Examples of such genes include ribosomal RNAs (especially the large subunit gene) (Tautz *et al.* 1986; Hancock and Dover 1988; Hancock 1995*b*), the TATA box-binding protein TBP (Gostout *et al.* 1993; Hancock 1993), the insect developmental gene *hunchback* (Tautz *et al.* 1987; Treier *et al.* 1989), and many other developmental genes and transcription factors (Wharton *et al.* 1985; Tautz 1989; Gostout *et al.* 1993; Green and Wang 1994; Künzler *et al.* 1995). It remains unclear whether these cryptically simple regions within genes are important for the function of the gene product and under selective constraint, or represent relatively weakly selected parts of the gene, although a population analysis of eleven genes on chromosome 2 of *Drosophila melanogaster* showed no evidence of selective constraints on trinucleotide variation (Michalakis and Veuille 1996). However there is evidence for effects of ribosomal RNA variable regions on ribosome function and RNA processing (Jeeninga *et al.* 1997; Van Nues *et al.* 1997), and changes in repeat numbers in the androgen receptor and huntingtin genes, which contain CAG repeats (encoding polyglutamine tracts) also appear to affect the interactions of the gene products (Chamberlain *et al.* 1994; Kazemi-Esfarjani *et al.* 1995; Li *et al.* 1995; Kalchman *et al.* 1997; Wanker *et al.* 1997; Rubinsztein, this volume), suggesting that such changes are not always completely selectively neutral. The recent discovery of diseases in humans caused by the expansion of tandem repeats within and near to genes (see chapter by Rubinsztein, this volume) suggests that inappropriately placed or sized repeats can be selectively disadvantageous.

## 1.3 Genomic distribution

Because of their use in genome mapping, the genomic distribution of microsatellites is best known in humans and mice. High resolution mapping suggests a more or less even (i.e. random) distribution over both genomes at the gross level (see Dib *et al.* 1996; Dietrich *et al.* 1996) although even the highest resolution maps contain some long gaps and low density regions, many near telomeres (see Fig. 1 of Dib *et al.* 1996, for example). There is also a suggestion of low concentrations of microsatellites on the mouse X chromosome (Dietrich *et al.* 1996), although this may reflect the experimental approach, which involved mapping known highly informative loci (Tautz and Schlötterer 1994). Higher microsatellite concentrations have been claimed for X chromosomes than for autosomes in *Drosophila* and *Anopheles gambiae* from *in situ* hybridization experiments (Huijser *et al.* 1987; Zheng *et al.* 1993).

At a more fine-grained level, simple sequences are rare within coding sequences (Hancock 1995a). This is particularly true for microsatellites not based on a repeat unit of three, such as poly(CA), as these can give rise to frameshifts if they mutate, a situation seen in some genetic diseases (see Chapter 7 of Cooper and Krawczak 1993). Some studies (in plants) have shown disparities in estimated frequencies of different microsatellites between genomic library screens, which search the entire genome, and sequence database screens, which are biased towards coding regions and their immediate vicinities (Broun and Tanksley 1996). This could be explained if some microsatellites were excluded from the immediate vicinity of coding regions, as well as from coding regions themselves.

Little is known in any detail about the factors that might restrict the distribution of microsatellites within non-coding regions. We might expect the use of particular intergenic (and intronic) segments as sites for transcription regulation (i.e. as binding sites for transcription factors) to restrict the locations of microsatellites somewhat, as both the sequences of binding sites and their spacings must be preserved. This is seen, for example, in the promoter regions of *hunchback* genes, in which repetitive sequences are relatively rare near the BICOID binding sites (J.M. Hancock, F. Bonneton, P.J. Shaw and G.A. Dover, manuscript in preparation). Sequence also affects the flexibility of DNA (el Hassan and Calladine 1996), which may be a factor in gene regulation (e.g. Chen *et al.* 1997) and the folding of sequences into chromatin (Sivolob and Khrapunov 1995), so these aspects may also constrain microsatellites somewhat. The locations of replication origins, although ill-defined in eukaryotes, may affect the organization of microsatellites in the genome, as direction of replication appears to influence the stability of microsatellites (Kang *et al.* 1995, 1996; Maurer *et al.* 1996; Freudenreich *et al.* 1997).

## 1.4 Microsatellite sequences

Poly(A)/poly(T) is the most common tandemly repeated sequence in the human genome (Beckmann and Weber 1992; Stallings 1992; Hancock 1996*b*). Despite its abundance, poly(A)/poly(T) is not suitable for mapping or population analysis because it is highly unstable during PCR, making sizing of alleles difficult or impossible. Database analyses of tandem repeats in genomic sequences by Beckmann and Weber (1992) showed that CA/TG repeats are the most common dinucleotide repeats, occurring about twice as frequently as AT repeats, and three times as often as AG/TC repeats. Amongst trinucleotide repeats (Stallings 1994) CAG and AAT repeats appeared to be most common, although the frequencies that were measured were low and likely to have been biased by the tendency of sequence databases to concentrate on coding sequences, which are relatively rich in CAG repeats (Stallings 1994; Hancock 1995*b*).

Other mammalian genomes appear to contain a similar complement of microsatellites to the human genome, although there appear to be subtly different frequencies even between rat and human (Beckmann and Weber 1992; Stallings 1992) while more divergent species may contain different mixtures (Stallings 1992). The yeast *Saccharomyces cerevisiae* is relatively richer in poly(AT) sequences than the human genome (Valle 1993; Hancock 1995*a*), while in plants GA and AT repeats are commoner than CA repeats (Stallings 1992; Lagercrantz *et al.* 1993). On the other hand, CA repeats form the most common microsatellites in *Drosophila melanogaster* (Schug *et al.* 1998). Bacteria contain very few tandemly repeated sequences, but again the commonest are poly(A)/poly(T) (Hancock 1995*a*, 1996*c*, unpublished observations; Field and Wills 1996).

The origins of these differences are not clear. Hancock (1995*a*) calculated that much of the variation in the frequencies of simple sequences in genomes could be explained on the basis of expected mechanistic biases in slippage, although a significant proportion could not, even in noncoding regions of the human genome. The abundance of poly(A)/poly(T) in the human genome has been suggested to reflect their spread by retrotransposons (Beckmann and Weber 1992; Arcot *et al.* 1995; Nadir *et al.* 1996; Yandava *et al.* 1997), poly(A) corresponding to the poly(A) tail of the reverse transcribed mRNA. The differences between species might be due purely to chance, or they might reflect particular mechanisms preferring particular sequences, or even selection acting on sequences that have adopted particular functions within a given evolutionary lineage.

It was mentioned earlier that cryptically simple sequences might arise by the accumulation of point mutations in microsatellites, but equally microsatellites could arise by point mutations within cryptically simple sequences, and this kind of process could give rise to different microsatellites in different evolutionary lineages (Messier *et al.* 1996; but see Gordon 1997). These two processes are not necessarily mutually exclusive, and indeed the death of microsatellites by point mutation might predispose to the subsequent birth of a new one. Evolutionary analyses of microsatellites have shown a wide variety of degrees of conservation. Only about 30 per cent of human microsatellites are conserved in rodents (Stallings *et al.* 1991) but other loci in other species have been conserved over very long periods of evolutionary time (FitzSimmons *et al.* 1995; Rico *et al.* 1996). It should be noted, however, that these findings are confounded with conservation of the primer sequence.

In other cases, particularly in disease-causing bacteria, changes in copy number of tandemly repeated sequences can affect the transcriptional activity of a promoter or the reading frame of a protein-coding region (Moxon *et al.* 1994). It is likely that in these bacterial cases selection has favoured the accumulation of repeats at particular loci. Variation in tandem repeats has also been suggested to contribute to quantitative genetic variation in eukaryotes (King 1994; Kashi *et al.* 1997). Proteins binding to a variety of tandem repeats have also been detected, and in some cases isolated (Aharoni *et al.* 1993; Richards *et al.* 1993; Timchenko *et al.* 1996; Epplen *et al.* 1996; Zhao *et al.* 1996). If these interactions are functionally significant, this would also be expected to affect the frequencies of particular microsatellites.

Another factor that may affect the frequencies at which particular sequence motifs become expanded into microsatellites is the propensity of particular sequences to form higher order structure during replication. It has been suggested that CAG/CTG repeats are particularly prone to expansion in triplet expansion disorders (see Rubinsztein, this volume) because they are particularly prone to form secondary structures during replication (Gacy *et al.* 1995). Differential abilities of different sequences to form such structures could influence their overall frequencies in the genome.

## 1.5 Mutational mechanisms of change in microsatellites

Rates of mutation of microsatellites are high compared to rates of point mutation, which are of the order of $10^{-9}$ to $10^{-10}$. Estimates of microsatellite mutation rates in *in vivo* systems are around $10^{-2}$ events per locus per replication in *E. coli* (Levinson and Gutman 1987b) and around $10^{-4}$ to $10^{-5}$ in yeast (Henderson and Petes 1992; Strand *et al.* 1993). Estimates from pedigree analysis in humans suggested a rate of around $10^{-3}$ events per locus per generation (Weber and Wong 1993), while studies in mice gave estimates of around $10^{-3}$ to $10^{-4}$ (Dallas 1992). In *Drosophila* mutation rates at microsatellites seem to be relatively low at around $6 \times 10^{-6}$ (Schug *et al.* 1997a). There are two types of model that can be invoked to explain these high rates of mutation. The first involves only a single DNA double helix and slipped strand mispairing (slippage) during DNA replication (Streisinger *et al.* 1966; Levinson and Gutman 1987a); the second involves recombination between DNA molecules (Smith 1976; Jeffreys *et al.* 1994).

Slippage during replication (see Levinson and Gutman 1987a) can take place when the nascent DNA strand dissociates from the template strand. When non-repetitive sequences are being replicated this does not pose a problem because there is only one way in which the nascent strand can reanneal precisely to the template strand before replication is recommenced, but if the replicated sequence is repetitive in nature the nascent strand may reanneal out-of-phase with the template strand. When replication is continued after such a misannealing, the eventual nascent strand will be longer or shorter than the template, depending whether the misannealing gives rise to looped-out bases in the template strand, in which case the product will be shorter, or the nascent strand, in which case it will be longer.

Recombination could potentially alter the lengths of microsatellites in two ways, by unequal crossing-over or by gene conversion. Unequal crossing-over involves crossing-over between chromosome strands (DNA molecules) which are misaligned. This occurs most easily for long, tandemly repeated sequences where the recombination machinery cannot easily determine the correct register between the two strands. Unequal crossing-over gives rise to a deletion in one DNA molecule and insertion in the other and can give rise to homogenization of variants within an array of a tandemly repeated family of sequences, as well as expansion and contraction of arrays (Smith 1976; Dover 1982). Unequal crossing-over can occur both between chromatids in the same chromosome (known as sister-chromatid exchange) and between chromosomes. Gene conversion, which involves unidirectional transfer of information by recombination, probably as a response to DNA damage, can transfer sequence in an out-of-phase manner from one allele to another, and has been suggested to generate diversity at minisatellite loci (Jeffreys et al. 1994; Armour, this volume), which are tandemly repeated arrays of basic motifs longer than those found in microsatellites. In this case the loci appear to acquire novel sequences in a polar manner (i.e. more so at one end of the locus than the other).

Perhaps the strongest evidence for the primary role of replication slippage in the generation of length mutation in microsatellites comes from genetic analysis of the process in yeast and E. coli. In both systems, length instability of tandem repeats is unaffected by mutants that greatly decrease recombination frequencies (Levinson and Gutman 1987b; Henderson and Petes 1992). On the other hand, mismatch repair mutants, which impair the ability of the replication machinery to detect bases looped out due to slippage, greatly destabilize microsatellites both in model systems (Levinson and Gutman 1987b; Strand et al. 1993, 1995) and in human cancers (reviewed by Kolodner 1996). Rates of microsatellite mutation are similar in mitotic and meiotic yeast cells, despite much higher recombination rates during meiosis (Strand et al. 1993), while mutation rates at tandem repeats are much higher than recombination rates (Ahn et al. 1988). Thus any role of recombination in microsatellite mutation would have to involve an as yet uncharacterized pathway (Sia et al. 1997a).

Mutations in the $\delta$ DNA polymerase, which is involved in replicative DNA synthesis, increase the frequency of mutations in $(GT)_n$ tracts approximately tenfold, which suggests a role for the proofreading function of this enzyme in correcting slippage errors (Strand *et al*. 1993). However, this effect is much less than the effect on point mutations, which increase 130 to 240-fold (Morrison *et al*. 1993) suggesting that δ polymerase proofreading is less important in correcting slippage errors than base mismatches. This may be because the δ polymerase proofreading function is most effective near to the site of polymerization, while bulged out bases due to slippage may form away from this site (Sia *et al*. 1997*a*). This argument is supported by a decrease in the contribution of proofreading activity to the replication fidelity of tandem repeats as array length increases (Kroutil *et al*. 1996).

A number of more circumstantial lines of evidence also support a primary role for slippage. The most direct of these is the observation that most length mutations at microsatellites represent gains or losses of single repeat units—recombination-based mutation would be expected to give rise to a wider range of novel mutants. A preponderance of small length changes is clearly seen in *E. coli* and yeast model systems (e.g. Levinson and Gutman 1987*b*; Sia *et al*. 1997*b*; Wierdl *et al*. 1997). It is difficult to obtain sufficiently large samples in humans, or in other complex eukaryotes, but clustering of mutant alleles around the parental allele length is also seen in family studies of triplet repeats associated with Huntington's disease in transmissions below the threshold for disease-causing instability (Goldberg *et al*. 1995; Chong *et al*. 1997). Because of the difficulty of obtaining direct data in complex eukaryotes, attempts have been made to fit population distributions of microsatellite lengths to models of microsatellite evolution that either assume the slippage model (the stepwise mutation model or variants thereof) or do not (infinite alleles models). These analyses are described in more detail by Estoup (this volume) but are generally supportive of the slippage model. Length distributions of microsatellites within the genome are also consistent with stepwise mutation (Bell and Jurka 1997).

A number of other properties of microsatellites are also consistent with the slippage model. *In vivo* studies confirm the expectation that microsatellite mutation rates increase with array length in both *E. coli* (Levinson and Gutman 1987*b*; Murphy *et al*. 1989) and yeast (Wierdl *et al*. 1997) as well as in T4 and T7 bacteriophages (Streisinger and Owen 1985; Kunkel *et al*. 1994). Length dependence is not seen in the growth of DNA fragments from pairs of complementary, unequal length oligonucleotides in the presence of DNA polymerases *in vitro* (Schlötterer and Tautz 1992), but it is not clear in this system that loop formation after partial strand displacement, rather than complete strand displacement and competitive reannealing, is the primary mechanism of fragment growth.

Informativeness of microsatellite loci might also be expected to increase with array length (Weber 1990), but analyses testing this prediction have not provided clear results (Valdes *et al*. 1993; Goldstein and Clark 1995; Schlötterer *et al*. 1997). This may be due to selective sweeps acting differently on the variation of microsatellites at different chromosomal loci (Schlötterer *et al*. 1997), although the relative contributions of selective sweeps and background selection to microsatellite evolution remain to be apportioned (Schug *et al.* 1997*b*; Schlötterer and Wiehe, this volume).

Microsatellite mutation rates also seem to show the expected dependence on repeat unit size. At least one early study suggested that there was not a simple decrease in microsatellite mutation rate with the length of the repeating unit (Weber and Wong 1993), but more recent studies *in vivo* (Sia *et al.* 1997*b*; see also Wells *et al.* 1967) suggest that this is in fact the case. Chakraborty *et al.* (1997) suggested from population analyses of a large number of di-, tri- and tetranucleotide loci that Weber and Wong's (1993) study was confounded by interlocus variation in mutation rate for microsatellites of the same basic repeat length, although their own study does not take into account non-neutral effects on the variance of allele lengths, such as possible size ceilings, or the variation in mutation rate associated with changes in allele length. Interruption of microsatellites results in reduced polymorphism (Weber 1990) and reduced mutation rates (Bichara *et al.* 1995; Chong *et al.* 1995; Kunst *et al.* 1997; Petes *et al.* 1997), consistent with the greater difficulty of forming slipped intermediates in the presence of sequence interruptions.

Do these lines of evidence completely eliminate the possibility of recombination acting at microsatellites? Perhaps not. There is *in vivo* evidence that recombination rates increase in the presence of microsatellites (Wahls *et al.* 1990; J.M. Hancock and M.F. Santibañez-Koref, unpublished observations). There are also some properties of microsatellites that are not readily explained by slippage alone, but which might be explicable by recombination. For example, it remains unclear what the mechanism of catastrophic expansion in non-coding triplet expansion diseases such as Fragile X syndrome, myotonic dystrophy and Friedreich's ataxia might be (see Rubinsztein, this volume). In yeast, long repeats have been shown to undergo an unexpectedly high number of large deletions, which might also have a recombinational basis (Wierdl *et al.* 1997). There is also evidence of mutational polarity across triplet repeats (Kunst and Warren 1994; Eichler *et al.* 1995*a*,*b*) which, by analogy with minisatellites (Jeffreys *et al.* 1994; see Armour, this volume), might reflect gene conversion at these loci. Finally, recent evidence in rodents suggests that rates of neutral point mutation are suppressed in the neighbourhood of long tandem repeats as a function of array length (J.M. Hancock and M.F. Santibañez-Koref, manuscript in preparation), which might reflect the action of gene conversion or recombinational repair. Low rates of divergence have also been reported to be associated with microsatellites in cetaceans (Schlötterer *et al.* 1991), although there is a suggestion that the opposite relationship may be true in alligators (Glenn *et al.* 1996). These lines of evidence represent only hints of a role for recombination in the genetics of microsatellites, but it seems likely this will become a major focus of research in the future.

# Acknowledgements

I thank Malcolm Schug for making his papers available to me before publication.

# 2 Functional roles of microsatellites and minisatellites

Yechezkel Kashi and Morris Soller

## Chapter contents

## Abstract

A number of lines of evidence are reviewed showing that microsatellite and minisatellite sequences serve as functional coding and regulatory elements in the eukaryote genome. These include the presence of homopolymeric amino acid tracts in numerous proteins, conservation of microsatellites in relation to coding sequences, and ability of microsatellites to bind proteins and to serve as enhancer elements in expression constructs. Variation in microsatellite repeat number is found to have pronounced effects on their regulatory and coding functions, and is associated with phenotypic variation at both the biochemical and organismic levels. The combination of high mutation rate and regulatory/coding function raises the possibility that microsatellites are a major source of eukaryotic genetic variation.

## 2.1 Introduction

As shown by the other chapters in this volume, microsatellites and minisatellites are most generally considered in terms of their roles as genetic markers for studies in population genetics, evolutionary relationships, and gene mapping. However, there is much accumulating evidence that microsatellite sequences also serve functional roles as coding or regulatory elements (Kunzler *et al.* 1995; Kashi *et al.* 1997). As regulatory sequences, microsatellites are found ubiquitously in upstream promoter regions of coding sequences, and in some instances have been found to be conserved in relation to coding sequences. Promoter regions containing microsatellites serve as enhancer elements in expression constructs, and deletion constructs lacking the microsatellite tracts have reduced enhancer activity. Microsatellites have also been shown to bind proteins, which is typical for upstream activating sequences. A number of studies show that the enhancing effect of microsatellites and their protein binding ability is a function of the number of tandem repeats in a specific microsatellite tract. As coding sequences, microsatellites have been found in numerous proteins, and variation in repeat number of homopolymeric amino acid tracts has associated functional effects. Finally, some recent studies show phenotypic effects of microsatellite length variation on physiology and development at the organismic level.

Viewed as functional elements of the genome, the special characteristics of microsatellites as mutational hotspots have led to the proposal that microsatellites may be a major source of quantitative genetic variation and evolutionary adaptation (Kashi *et al.* 1990, 1997; King 1994), enabling a population to replenish genetic variation lost through drift or selection, and serving as adjustable tuning knobs through which specific genes are able to rapidly adjust the norm of reaction to minor or major shifts in evolutionary demands (King *et al.* 1997; King and Soller 1998). These proposals will not be pursued in the present essay, which will limit itself to reviewing the evidence supporting a functional role for microsatellites.

## 2.2 Microsatellites are conserved in relation to coding sequences

In numerous instances microsatellites are found in upstream promoter regions of coding sequences. Many of these will be referred to in the following sections of this review. In a number of instances microsatellite tracts have been found in corresponding upstream locations of the same gene in different species. These include: poly(TG) tracts upstream of the rRNA transcription unit in mouse, rat and man (Braaten *et al.* 1988), and upstream of the rat and human somatostatin genes (Shen and Rutter 1984; Hayes and Dixon 1985); poly(GGAAG) upstream of the chicken ovotransferrin gene with a related motif, poly(GGAAA), in the corresponding region of the pheasant ovotransferrin gene (Maroteaux *et al.* 1983); poly(CTGA) upstream of the same two genetic loci (*Gld, Ted*) in each of three *Drosophila* species (Cavener *et al.* 1988); and the motif $(TCCC)_2 \cdots (TCCC)_2$ found in the upstream promoter region of both the chicken and mouse $\alpha 2(I)$ collagen genes (McKeon *et al.* 1984). A highly polymorphic pentanucleotide repeat (CCTTT) within the $5'$ promoter region of the human inducible nitric oxide synthase gene (*iNOS, NOS2*) was also found in the corresponding $5'$ region of chimpanzee, gorilla, orang-utan and macaque *iNOS* genes (Xu *et al.* 1997). The CGG repeat in the $5'$ promoter region of the *FMR1* gene, involved in Fragile X syndrome (FRAXA) was found in each of 44 mammalian species ranging from bat to dolphin, and covering 150 million years of evolution, providing strong evidence of its functionality (Eichler *et al.* 1995*b*). Similarly, the core motif of the microsatellite found in the first intron of the human *TH* gene is conserved in several non-human primate genera (Meyer *et al.* 1995). Numerous other studies document conservation of microsatellites in different mammalian genomes (Moore *et al.* 1991; Stallings *et al.* 1991; Hino *et al.* 1993; Deka *et al.* 1994; Stallings 1994, 1995). Conservation of sequences across species is often an indication of biological function.

## 2.3 Promoter regions containing microsatellites serve as enhancer elements in expression constructs

Microsatellites are found in a number of promoter regions that can supply enhancer function to expression vectors. These include poly(ACAT) (Gilmour *et al.* 1989) and poly(GGA) (Ishii *et al.* 1987; Tal *et al.* 1987; Suen and Hung, 1990). A number of studies show enhancer effects of minisatellite sequences within viral genomes; the minisatellite sequence (28 bp repeat motif) located 1 kb downstream from the c-Ha-rasI gene, *HRAS1*, also shows moderate enhancer activity (reviewed in Trepicchio and Krontiris 1992).

In one instance, promoter activity was obtained with a construct containing a promoter fragment with poly(GGA) and additional activity with a larger fragment containing poly(GCAA) as well (Suen and Hung 1990). It should be noted, however, that not all microsatellites present in a promoter region can be shown to have enhancer activity. Thus, the promoter region of the human insulin receptor gene contains seven repeats of potential Sp1 binding sites in the form of poly(GGGCGG)$_4$ distally, and poly(CCGCCC)$_3$ proximally, as well as a poly(TCC) tract. The fragment containing the proximal three Sp1 binding sites alone had full promoter activity (Araki *et al.* 1987). Similarly, a fragment from the upstream sequence of the mouse E$\beta$ immunoglobulin gene, containing a variety of microsatellite tracts, served as a positive enhancer; a subfragment containing most, but not all, of these tracts was equally active (Gillies *et al.* 1984). Also, a long element from the upstream region of the chicken $\alpha 2(VI)$

collagen gene containing numerous microsatellite tracts—$(GGAA)_{25}$, $(GGGA)_6$, $(GGGAGGAA)_4$, $(GGGGGGAA)_2$—was found to have negligible effect on promoter activity (Koller *et al.* 1991). It is possible, however, that in these instances the apparently inactive microsatellite tracts do function under *in vivo* conditions.

### 2.3.1 Microsatellites *per se* as enhancer elements in constructs

In some instances, not only do promoter sequences containing microsatellite motifs enhance transcription, but deletion constructs lacking the microsatellite sequences markedly reduce transcriptional activity of the promoter. Examples include: deletion of $(GA)_2$ and $(GA)_3$ motifs from the upstream promoter sequence of the *Drosophila Ubx* gene (Biggin and Tjian 1988); deletion of two separated repeats of $(TCCC)_2$ from the upstream promoter region of the chicken $\alpha2(I)$ collagen gene (McKeon *et al.* 1984); deletion of two tracts of poly$(TCC)_3$ in the promoter region of the epidermal growth factor (EGF) receptor (Johnson *et al.* 1988); and deletion of a $(TCCC)_7$ tract from the promoter region of c-KI-ras (Hoffman *et al.* 1990). Deletion mapping of the promoter region of the *TGF-β3* gene in a CAT expression system showed that promoter activity is attributable to a 60 bp region containing a $(TCCC)_3$ sequence (Lafyatis *et al.* 1991). Deletion or point mutations in the $(TCCC)_3$ sequence dramatically decreased promoter activity. Deletion mapping of the promoter region of the *Drosophila hsp26* (heat shock protein) gene showed that a degenerate poly$(TC)$ sequence was required for efficient transcription in extracts from unstressed embryos (Sandaltzopoulos *et al.* 1995). Degenerate poly(TC) sequences also serve as transcription elements in *Aspergillus* (Punt *et al.* 1990) and *Phytophthora infestans* (Chen and Roxby 1997). Deletion mapping of the upstream promoter regions of the *D. discoideum* spore germination-specific protein *cel A* showed that an 81 bp sequence, consisting primarily of poly$(A_n T_m)$, was required for proper temporal transcription of this gene (Ramalingam *et al.* 1995). In one case, a reverse effect was found: a promoter region from the rat prolactin gene, containing a poly(TG) tract 170 bp in length, was found to exert a negative effect on transcription in two different expression vectors (Naylor and Clark 1990). Deletion of the poly(TG) tract eliminated this effect.

In yeast (*S. cerevisiae*), deletion analysis showed that a poly$(T)_{15}$ tract found between the two constitutively expressed genes *HIS3* and *PET56* serves as an upstream promoter element for *HIS3*, and a downstream promoter element for *PET56*. Similarly, deletion of a poly(T)-containing T-rich tract from the promoter region of the constitutive *DED1* gene reduced transcription below the wild-type level (Struhl 1985). Poly(T) sequences are found upstream of many other yeast genes. Thus, poly(T) sequences may be generally used as upstream promoter elements for the constitutive transcription of yeast genes (Struhl 1985).

Insertion of genomic or synthetic sequences carrying microsatellites can also serve as positive promoter elements. A 48 bp synthetic oligonucleotide containing only the poly(T)-rich portion of the *DED1* promoter sequence stimulated transcription in an *in vitro* expression system (Lue *et al.* 1989). Moving the oligonucleotide 103 bp further upstream diminished the level of transcription two- to fourfold. Similarly, when human genomic DNA fragments (1 to 1.5 kbp in length) containing a poly(TG)$_n$ tract ($n = 15$, 20 or 25), or when synthetic poly(TG) oligonucleotides (50 to 130 bp) were inserted in an expression vector, constructs carrying the TG tracts were two to ten times more active than controls, depending on the location of the TG tract (Hamada *et al.* 1984). Furthermore, the TG tract was more effective when closer to the promoter, and its orientation was not crucial. Thus, in both systems the microsatellite elements appeared to have characteristics similar to those of viral enhancers. The functional role of a tetranucleotide repeat (TCAT)$_{10}$ located in the first intron of the gene encoding tyrosine hydroxylase (a candidate gene for neuropsychiatric diseases) was investigated in a construction made by linking the repeat to the luciferase reporter gene under the control of thymidine kinase minimal promoter (Meloni *et al.* 1998). In a variety of cell lines, these sequences increased the basal transcription up to ninefold. Here, too, the effect was independent of sequence orientation, as is typical of enhancer elements.

It should be noted that the enhancer effect of microsatellite sequences is generally rather mild. Thus, when the entire *DED1* promoter region was added to an expression construct, activity was increased 150-fold (Lue *et al.* 1989). Similarly, enhancer activity of poly(TG) was much weaker than that of the simian virus 40 enhancer (Hamada *et al.* 1984).

## 2.3.2 Microsatellites and minisatellites bind proteins

Short nucleotide motifs, found in upstream activation sequences, serve as binding sites for a variety of regulatory proteins (reviewed in Ptashne 1988; Lue *et al.* 1989). Microsatellite-containing promoter regions have also been shown to bind specific proteins. The level of transcription obtained in constructs containing the poly(T)-rich portion of the *DED1* promoter was diminished in the presence of synthetic oligonucleotides having this sequence, indicating the presence of a transcription factor that specifically binds the poly(T) rich sequence (Lue *et al.* 1989). Using a competition band-shift assay with yeast extracts, a protein (datin) was detected which was shown to bind only to poly(T) tracts in duplex DNA. Deletion of the gene coding for datin eliminated poly(T) binding activity (Winter and Varshavsky 1989). Poly(T) binding proteins are found in other species as well (Garreau and Williams 1983; Solomon *et al.* 1986). A protein specifically binding to single strand poly(C) was cloned from a cDNA rat hepatoma cell line (Ito *et al.* 1994).

In Southwestern analysis of nuclear proteins from *D. melanogaster* embryos, several proteins were found which specifically bind poly(TG). When the filters were incubated with poly(TC), a distinctly different pattern of labelled DNA-protein complexes was obtained (Vashakidze *et al.* 1988). Using similar methods with mouse nuclear proteins, a protein doublet specifically binding to a poly(TC) probe was found. Single-stranded poly(TC) acted as an effective competitor. Poly(TC)-binding proteins were also found in cell lines of Chinese muntjac, chimpanzee, African green monkey, and human (Yee *et al.* 1991). A protein which binds to both duplex poly(CT) elements and to single-stranded poly(CT) elements was cloned and shown to have separable binding activities, as demonstrated by mutations that disrupted single-strand binding but did not disrupt duplex poly(CT) binding (Kolluri *et al.* 1992). Using a DNA affinity resin containing GAGA elements [i.e., poly(CT)], a protein was isolated from nuclear extracts of *Drosophila* embryos which strongly stimulated transcription of the expression construct containing $(GA)_2$ and $(GA)_3$ motifs from the upstream promoter region of *Ubx*, but did not affect other promoter regions lacking such sequences (Biggin and Tjian 1988). A similar, perhaps identical protein was isolated from nuclear extract of *Drosophila* embryos (Gilmour *et al.* 1989). This protein, later termed GAGA factor, was also shown to interact with poly(GA) tracts in the promoter region of *Drosophila hsp26* gene (Sandaltzopolos *et al.* 1995). Transcription factors binding to poly(CT) elements have also been identified in *Aspergillus* (Punt *et al.* 1990) and *Phytophthora infestans* (Chen and Roxby 1997). Single-stranded poly(GA)- and poly(GT)-binding proteins have been identified in human fibroblasts (Aharoni *et al.* 1993). The poly($A_n T_m$) sequence of the *D. discoideum cel A* gene promoter region specifically bound factors in an extract prepared from germinating spores (Ramalingam *et al.* 1995).

A number of studies have identified poly(TCC) binding proteins. The proximal fragment of the promoter region of rat *neu* gene containing poly(TCC)$_7$ formed protein-DNA complexes when incubated with crude nuclear extract in gel retardation experiments (Suen and Hung 1990). Two of the poly(TCC)-containing regions of the promoter region of the *EGF* receptor gene were found to bind nuclear proteins (Ishii *et al.* 1987). Mobility shift assays using an oligonucleotide composed of poly(TCC) sequences from the c-Ki-ras TCC motif detected three discrete complexes (Hoffman *et al.* 1990). Competition studies involving oligonucleotides representative of poly(TCC) motifs located in the 5′ regions of the *EGF-R* and *I-R* genes inhibited formation of one complex and reduced the formation of a second. These results indicate that one or more protein factors can bind to the c-Ki-ras poly(TCC) domain, and at least one of these can also interact with similar domains found in the 5′ flanking regions of the *EGF-R* and *I-R* genes. In a number of cell lines, a nuclear protein was identified that binds to single-stranded poly(AGG) elements found in the promoter regions of the *CFTR* (cystic fibrosis transmembrane conductance regulator) and *MUC1* genes (Hollingsworth *et al.* 1994). A poly(CCT) binding protein has also been extracted from mouse tumour cells (Muraiso *et al.* 1992).

Involvement of poly(CAG) and poly(CGG) repeat expansion in hereditary neurodegenerative diseases, and of poly(CTG) in myotonic dystrophy (see later), led to investigation of protein-binding ability of these sequences. Two poly(CAG) binding proteins were purified from mouse adult brain, and were found to bind to poly(AGC), poly(AGT), poly(GGC) and poly(GGT), but not to various other trinucleotide repeats (Yano-Yanagisawa *et al.* 1995). Poly(CCG)- and poly(CGG)-binding proteins have also been reported in HeLa cell nuclear extracts (Richards *et al.* 1993). Bandshift analysis uncovered proteins in human nuclear extracts binding to poly(CTG) in DNA and poly(CUG) in RNA (Timchenko *et al.* 1996); it is the poly(CTG) sequence located in the 3′ non-protein coding region of the myotonin kinase gene at which triplet expansion leads to myotonic dystrophy.

The (TCCC)$_3$ sequence found in the *TGF-β3* promoter region binds a sequence-specific DNA-binding protein which may regulate aspects of *TGF-β3* expression (Lafyatis *et al.* 1991). Protein binding is inhibited by point mutation changes in the (TCCC)$_3$ repeat. The (TCAT)$_{10}$ sequence having enhancer qualities, originally noted in the first intron of the tyrosine hydroxylase gene, was also found to bind HeLa cell nuclear protein extracts, forming two distinct bands (Meloni *et al.* 1998).

Proteins binding to specific minisatellite sequences are also present in nuclear extracts. Mammalian testis nuclear extract was screened by Southwestern analysis, using a synthetic double-stranded minisatellite containing five repeats of the Jeffreys minisatellite core sequence. A single major DNA-binding protein, termed MSBP-1, was revealed. This protein was also found in nuclear extracts of mouse brain and other tissues, and in nuclear extracts from rabbit, *Xenopus*, *Drosophila*, human, rat, dog, and trout (Collick and Jeffreys 1990). In nuclear extracts from a variety of human cell lines, the minisatellite sequence (28 bp repeat motif) located downstream of *HRAS1*, bound four proteins, apparently belonging to the rel/NF-$_\chi$B family of transcriptional regulatory factors (Trepicchio and Krontiris 1992). Similarly, the minisatellite sequence (50 bp repeat motif) within the *DH-JH* interval of the human immunoglobulin heavy chain gene bound nuclear transcription factors belonging to the myc/helix-loop-helix family (Trepicchio and Krontiris 1993). In both of these cases, protein binding can be attributed to sequences embedded within the minisatellite motif that are known to interact with the corresponding transcription factors.

## 2.4 The number of repeat elements in a microsatellite or minisatellite tract affects expression and protein binding

Differences in the number of protein-binding sequences in upstream promoter regions have been shown to affect transcription. When synthetic oligonucleotides composed of protein-binding sequences or of microsatellites are inserted into promoter regions of expression vectors, constructs containing additional repeats often show increased activity. In some instances, repeat copy number of microsatellites has been shown to affect protein binding directly.

## 2.4.1 Effect of number of protein-binding sites on transcription in expression constructs

When tandemly repeated ATGATT sequences (a protein-binding element found in the upstream promoter region of the yeast DNA ligase gene) were inserted in the promoter region of an expression vector, there was a clear positive correlation between the number of correctly oriented repeats and transcription activity (White *et al.* 1991). When the same vector was used to study the activity of the CAGGAAAA sequence (a protein-binding element found in the upstream promoter region of the *HO* endonuclease involved in mating-type switching in yeast), one repeat had relatively little effect, while the addition of a second and third repeat had far greater effects, with increases of much more than two- and threefold, respectively (Nasmyth 1985). *HO*, itself, is transcribed powerfully, but only transiently, during late G1 phase of the cell cycle. Deletion mutants lacking GAGGAAAA sequences produce *HO* transcripts early in G1, and do not show increased activity in late G1. Deletion mutants in which copies of the CAGGAAAA sequence have been inserted mimic wild-type in that they delay transcription until late in G1, but reach peak transcription levels several-fold higher than the original deletion mutant. Peak transcription levels are proportional to the number of CAGGAAAA sequences inserted, increasing from 1 to 3 to 5. Similarly, the negative effect of the GAGGAAAA sequence on transcription early in G1 is proportional to the number of sequences, showing less effect with two copies than with three, and less still with only one copy (Nasmyth 1985). Again, as more of the pair of $(TCC)_3$ sequences upstream of the EGF receptor were deleted, enhancer activity decreased steadily. When both pairs of repeats were deleted, activity dropped by about fourfold (Johnson *et al.* 1988).

Effects of copy number on transcription are not necessarily linear. Thus, in an expression construct, it was shown that a synthetic oligonucleotide containing a pair of low affinity GAL4 binding sites gave 15 times the stimulation given by each site individually, while a pair of high affinity binding sites gave twice the stimulation of a single site (Giniger and Ptashne 1988).

## 2.4.2 Effects of repeat number of microsatellite tracts on transcription

The first example of a relation between length of microsatellite tract and expression was given by the *ADR2* locus of yeast. This locus is repressed by glucose and derepressed when grown on a non-fermentable carbon source. Two constitutive mutants were found which gave *ADR2* levels up to five times those found in wild-type cells when grown on non-fermentable carbon sources. Promoter regions of these mutants showed them to have $poly(T)_{53}$ and $poly(T)_{54}$ tracts upstream of the coding sequence, instead of the usual $poly(T)_{20}$ (Russell *et al.* 1983). The longer stretch of poly(T) associated with *DED1* also appears to be a more effective promoter element than the short stretch associated with *HIS3* and *PET56*, since constitutive *DED1* levels are about five times higher than those of *HIS3* or *PET56* (Struhl 1985). Also, deletion of nine T residues from the 3′ end of the synthetic *DED1* poly(T)-rich oligonucleotide abolished its activating effect, while deletion of nine residues from the 5′ end reduced expression 20-fold. Extending the uninterrupted stretch of Ts in the poly(T)-rich element by converting all G and C residues to Ts increased the stimulatory effect (Lue *et al.* 1989). Similar results were obtained in studies involving a poly(T) tract in the promoter region of the *actin 15* gene of *D. discoideum* (Hori and Firtel 1994). Creating an internal deletion that removed 25 of the 45 consecutive dT residues from the middle of the homopolymeric dT region resulted in a 25-fold decrease in a transient infection expression assay. In a second series of experiments, a complex selection scheme was used to construct actin promoter–NPTI expression vectors carrying poly(T) elements of various lengths. Expression was monitored by G418 resistance. The original 45-repeat element provided resistance to 8000 µg/ml G418; while constructs with only 19 and 11-repeats were able to provide resistance to concentration of 1000 and 200 µg/ml, respectively.

The human *MxA* gene is one of several interferon-inducible genes that have recently been isolated from a number of different species; some *Mx* genes have been shown to confer resistance against certain viruses. A 19 bp repressor element, consisting of a 9 bp direct repeat, was identified in the *MxA* promoter region. Deletion of the repressor element resulted in high constitutive expression in a transient infection CAT reporter system (Chang *et al.* 1992). Remarkably, constructs containing multimers of the repressor element resulted in expression levels much higher than those of the deletion construct. Multimers of the repression element were also shown to have enhancer-like characteristics in a CAT reporter system containing an enhancerless SV40 promoter; monomeric elements did not show enhancer activity.

In a direct experiment to test the effect of microsatellite length on enhancer effect, enhancer activities of various lengths of poly(TG) were compared in an expression construct. Maximum enhancement was obtained with 30–40 bp of poly(TG), although the activity of shorter poly(TG) tracts was not determined. As the length of the poly(TG) tract increased from 40 to 130 bp, enhancer activity fell. A length of 130 bp of poly(TG), for example, was fivefold less active than 50 bp. Interestingly, the size range of the bulk of poly(TG) elements found in the genome is between 20 and 60 bp, which is the size that seems to have maximum enhancer activity in this system (Hamada *et al.* 1984).

### 2.4.3 Effect of repeat copy number on protein binding by microsatellites

The number of repeats of a protein-binding microsatellite sequence has been shown to affect the degree of protein binding. The effects at this level are often non-linear. Thus, datin binds to poly(T) tracts 19, 15, and 11 bp long with comparable affinities, but does not bind to poly(T) tracts 3, 5, 7, or 8 bp long (Winter and Varshavsky 1989). The African green monkey $\alpha$-satellite-binding protein was found to bind with approximately equal affinity to any run of six or more A's, to many if not all runs of five A's, but rarely, if at all, to runs of four A's (Solomon *et al.* 1986). MSBP-1, the minisatellite-binding protein identified by Collick and Jeffreys (1990), did not bind to oligonucleotide probes consisting of a single minisatellite repeat, but binding of MSBP-1 to DNA increased with increase in probe repeat copy number, with a tenfold increase in binding efficiency of a five-copy probe compared with a four-copy probe (Collick and Jeffreys 1990).

## 2.5 Repeat number variation of microsatellite tracts in coding regions in relation to gene expression and function

Tracts of polyglutamine (20 or more residues) or polyproline (more than ten residues) have been found in at least 67 transcription factors, including human TATA-binding protein, and the androgen receptor from several species, the proteins encoded by the yeast gene *GAL11* and *SNF5*, and by the *Drosophila* genes *antennapedia*, *engrailed* and *Notch* (reviewed in Gerber *et al.* 1994). A survey of poly(CAG) and poly(CTG) repeats in human cDNAs uncovered 72 unique clones representing new genes (Neri *et al.* 1996). Screening of protein-sequence databases for human, mouse, *Drosophila*, yeast and *E. coli* uncovered multiple long homopolymeric amino acid tracts in 12 per cent of *Drosophila* proteins, about 1.7 per cent of human, mouse and yeast proteins, and none among *E. coli* proteins. A large majority of the identified *Drosophila* proteins were essential developmental proteins, many playing a role in central nervous system development. Almost half of the human and mouse proteins identified were homeotic homologues (Karlin and Burge 1996).

Repeat-number variation also affects the function of polyglutamine tracts in proteins. Using a CAT reporter system carrying an androgen response element in the presence of dihydrostestosterone, it was shown that deletion of CAG repeats in the rat or human androgen receptor resulted in higher transactivation, while combining human and rat glutamine tracts significantly lowered transactivation function. In the same system, expansion mutants, ranging from 25 to 77 repeats, showed a progressive decrease in transcriptional transactivation with increasing CAG repeat length (Chamberlain *et al.* 1994). Similar results were obtained using a somewhat different reporter system and androgen receptor poly-Gln tract length ranging from zero to 50 repeats (Kazemi-Esfarjani *et al.* 1995).

Transfection analysis of rat glucocorticoid receptor, which also contains a CAG repeat encoding poly-Gln, showed that transcription activation was slightly reduced by increasing the length of the poly-Gln tract. Changing a short length of the poly-Gln tract to a poly-Ser tract by a frameshift substitution slightly increased transactivation, but transactivation was progressively and strongly reduced by frameshift substitution of longer lengths of poly-Ser. Finally, transactivation was unaffected by frameshift substitution of poly-Ala lengths up to 23 repeats; but abolished by frameshift substitutions of somewhat longer lengths of poly-Ala (Lanz *et al.* 1995). Gerber *et al.* (1994) compared the transactivation activity of poly-Gln or poly-Pro in a GAL4 reporter system. They showed progressive increase of *in vitro* activity with HeLa nuclear extract of constructs with polymer length up to 90 glutamines or 50 prolines. *In vivo*, transient coinfection studies gave maximum activity at ten to 30 glutamines and about ten prolines.

## 2.6 Phenotypic effects of microsatellite repeat-number variation

It is intrinsically plausible that variation at a biochemical level will have phenotypic effects at the level of organismic physiology and development. However, demonstration of quantitative phenotypic effects associated with DNA-level variation is exceedingly difficult. At present, the most direct evidence that functional variation due to microsatellite length variation does have phenotypic consequences comes from studies of human genetic disorders that result from trinucleotide-repeat expansion, and in particular from the observed strong correlations between repeat length and age of onset, and (in some cases) severity of disease (reviewed in Ashley and Warren 1995; Sutherland and Richards 1995). The first repeat expansion disease locus was identified in 1991 in association with fragile X syndrome, the most common form of familial mental retardation. By 1995, the total had reached 11 loci and nine genetic diseases: five fragile sites involving poly(CGG) expansion to ten to 1000 times normal length; and six neurodegenerative disorders involving exonic poly(CAG) expansion to two to three times normal length. Since then the list has grown to include: Synpolydactyly, due to exonic poly(CGG) expansion; Friedreich's ataxia, due to intronic poly(GAA) expansion (Campuzano *et al.* 1996); Autosomal Dominant Pure Spastic Paraplegia, due to exonic poly(CAG) expansion (Nielsen *et al.* 1997); and Autosomal Dominant Oculopharyngeal Muscular Dystrophy, due to exonic poly(GCG) expansion (Brais *et al.* 1998).

However, these diseases are associated with extreme increases in repeat numbers, rather than with typical variation in microsatellite repeat number, which are of much smaller extent. Several other cases might thus be more relevant. When the Y chromosomes from certain *Mus musculus domesticus* wild strains are introduced into the laboratory mouse strain C57BL/6J (B6), hermaphroditic or fully sex-reversed female XY progeny are produced. The mouse Y chromosome sex-determining gene, *Sry*, encodes a polyglutamine sequence, thought to act as a transcription-activating domain, controlling the expression of other genes involved in gonadal development. *Sry* clones from normal mice and from five wild mouse strains that do not cause sex reversal, all code for 12 glutamines at this site. Two strains that caused complete sex reversal, and four strains that caused partial sex reversal, coded for 11 and 13 glutamines, respectively. Thus, the number of glutamine residues appears to be critical for proper functioning of the *Sry* gene (Coward *et al.* 1994). Using a generalized method for repeat expansion detection, a significant shift toward larger CAG and CTG products compared with controls was observed in individuals suffering from schizophrenia or bipolar affective disorders (O'Donovan *et al.* 1995). Expansion of the poly(CAG) tract in the human androgen receptor (*AR*), to $n > 40$, causes spinobulbar muscular atrophy (SBMA), an adult-onset selective motor neuronopathy that is often associated with a mild form of androgen insensitivity (Kennedy syndrome), which causes impaired androgen-mediated feedback inhibition of pituitary, hypothalamic, and testicular hormones. As androgen influences prostate cancer growth and spermatogenesis, effects of normal range polymorphism for CAG repeat number on age of onset of prostate cancer were investigated. A significant correlation between CAG repeat length and age at onset was observed, with shorter repeat length associated with earlier development and greater risk of prostate cancer (Anon 1996; Hardy *et al.* 1996), while longer CAG tracts were associated with a fourfold increased risk of impaired spermatogenesis (Tut *et al.* 1997). Both observations support decreased functional competence of *AR* with longer glutamine tracts.

The most direct association of a microsatellite sequence with function is provided by the threonine-glycine repeat with the *Drosophila period* clock gene, *per*. The two major variants (Thr-Gly)$_{17}$ and (Thr-Gly)$_{20}$ are distributed as a highly significant latitudinal cline, with high frequencies of the shorter variant in the southern Mediterranean, and of the longer variant in northern Europe. When tested at 18 °C and 29 °C, twelve lines carrying the 17-repeat variant showed a period closer to 24 h at the warmer temperature, but the period lengthened significantly at the colder temperature; eleven lines carrying the 20-repeat variant showed no overall significant difference between the two temperatures. To test the validity of these differences, *per* transgenes in which internal deletions of the repetitive tract from a cloned (Thr-Gly)$_{20}$ *per* gene were made. In conformity with the evidence from the natural variants, periods for two zero-repeat, three one-repeat, and two 17-repeat transformed lines were all significantly greater at 29 °C than at 18 °C; this was the case for only two of four 20-repeat transformants (Sawyer *et al.* 1997).

A number of studies also show effects of minisatellite repeat-number variation on expressed phenotype. Expansion of a minisatellite with a repeated 15 to 18-mer unit located in the putative promoter region of the gene encoding cystatin B (*CST6*, a cysteine protease inhibitor) was found to account for most individuals afflicted with progressive myoclonus epilepsy, an autosomal recessive disorder. Presence of the expanded minisatellite in the promoter regions is associated with reduction in *CST6* mRNA levels (Virtaneva *et al.* 1997). A fascinating study involved Exon III of the D4 dopamine receptor gene (*D4DR*), which contains a polymorphic 16 amino acid repeat region; the most frequently observed alleles are the 4-repeat and 7-repeat variants. It has been hypothesized that individual variations in the Novelty Seeking trait defined by the Cloninger tridimensional personality questionnaire (TPQ) are mediated by genetic variability in dopamine transmission. It has been shown that the number of repeats of the Exon III repeat region can affect ligand binding by the receptor (Van Tol *et al.* 1992; Asghari *et al.* 1994). This led to a study of 124 normal adult male and female volunteers, in which it was found that subjects with the 7-repeat allele exhibited significantly elevated Novelty Seeking scores, in comparison to those lacking the 7-repeat allele (Ebstein *et al.* 1996). A similar study based on 315 mostly male siblings, and using the NEO personality inventory, showed Extraversion and Novelty Seeking scores to be significantly higher for subjects with 6 to 8-repeat alleles, while Conscientiousness score was significantly less (Benjamin *et al.* 1996).

A polygenic component of genetic susceptibility to insulin-dependent diabetes mellitus (IDDM) was mapped to a minisatellite containing polymorphic region upstream of the insulin gene (Bell *et al.* 1984). By cross-match haplotype analysis and linkage disequilibrium mapping, the causative mutation was mapped to the minisatellite itself (Bennett *et al.* 1995). By means of transient transfection studies in pancreatic B-cells, using constructs containing native human insulin promoters, it was shown that constructs containing a short form of the minisatellite had a relatively small increase in transcriptional activity compared with a construct lacking the minisatellite, while constructs containing the long form of the minisatellite had twice the transcriptional activity of the short form (Kennedy *et al.* 1995). The difference in activity was also observed with a heterologous promoter. The transcription factor PUR-1 (a GAGA box-binding protein) was found to bind to the minisatellite with high affinity, and to bind with different affinity to different repeats (which differ in a minor way in nucleotide sequence) within the minisatellite. This strengthens the hypothesis that minisatellites may affect transcription of nearby genes, as a result of regulatory sequences embedded in the repeated motif. Along these lines a minisatellite located 1 kb downstream from the human proto-oncogene *HRAS1* showed an association with the risk of cancer (Krontiris *et al.* 1993; Phelan *et al.* 1996) and was found to affect transcriptional activity of reporter gene constructs (Green and Krontiris 1993).

## 2.7 Conclusion

The present review documents accumulating evidence that microsatellites and minisatellites code as functional elements of protein molecules and serve as regulatory elements of transcription. It was further shown that changes in microsatellite repeat number can cause quantitative variation in protein function and gene activity, and can affect organismal physiology and development. Together with the well known high rate of spontaneous mutation in repeat number, these observations support the hypothesis that microsatellites are an important source of observed quantitative genetic variation, and a substrate for evolutionary change (Kashi *et al.* 1997; King *et al.* 1997). We raise elsewhere the possibility that variation in microsatellite repeat number may be evolutionarily programmed to provide a copious source of quantitative genetic variation to replenish that lost by drift and selection. This is needed to enable a population to track a moving optimum and exploit fully the variety of micro-ecological niches available to it (King *et al.* 1997; King and Soller 1998).

# Acknowledgement

This review was supported by the United States–Israel Binational Science Foundation (BSF) and the United States–Israel Binational Agricultural Research and Development Fund (BARD).

# 3 Minisatellites and mutation processes in tandemly repetitive DNA

John A.L. Armour, Santos Alonso Alegre, Sue Miles, Louise J. Williams and Richard M. Badge

## Chapter contents

# Abstract

Detailed analysis of mutational processes at some human minisatellite loci is possible because of the very high rates of germline mutation and the presence of internal structural variation. This variation in the distribution of variant repeat units can be used both to survey allelic variation across populations and to define the structural basis of length change mutations. More recently, the fidelity with which minisatellite alleles can be amplified by PCR (even from the single molecule level) has allowed direct assays of germline mutation from sperm DNA samples. The unexpected features of minisatellite mutation include polarity, recombinational exchanges between alleles, and mutation rate variation between alleles. The implications of these studies for our view of microsatellite mutation and evolution will be discussed.

## 3.1 Satellites, minisatellites and microsatellites

### 3.1.1 General properties

The tandemly-repeated component of complex genomes is generally classified into three main divisions: (a) major *satellite* arrays, in which a single repeat sequence family can constitute several percent of the total genome, and can occur in individual repeat arrays as large as 5 Mb; satellites (and in particular the alphoid family) are often preferentially associated with centromeres; (b) *minisatellites*, which may be present at hundreds or thousands of different loci per genome, in which a repeat unit sequence long enough ($> 10$ bp) to be locus-specific is repeated to give repeat blocks of intermediate size (0.5–30 kb); and (c) *microsatellites*, which can be extremely abundant, in which short repeat sequences (2–8 bp) are reiterated to give short arrays (20–100 bp) at each locus.

All three major classes of tandemly repeated DNA can show extensive polymorphism in length. The intrinsic properties of minisatellites (repeat unit variation, intermediate size, and high mutation rate) have meant that our knowledge of their mutation processes is more detailed than for other classes of repeat. After discussing genotyping methods, we will review what is known about mutation at minisatellite loci before making comparisons with microsatellite loci.

### 3.1.2 Genotyping and information content

#### Satellites

Major satellites are only infrequently used to genotype individuals, but have been useful in human genome mapping in providing genetic markers anchored at centromeres. They can be typed either by Southern blot hybridization after pulsed-field gel electrophoresis to measure the total size of the array (Mahtani and Willard 1990; Oakey and Tyler-Smith 1990), or using restriction digests or PCR primers which detect locus-specific repeat unit variants (Warburton and Willard 1996). Detailed analyses have provided interesting insights into the turnover of repeats at these large arrays, but their application to more general genetic analysis has been limited.

### Minisatellites: multi-locus 'fingerprinting'

Some tandemly repeated probes can be used in low-stringency hybridizations to detect a large number of tandemly repeated loci, many of which are hypervariable (Jeffreys *et al.* 1985*a,b*). Some probes have been shown to give individual-specific profiles in humans, which are therefore called 'DNA fingerprints'. They have had wide applications to problems requiring the determination of family relationships and the origin of biological samples (including forensic casework).

Probes consisting of simple repeats [especially $(CAC)_n$] have also been successful in detecting fingerprint patterns in a variety of species (Ali *et al.* 1986); analysis of cloned loci shows that the loci detected are generally similar in structure to minisatellite loci detected using other probes, with long arrays of large (e.g. 15 bp) higher-order repeat units: they do not seem to be simply very large arrays of triplet repeats (Zischler *et al.* 1992).

The advantages of this type of multi-locus fingerprinting are that a large amount of information can be obtained from a single test, and that many probes work well in a wide variety of species: there is often no need to discover a new fingerprinting probe to start work in a particular species. The disadvantages are that a relatively large amount of genomic DNA (about 1 μg or more) is needed for high-quality results, that the same loci are not identifiable between different families, and that the bands observed are not necessarily all from different loci, complicating population genetic inference: initial work using pedigrees of undisputed provenance is required to verify the independence (or otherwise) of the bands detected.

### Minisatellites: single-locus genotyping

DNA fingerprinting probes can be used to screen libraries of genomic DNA to identify clones from individual hypervariable loci (Wong *et al.* 1986, 1987; Armour *et al.* 1990; Vergnaud *et al.* 1991). DNA from such a clone can in turn be used as a high-stringency probe to detect a single hypervariable locus.

*Single-locus probes* are generally more sensitive, thus allowing genotyping from smaller DNA samples, and some of the shorter loci can be typed by PCR. In general, since each probe only detects a single locus, a panel of hypervariable probes is required to achieve high levels of statistical power. The main barrier to their use is the initial work required to develop locus-specific probes for individual species: cloning hypervariable loci from genomic DNA has its own technical difficulties (Armour *et al.* 1990).

In population genetic work assaying length at several independent hypervariable loci, minisatellites have been analysed primarily in humans for validating forensic work in different populations (see, for example, Balazs *et al.* 1989). For work using allele frequencies to determine genetic distance measures they have generally been superseded by microsatellites. Detailed information on the evolutionary histories of chromosomal lineages, however, may be obtained from the analysis of internal structure (by MVR-PCR, see below) at individual minisatellite loci (Armour *et al.* 1996*b*).

## Microsatellites

Genetic information from multi-allelic loci is also available by typing *microsatellites*. Although individually they do not approach the levels of variability exhibited by the best minisatellite loci, this is more than compensated for by their abundance and random spacing in the genomes of a wide variety of species. They are also much simpler to isolate than minisatellite loci. Individual loci must be sequenced before they can be used to genotype samples by PCR, but this brings the collateral advantage of portability: clones do not have to be posted between laboratories.

As with single-locus minisatellite probes, the information from each locus is limited, and genotypes from several unlinked loci are generally required to give sufficient statistical power for the resolution of (for example) parentage (Blouin *et al.* 1996). Amplification of microsatellites by PCR is generally simple and reliable, but the generation of additional products during PCR ('stutter bands') is a particular problem in analysis of (for example) mixed DNA samples. In this respect loci with longer repeat units, such as *tetranucleotides*, seem generally superior to the more commonly used dinucleotide repeat loci, although there is less work on their mutation processes.

# 3.2 Minisatellite mutation

## 3.2.1 Direct observation

### Pedigree analysis

The extreme levels of diversity observed at some human minisatellite loci can be most simply explained by very high *germline mutation* rates to new length alleles. This prediction was fulfilled by the direct observation of length-change mutations in pedigrees (Jeffreys *et al.* 1988). These observations allowed the estimation of germline mutation rates up to 5.2 per cent, and the definition of simple general properties such as the distribution between male and female germ-lines. While the initial study found no overall excess from either sex, in retrospect this may be a consequence of the set of loci examined. Subsequent studies have shown that other loci do have strong biases towards mutation in the male germline (Vergnaud *et al.* 1991; Jeffreys *et al.* 1994). Other loci, with apparently similar rates in male and female germ-lines, appear to undergo different types of processes in the different sexes (Jeffreys *et al.* 1994).

A clear feature of all these studies is that small changes are the most frequent. At some loci with small repeat units and long alleles (such as D1S7, with a 9 bp repeat and alleles as long as 20 kb) a single repeat unit change would be very hard to demonstrate, and thus the true rate of mutation may be higher than directly observed; nevertheless, at loci containing fewer repeats, observed mutations are by no means restricted to single repeat unit changes.

## Small-pool PCR

Even with a germline mutation rate of 1 per cent, large numbers of samples need to be typed to find a representative selection of mutations in pedigrees. Furthermore, population-based screening gives no indication of heterogeneity in mutation rates and processes between alleles at a locus.

*Small-pool PCR* (SP-PCR) addresses both of these problems. It uses PCR to amplify minisatellite alleles from small pools (generally 20–100 molecules) of DNA; using pools of this size, any mutations should be clearly visible as a detectable fraction of the total signal, and several thousand molecules from the same individual can be screened in a single experiment. The method therefore allows the recovery of mutants from the same source, limited only by the supply of DNA (Jeffreys *et al.* 1994; Monckton *et al.* 1994; May *et al.* 1996).

The method relies on the amplification of single mutant molecules from the pools, and therefore requires strict regimes to prevent contamination. It is generally applicable to loci at which single molecules can be amplified with fidelity, and at which length change mutations occur at a high enough rate ($> 10^{-3}$, see below). It has been used to investigate the detailed mechanisms of mutation at minisatellites, by recovering numerous mutations from the same source.

*SP-PCR* systems have been successfully developed for MS32 (D1S8) and MS205 (Jeffreys *et al.* 1994; May *et al.* 1996). At both loci, the authenticity of at least the majority of mutants detected is verified by (a) the appearance of mutants at high frequency in germline (sperm) but not somatic DNA, and (b) the MVR-PCR analysis of the mutants, showing the predicted polar location of rearrangements expected for genuine mutants (see below). Furthermore, the signal from putative mutants matches that expected from single target molecules, and the mean apparent rates of mutation from SP-PCR are consistent with rates found by pedigree-based surveys.

SP-PCR analysis at both MS32 and MS205 demonstrates allelic variation in mutation rate (see below), as well as a bias towards size increases. This bias is smaller at MS205, and may not be large enough to lead to a deterministic growth in the mean size of the locus in real populations. However, the existence of a greater bias at MS32 suggests that some minisatellites may have an intrinsic tendency to expand by mutation: in longer-term evolution, this tendency to steady expansion may be counteracted by other forces such as selection against some classes of alleles (Krontiris *et al.* 1993; Bennett *et al.* 1995; Kennedy *et al.* 1995; Armour 1996) or repeat sequence degeneration in large alleles. Alternatively, since analysis of the primate counterparts of hypervariable human minisatellites has suggested that the 'hyperactive' phase of evolution at any one minisatellite locus may be very short (Gray and Jeffreys 1991), the time during which bias towards expansion operates may be too short to lead to major growth of the locus.

## 3.2.2 Internal structural analysis: MVR-mapping

Many hypervariable minisatellites show some variation in the precise sequence of repeat units. A typical allele at a hypervariable minisatellite in humans is not a perfectly repeated array, but an interspersed mixture of two or more repeat unit sequences. These repeat unit sequence variants ('Minisatellite Variant Repeats' or MVRs) can be mapped either by restriction analysis (Jeffreys *et al.* 1990) or by PCR methods (Jeffreys *et al.* 1991*a*; Armour *et al.* 1993; Neil and Jeffreys 1993; Buard and Vergnaud 1994). In addition to these studies, in which MVR mapping of hypervariable loci has contributed to our understanding of minisatellite mutation, MVR analysis has been used to add information at less variable loci of medical importance such as H-*ras* (Conway *et al.* 1996) and ApoB (Ellsworth *et al.* 1995).

The resulting profile of the interspersion pattern of repeat units (the 'MVR map') gives much more information than the simple measurement of allele length. Indeed, pairs of alleles can be found which contain exactly the same number of repeat units, but which have clearly different internal structures: they have arrived at the same length state via two very different evolutionary pathways ('length homoplasy'; see the chapter by Estoup *et al.* this volume) (Fig. 3.1). Combined with the polar nature of the length change mutation process, MVR mapping of minisatellites therefore gives information on the ancestry of an allele, not just its present state.

MVR mapping has been used to examine the mutation process at human minisatellites in two ways. Firstly, at some loci, randomly sampled alleles from a population can demonstrate features of the overall pattern of mutation. Secondly, direct evidence for structural rearrangements can be obtained from the analysis of new mutant alleles observed in pedigrees.

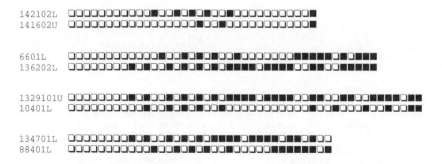

**Fig. 3.1** MVR maps of alleles at MS205 (D16S309), showing four pairs of alleles containing the same numbers of repeat units but with different internal structures ('isoalleles', Monckton and Jeffreys 1991). These pairs of alleles have coincidentally arrived at the same length via different evolutionary pathways. Black and white boxes represent the two different repeat unit variants assayed by MVR-PCR at this locus.

### 3.2.3 Unexpected features of minisatellite mutation

#### Polarity

At three loci studied to date, both population and pedigree analysis provide strong evidence for mutational polarity; that is, most mutations are restricted to a small region at one extremity of the locus (Armour *et al.* 1993; Neil and Jeffreys 1993; Jeffreys *et al.* 1994). At a highly polar locus such as MS205 (D16S309), all mutant alleles discovered in pedigrees involve rearrangements of repeat units at one extremity of the array, shown on the right in Fig. 3.2. Similarly, alignment of alleles sampled from a population shows many examples in which alleles differ only by small changes at the 'active' end of the locus (Fig. 3.3). Both of these observations suggest that the mutation process acts to rearrange repeat unit structure preferentially at this active end, but is only infrequently directed towards other parts of the locus. The recent evolution of groups of related alleles can thus be partially reconstructed by aligning MVR-maps according to their similarity at the relatively stable end of the array (Armour *et al.* 1996*b*).

**Fig. 3.2** Examples of new mutant alleles at MS205 (D16S309). In each case, the two parental alleles (P1 and P2) are shown above the new mutant allele M. The rearrangements occurring to produce the new mutant M (in these cases via intra-allelic modification of P2) cannot be specified exactly, but must involve changes in a block of repeats near one extremity (the right-hand end) of the progenitor allele.

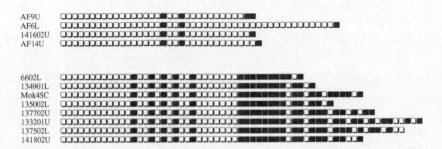

**Fig. 3.3** Selected groups of alleles at MS205 (D16S309). Each group contains alleles differing only by changes at the 'active' end, suggesting that these alleles are recently descended from a common ancestor, i.e. are monophyletic.

Although MS32, MS205 and MS31 have clear mutational polarity, it is not a universal feature of mutation at minisatellites in humans. The extremely unstable CEB1 (D2S90) minisatellite appears to undergo mutation along its entire length (Buard and Vergnaud 1994), and MS621 (D5S110) may also undergo non-polar mutation (Armour *et al.* 1996*a*).

## Inter-allelic exchanges

Analysis of mutations in pedigrees by MVR-PCR (see below) demonstrates that while many mutations involve the rearrangement of repeat units from only one parental allele, other events could only be satisfactorily explained as the result of exchanges of repeat units between both parental alleles (Buard and Vergnaud 1994; Jeffreys *et al.* 1994; Monckton *et al.* 1994; May *et al.* 1996). Thus while entirely intra-allelic events could be explained by simple mechanisms such as replication slippage or unequal sister chromatid exchange, inter-allelic events require the involvement of a recombinational mechanism. In those events for which flanking markers are informative, no exchange of flanking markers has been observed, suggesting that these recombinational events are—at least in the majority of cases—not simple unequal crossovers but instead 'patches' of exchange akin to a gene conversion event.

## Mutation rate variation

Small-pool PCR (SP-PCR) analysis of sperm DNA has changed the way in which mutation can be analysed: instead of screening the offspring of many different parents in a population survey, direct analysis of germline DNA allows numerous mutations to be discovered in DNA from the same individual. These studies have revealed that there is considerable variation in the mutation rates attributable to individual alleles at two different minisatellite loci: MS32 and MS205 (Jeffreys *et al.* 1994; Monckton *et al.* 1994; May *et al.* 1996). Population-based studies have suggested that mutation rates at other minisatellites may also be allele-specific (Andreassen *et al.* 1996). Variation in mutation rate appears to be allele-specific: two alleles in one sperm sample can exhibit very different rates of mutation (Monckton *et al.* 1994; May *et al.* 1996). Consistent with this observation is the correlation between the mutation rate and the population variability: those alleles which have reduced mutation rates as assayed by small pool PCR are relatively common in the population. Since rapid mutation diversifies most alleles (so that each form is infrequent enough to be only encountered once even in large surveys of unrelated individuals), alleles which mutate relatively slowly can become frequent enough to be sampled more than once.

Mutation rate shows no simple relationship with allele length. At the MS32 (D1S8) minisatellite, low mutation rates correlate with the presence of substitutional variants in the DNA immediately flanking the 'active' end of the locus. Taken together with the length-independence and polarity of mutation, the correlation of mutation rate with flanking genotype suggests that mutation to new length alleles is not an intrinsic property of the tandem repeats, but is instead directed by sequences flanking the active part of each locus. Recent work on extending the detection limits of sperm DNA analysis for mutation at MS32 by combining SP-PCR with size selection suggests that the detection limit in unfractionated DNA may be as high as $10^{-3}$. Furthermore, the effect of flanking variants is not simply to 'switch off' a high frequency germline mutation process, leaving only the products of a less frequent somatic process: instead, analysis of mutational spectra shows that alleles with low mutation rates also undergo germline-specific mutation processes, but at a lower rate (Jeffreys and Neumann 1997).

## 3.3 Comparisons with microsatellites

### 3.3.1 Information content

#### Mutation rate

Analysis of mutation rates and mechanisms is most rewarding when there are many mutants to examine. In general, the most unstable minisatellites investigated have high rates of germline mutation (0.4–7 per cent per gamete), whereas only a very small number of microsatellite loci have been shown to have mutation frequencies in this range (Mahtani and Willard 1993; Weber and Wong 1993; Talbot *et al.* 1995).

#### Variant repeats

The apparent dependence of microsatellite mutation on blocks of perfect repeats removes an extremely important source of mechanistic information. Indeed, nearly all our detailed knowledge of minisatellite biology depends on MVR analysis, and in the absence of internal structural analysis, characterization of mutation events is reduced to length measurement and analysis of flanking polymorphic markers. For this reason, it is not strictly possible to exclude more complex mechanisms (such as unequal gene conversion) in microsatellite mutation.

However, a number of features suggest that mutation processes at minisatellite and microsatellite loci may differ in important respects. The most obvious of these is the comparison of the dependence of microsatellite mutation on length, and in particular the length of perfect repeats, with the independence of minisatellite instability from either length or perfect repetition. Minisatellite mutation appears to conform to a model in which instability is not an intrinsic property of the tandem repeats (see above); by contrast, everything we know about microsatellite mutation suggests that it comes from the repeats themselves, and may therefore involve simple processes dependent on repeat-repeat annealing such as replication slippage. Although detailed analysis of microsatellites has generally looked at allele frequencies rather than mutations, the results appear most simply consistent with a mutation model in which most events involve single repeat unit changes (Shriver *et al.* 1993; Valdes *et al.* 1993; Weber and Wong 1993).

### 3.3.2 Using mutation data to build realistic models

#### Ascertainment bias in choice of loci and species?

Given these major differences, how can we use empirical observations to build realistic models of microsatellite mutation? Before considering how we might work towards better models of microsatellite instability, it is useful to sound a note of caution about locus ascertainment. Our choice of model systems is not a random one: at both minisatellites and microsatellites, most work on mutation has naturally come from those loci with the highest rates of mutation. The vast majority of loci, including some of considerable interest (Morral *et al.* 1994), have rates too low to measure by direct observation. Does this reliance on information from highly unstable systems distort our view? In particular, does our reliance on 'fast moving' loci for information on minisatellite or microsatellite mutation mainly tell us about the properties of a minority of highly unstable loci, or can their behaviour truly be extrapolated to a 'silent majority' of loci, which behave the same way but at a slower rate?

#### Allelic rate variation

The precedent from minisatellites suggests that it may be important to consider the rates and processes of mutation for each of the different length alleles at microsatellites separately, rather than lump them all together. This would of course compound the difficulty of getting direct observational evidence about mutation at any one locus: germline mutations are rare enough, without stipulating mutation of a particular allele. The difficulty of amplifying microsatellites with the fidelity needed to detect small admixtures of mutant alleles is currently a formidable technical barrier to SP-PCR at microsatellite loci.

That is not to say, however, that empirical evidence is a hopeless cause, and recent work at HLA-DQβ shows that it is possible to make useful if indirect inferences about the mutational properties of lineages within a locus. That study used closely linked (highly variable) HLA sequences to examine diversity at a dinucleotide array; some lineages appeared to generate more variation than others, and the simplest correlation was between diversification and allele length at the dinucleotide array (Jin *et al.* 1996). More generally, this method of lineage-specific analysis may be used to provide empirical support for more realistic models of microsatellite mutation for loci at which mutation rates are too low to allow direct observation of mutation events in pedigrees.

## Acknowledgements

Work in our laboratory is supported by the Wellcome Trust (Grants 047113/Z/96/Z and 047697/Z/96/Z) and the University of Nottingham.

# 4 Mechanistic basis for microsatellite instability

Jonathan A. Eisen

## Chapter contents

## Abstract

The inherent instability of microsatellite loci makes them exceptionally useful for evolutionary and genetic studies. This instability is predominantly due to changes in the number of copies of the microsatellite repeat. Most copy number changes at microsatellites are caused by slip-strand mispairing errors during DNA replication. Some of these errors are corrected by exonucleolytic proofreading and mismatch repair, but many escape repair and become mutations. Thus microsatellite instability can be considered to be a balance between the generation of replication errors by slip-strand mispairing and the correction of some of these errors by exonucleolytic proofreading and mismatch repair. The factors that cause this process to occur much more frequently in microsatellites that in non-repeat containing DNA are discussed. However, not all microsatellites are equally unstable because not all are equally prone to this mutation process.

The mechanisms by which a variety of factors cause this variation in stability among microsatellites are discussed.

## 4.1 Introduction

The characteristic that makes loci that contain microsatellite repeats particularly useful for evolutionary and genetic studies is their inherent instability. The mutation rates at most microsatellite loci are usually orders of magnitude higher than mutation rates at other loci within the same genome. Although many types of mutations occur at microsatellite loci, the elevated mutation rate is primarily caused by an elevated rate of one particular class of mutations: changes in the length of the repeat tract. Thus the term 'microsatellite instability' is frequently used to refer specifically to these tract-length changes. Since most of these tract-length changes result from changes in the integral number of copies of the repeat, they are also frequently referred to as copy number changes.

Ever since it was recognized that microsatellites are so prone to changes in tract length, researchers have been trying to determine why. A variety of approaches have been useful for this purpose. Evolutionary and population genetic comparisons have been used to document the patterns of tract-length variation at microsatellites, and to test the robustness of different mutation models when averaged over long time-scales. Biochemical experiments with purified proteins or cell extracts have been used to characterize each step in the mutation process, and to determine the factors that control that step. Genetic studies have given insight into the genes that control microsatellite stability, and have allowed the accurate quantification of the stability of different microsatellites in controlled genetic backgrounds. Only by combining the results of these different types of studies has the mechanism of the mutation process become well characterized. Since evolutionary studies of the mutation mechanism are described in detail elsewhere in this book, I focus here on the biochemical and genetic studies.

To have a complete understanding of the mechanism of microsatellite instability one must also explain why stability varies both within and between species. Clues to the cause of this variation have come from the identification of factors that correlate with the level of microsatellite stability. Such factors include size of the repeat unit, number of copies of the repeat, presence of variant repeats, and amount of transcription in the region of DNA containing the repeat. Many studies that use data on microsatellite variation use models of the mutation process to enhance the analysis being done. Such studies should be improved by a better understanding of the mechanism underlying microsatellite instability as well as the causes of differences in stability among microsatellites. In this chapter, I summarize what is known about the mechanism underlying microsatellite instability and discuss some of the factors that cause variation in stability within and between species.

## 4.2 Microsatellite mutation models

The central debate about the mechanism of microsatellite instability has focused on two competing but not necessarily mutually exclusive models. One model proposes that microsatellite instability is caused by an elevated rate of unequal crossing-over (UCO) within microsatellite repeats. Unequal crossing-over is the result of recombination between homologous chromosomes that are imperfectly aligned. The UCO microsatellite instability model suggests that UCO occurs at an elevated rate in microsatellites because the presence of repeats increases the likelihood of misalignment between homologues. A similar proposal has been made to explain the high rates of copy number changes observed in tandemly repeated genes (Smith 1973). The alternative model proposes that microsatellite instability is caused by an elevated rate of slip-strand mispairing (SSM) errors during DNA replication. The SSM process, which was first proposed to explain frameshift mutations in any type of DNA (Fresco and Alberts 1960), begins with the DNA polymerase 'slipping' during replication, causing the template and newly replicated strands to become temporarily unaligned. For replication to continue, the strands must realign. Mutations will be generated if this realignment is imperfect. The SSM microsatellite instability model proposes that SSM occurs at an elevated rate in microsatellites because the presence of repeats increases the likelihood of misalignment after slippage (since repeats can easily be looped out of the DNA double-helix) (Streisinger *et al.* 1966).

The results of many studies indicate that an elevated rate of SSM is the main cause of microsatellite instability. The key evidence that supports the SSM model against the UCO model is summarized below (see Sia *et al.* 1997*a*, for a review):

- Microsatellite stability is unaffected by defects in genes with major roles in recombination, such as *recA* in *Escherichia coli* (Levinson and Gutman 1987*b*), and *rad52* in *Saccharomyces cerevisiae* (Henderson and Petes 1992). This suggests against the UCO model since mutations are dependent on recombination in this model.

- In humans, copy number changes at microsatellites can be generated without exchange of flanking genetic markers (and thus probably without recombination) (Morral *et al.* 1991).

- In *S. cerevisiae*, microsatellite stability is similar in mitotic and meiotic cells (Strand *et al.* 1993). Since recombination occurs more frequently in meiosis than mitosis, if the UCO model were correct, microsatellites should be more unstable during meiosis.

- Microsatellite stability is reduced by defects in genes involved in DNA replication error correction pathways. This is consistent with the SSM model since this model requires DNA replication to occur. In addition, genetic and biochemical experiments show that these error-correction pathways can recognize and repair the types of DNA loops that would be created by SSM (Bishop *et al.* 1989; Parker and Marinus 1992).

- The orientation of a microsatellite relative to the leading and lagging strands of replication influences its stability (Freudenreich *et al.* 1997). This is not expected by UCO model but is consistent with the SSM model since the leading and lagging strands have somewhat different mechanisms of replication.

These and other results show that SSM is an integral component of the mutation process leading to microsatellite instability. However, SSM alone does not provide a full picture of this mutation process. As suggested above, not all SSM errors become mutations—some are 'repaired' by error-correction mechanisms. The two error-correction pathways that have been shown to be important in repairing SSM errors are exonucleolytic proofreading and post-replication mismatch repair. Thus a complete description of the mutation process must include both the generation of replication errors by SSM and the correction of some of these errors by mismatch repair and proofreading (see Fig. 4.1). In the following sections I discuss each of the steps in the microsatellite instability mutation process, providing some details about the mechanism of each step and the methods used to study those mechanisms. In addition, I discuss how variability in each step contributes to variation in microsatellite stability within and between species (see Table 4.1).

## 4.3 Slip-strand mispairing replication errors

To study the mechanism of the SSM process, one must functionally isolate SSM from the downstream error-correction steps. One approach to achieve such functional separation is to study the replication of DNA *in vitro* (Kunkel 1986, 1990; Schlötterer and Tautz 1992). *In vitro* studies allow straightforward comparisons of replication errors by different polymerases, as well as comparisons of errors by the same polymerase using different templates. However, *in vitro* studies are limited because they may not accurately reflect what occurs during intracellular replication conditions. To study SSM errors *in vivo*, researchers have used strains with defects in either exonucleolytic proofreading or mismatch repair or both. In such strains, since SSM errors are not corrected, SSM error rates and patterns can be inferred directly from observed mutations (e.g., Wierdl *et al.* 1997). Results from many such *in vitro* and *in vivo* studies show that the SSM process can be subdivided into three distinct steps: slippage of the DNA polymerase during replication, mis-realignment of the template and newly replicated DNA strands, and continuation of replication from a misaligned template (see Fig. 4.1).

These studies confirm the prediction of the SSM model that SSM errors are more likely to occur in microsatellite repeats than in 'normal' DNA. However it has not been determined which step of SSM is most affected by the presence of repeats: slippage, misalignment, or extension. It is almost certain that misalignment is more common in repeat regions than in 'normal' DNA. Loops generated by misalignment will be more stable in microsatellites than in non-repeat regions since base pairing is not significantly changed when one or more copies of a

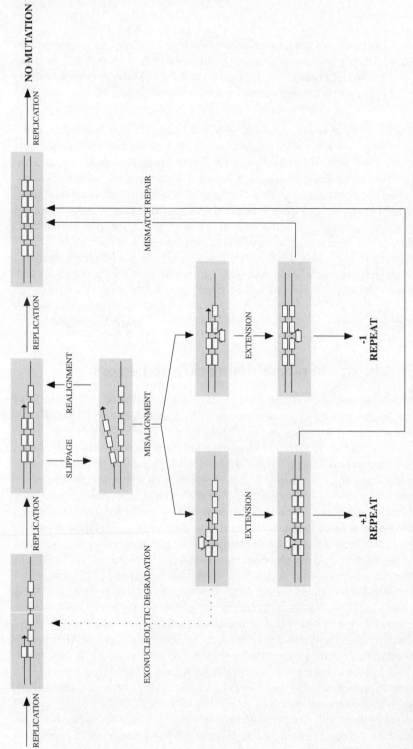

**Fig. 4.1** Model of the mutation process at microsatellite loci. Cartoons of double-stranded DNA containing a microsatellite repeat are shown at different stages of the replication and mutation process. In the cartoons, DNA strands are represented by thin lines, microsatellite repeats by small boxes, and ongoing replication by small arrows. Flow arrows point down for steps that lead to mutations, up for steps that prevent mutations from occurring, and to the right for steps in the DNA replication process. The exonuclease step is shown with a dashed line since it has only a limited role in regulating microsatellite mutations. Details about each step are provided in the main text.

repeat are in a loop (see Fig. 4.1). However, there is also reason to believe that slippage occurs more frequently in microsatellite repeats than in normal DNA. *In vitro* studies show that DNA containing microsatellite repeats is particularly prone to the formation of unusual DNA structures. Such structures likely interfere with the replication process, which could lead to slippage by the polymerase (Kang *et al.* 1995; Samadashwily *et al.* 1997). Thus, the elevated SSM rates at microsatellites relative to normal DNA may be caused by an increased likelihood of both slippage and misalignment.

## 4.3.1 SSM variation: effects of the nature of the microsatellite

Although in general SSM errors are more frequent in microsatellite-containing regions than in other regions of the genome, the rate and type of such errors are not equal for all microsatellites. The nature of the microsatellite itself has a large impact on SSM. For example, the likelihood of SSM for a particular microsatellite is correlated with the number of copies of the repeat. The most detailed study of this copy number effect is that of Wierdl *et al.* (1997), in which the stability of five microsatellites with different numbers of copies of a GT repeat was analysed. The mutation *rate* was found to increase with more repeats (as is expected since there are more places to slip and misalign) but the increase was greater than expected (more than two orders of magnitude between loci with 7.5 and 52.5 repeats). The *types* of mutations also differed between the microsatellites with different numbers of the repeat. The long tracts (those with more repeats) were more likely to have large, multi-repeat deletions than short tracts. In addition, the mutations that resulted in single-repeat changes (plus or minus one repeat) were different between long and short tracts. The single copy changes in long tracts were mostly additions while those in short tracts included roughly equal numbers of additions and deletions. Wierdl *et al.* showed that these copy number effects were not due to biases in mismatch repair, since the effects were seen in mismatch repair mutants. Therefore, they concluded that the copy number effects were probably caused by differences in SSM between microsatellites with different numbers of repeats. However, they were not able to determine the step of SSM that was influenced by copy number. One possibility is that the unusual DNA structures, discussed above as a potential cause of increased slippage in microsatellite repeats, may be even more likely to occur as the number of repeats increases. Regardless of the exact mechanism, the details of the effects of copy number on SSM (and thus on microsatellite stability) help to explain why the number of repeats at a particular microsatellite is somewhat stable over evolutionary time. Long tracts may be biased towards getting shorter (due to the large deletions) and short tracts may be biased towards getting longer (because of a slight bias in additions over deletions). An effect of copy number may also explain why certain microsatellites (e.g. those associated with some human diseases) become particularly unstable after they cross a threshold number of copies of the repeat (see chapter by Rubinsztein, this volume).

**Table 4.1**
Factors that lead to variation in mutation rates and patterns at microsatellite loci[1]

| Step in mutation process affected by factor | Nature of the microsatellite | | | | |
|---|---|---|---|---|---|
| | Repeat unit size | Number of repeats | Type of repeat | Variant repeats | Replication orientation |
| SSM (any step) | ± | + | ± | + | ± |
|   Replication slippage | ± | ± | ± | ± | ± |
|   Misalignment | ± | ± | ± | ± | ? |
|   Extension[5] | ? | ? | ± | ? | ± |
| Exonuclease | + | + | ± | ? | ± |
| Mismatch repair | +[6] | ? | + | ? | + |

Another aspect of the microsatellite that influences the likelihood of SSM is the presence of variant repeats. Evolutionary and genetic studies have shown that the presence of variant repeats is correlated with the stability of a microsatellite (e.g. Goldstein and Clark 1995). Petes *et al.* (1997) have studied this effect in controlled laboratory conditions in *S. cerevisiae* to try to determine the underlying mechanism. This study showed that the presence of variant repeats leads to an approximately fivefold stabilization of GT repeats. Since this stabilizing effect was also seen in mismatch-repair mutants, the authors suggested that the variant repeats exerted their effect by reducing the likelihood of SSM errors. However, as with the copy-number effect described above, it has not been possible to determine what step of SSM was most 'stabilized' by variant repeats.

### 4.3.2 SSM variation: effects of external factors

There are many reasons to believe that external factors (i.e. factors other than characteristics of the microsatellite) can influence SSM error rates and patterns. For example, base misincorporation error rates and patterns are influenced by many external factors. Since base misincorporation and SSM are both forms of polymerase error, it is likely that these factors will also influence the SSM process. External factors that influence misincorporation errors include local DNA sequence (e.g. the GC content or the ability to form secondary structures), genome position (e.g. proximity to replication origins or chromosome ends), and even the chromosome in which a sequence is found (e.g. nuclear, organellar, plasmid) (Wolfe *et al.* 1987, 1989; Kunkel 1992; Hess *et al.* 1994). In addition, misincorporation error rates are dependent on many conditional factors including methylation state, amount of chromosome packaging, temperature, phase of the cell cycle during which a particular section of DNA is replicated, and amount of DNA damage and repair prior to replication. Future studies of microsatellite mutation mechanisms would benefit by examining whether some of these factors influence SSM errors.

**Table 4.1**
*Continued*

| Neighbouring DNA[1] | | Cellular conditions | | | | Organismal differences | |
|---|---|---|---|---|---|---|---|
| GC content | Sequence content | Transcrip-tion | Methyla-tion state | Cell cycle stage | Pathway used[2] | Pathway presence[3] | Pathway biases[4] |
| + | + | ± | ± | + | + | ? | ± |
| ± | ± | ± | ± | ± | + | ? | ± |
| ± | ± | ? | ? | ± | + | ? | ± |
| ± | ± | ? | ? | ± | + | ? | ± |
| + | + | ? | ? | ± | + | + | + |
| + | + | ± | + | + | + | + | + |

[1] Some of these effects have only been shown for base-misincorporation errors.
[2] For example, different polymerases are used for chromosome replication and DNA repair replication.
[3] Mismatch repair is absent in many strains and species and not all polymerases have associated exonucleases.
[4] For example, the ability to recognize loops for mismatch repair varies greatly between repair systems in different species.
[5] The extension and exonuclease steps are related in that they both work with the same substrate (see Fig. 4.1) but they can be functionally separated.
[6] Mismatch repair is affected by the total size of the loop, thus both number and size of repeats are important.

### 4.3.3 SSM variation: differences between individuals or species

Although the SSM mechanism and its role in causing microsatellite instability is conserved between species, it is likely that the specific rates and patterns of SSM differ greatly between species. For example, polymerases from different species have significantly different base misincorporation error rates (Kunkel 1992) and thus likely also have different SSM rates at microsatellites. In addition, many of the factors described above as influencing SSM errors within a species differ greatly between species (e.g. GC content, temperature, methylation). Thus it remains to be seen whether all species are affected by copy number and variant repeats in the same ways as described above.

## 4.4 Correcting SSM errors I: exonucleolytic proofreading

Exonucleolytic proofreading is a process in which DNA that has been recently synthesized is examined for errors made by the DNA polymerase. If errors are found, the exonuclease will degrade the newly replicated strand, the DNA polymerase will back up, and the strand will be recopied. Thus many errors made by the DNA polymerase will not become mutations because they will be 'erased' by proofreading. Proofreading was originally characterized for its role in limiting mutations due to base misincorporation errors. The role of proofreading in regulating microsatellite stability has been determined by methods that are similar to those used to study SSM. *In vitro* studies have been used to compare the error rates and patterns of polymerases with and without associated exonucleases, and to determine the types of substrates that the proofreading exonucleases will degrade. *In vivo* studies have allowed the determination of errors with and without exonucleases under realistic cellular conditions. In such *in vivo* studies, it has been helpful to use strains with defects in mismatch repair, so that the role of the proofreading step is clear.

Studies such as the ones described above have shown that proofreading is involved in regulating the stability of microsatellites, but the extent of this role is limited in two ways. First, proofreading only significantly influences the stability of a subset of microsatellites: those with both small unit size (mostly mono- and di-nucleotide repeats) (Kroutil *et al.* 1996; Sia *et al.* 1997a; Strauss *et al.* 1997) and few copies of the repeat (Streisinger and Owen 1985; Kroutil *et al.* 1996; Tran *et al.* 1997). In addition, even for this subset of microsatellites, the impact of proofreading is limited—the stability of such microsatellites only decreases by about five- to tenfold in exonuclease mutants.

The details of the mechanism of proofreading help to explain why this process has only a limited role in regulating microsatellite stability (for reviews see Echols and Goodman 1991; Kunkel 1992). Proofreading exonucleases detect errors by monitoring the DNA that has just been replicated to determine whether it forms normal double-helical DNA structures with the template strand. Abnormal DNA structures trigger the exonuclease activity. This is how proofreading prevents many base misincorporation errors from becoming mutations. A base misincorporation error will lead to a base : base mismatch between the newly replicated and template DNA strands and many such mismatches will be recognized by proofreading exonucleases. However, proofreading exonucleases are only able to monitor the DNA within a few bases of the active site of the polymerase. This proximity effect explains why proofreading has at most a small impact on microsatellite stability. Most loops generated by SSM will be too far from the replication fork to be recognized by proofreading exonucleases. The lack of a role of exonucleases in repairing most SSM errors at microsatellites helps to explain the high rate of microsatellite copy-number changes relative to point mutation rates.

## 4.4.1 Proofreading variation

While the impact of proofreading on microsatellite stability is limited, variation in proofreading can account for some of the variation in stability of microsatellites. As with SSM, the nature of the microsatellite has a profound impact on proofreading. The best example of this was described above—proofreading only works on microsatellites that are short and in which the repeat unit size is small. The mechanism of both of these biases is directly related to the proximity effect described above. As the number of copies of a repeat increases, the impact of proofreading decreases because those loops that are generated by SSM will be even more likely to be far from the replication fork. In addition, in microsatellites with repeats of large unit size (e.g. 5 bp repeats), a loop just one repeat away from the replication fork may be too far away to be proofread (the base pairing of one repeat may be enough to stabilize the DNA structure at the fork). Proofreading is also likely to be affected by many external factors. For example, the efficiency of some exonucleases is affected by both GC content and sequence context (Kunkel 1992). Thus the sequence around a mononucleotide repeat may influence its mutation rate by altering the efficiency of proofreading. Finally, the impact of proofreading on microsatellite stability is also likely to vary greatly between species. For example, some species do not even have proofreading exonucleases associated with their DNA polymerases. Microsatellites with short mono- and dinucleotide tracts should be more unstable in species without proofreading than in species with proofreading.

## 4.5 Correcting SSM errors II: mismatch repair

Mismatch repair was named based on its role in recognizing and repairing base : base mismatches that arise due to base misincorporation errors. It is now clear that the same process can repair DNA containing loops such as those generated by SSM at a microsatellite (see Fig. 4.1). Mismatch repair has a much more significant impact on microsatellite stability than proofreading. Defects in mismatch repair can cause microsatellite instability to increase by many orders of magnitude (see below for more details). Since mismatch repair plays such a key role in regulating microsatellite stability, differences in the repair of loops by mismatch repair could account for a great deal of the variation in microsatellite stability within and between species.

Before discussing the specifics of loop repair and how it varies within and between species, it is useful to review some details about the general mechanism of mismatch repair. Mismatch repair has been found in a variety of species from bacteria to humans. It has been characterized in the most detail in *E. coli*. In the other species in which it has been characterized, the overall scheme of mismatch repair works in much the same way as in *E. coli*. Thus the *E. coli* system has served as a useful model for mismatch repair in all species. The first critical step in mismatch repair in *E. coli* is the recognition of mismatched DNA by the MutS protein (see Modrich 1991 for a review). Specifically, a dimer of MutS (two MutS proteins bound together) binds to the site of a mismatch in double-stranded DNA. Subsequently, through an interaction between the MutS dimer, a dimer of the MutL protein, and a single MutH protein, a section of one of the DNA strands at that location is targeted for removal. Other proteins complete the repair process: the section of DNA that has been targeted is removed and degraded, a patch is synthesized using the complementary strand as a template, and the patch is ligated into place resulting in a repaired section of double-stranded DNA without mismatches.

The evidence that mismatch repair is involved in repairing SSM errors at microsatellites comes from three types of studies. First, defects in mismatch repair cause decreases in microsatellite stability (anywhere from 10- to 5000-fold, depending on the species and the microsatellite). In addition, when DNA containing loops is transformed into cells, the loops can be repaired, but only if the cells have functional mismatch repair (Dohet *et al.* 1986; Bishop *et al.* 1989; Parker and Marinus 1992). Finally, *in vitro* studies have shown that repair of loops can be carried out by purified mismatch repair proteins (Learn and Grafstrom 1989; Parker and Marinus 1992). Each of these results has been found in a variety of species, showing that the role of mismatch repair in repairing loops at microsatellites is highly conserved. Incidentally, this is what led to the discovery that mismatch repair genes are defective in hereditary non-polyposis colon cancer in humans—cells from patients with this disease showed high levels of microsatellite instability. In summary, these studies show that the repair of loops is very similar to the repair of mismatches.

### 4.5.1 Mismatch repair variation: effects of the nature of the microsatellites

Perhaps the most important cause of variation in mismatch repair is the nature of the microsatellite. Loops are not all recognized equally by mismatch repair system, and this specificity varies between species. One factor that is very important to the recognition step is the size of loop. For example, in *E. coli*, transformation studies have shown that loops of 1–3 bases are repaired well, those of 4 bases are repaired poorly, and those greater than 4 bases are not repaired at all. *In vitro* studies of purified mismatch repair proteins show that this is due to inability of MutS to recognize loops larger than 4 bases in size (Learn and Grafstrom 1989; Parker and Marinus 1992). Thus in *E. coli*, microsatellites in which the repeat unit size is 4 bp or greater have especially high rates of instability, since SSM errors in such regions are not repaired well. Mismatch recognition is also biased by loop size in many other species, although the specific size preferences are not completely conserved. For example, the yeast mismatch repair system appears to be able to recognize and repair loops up to 6 bp well (and possibly even up to 14 bp, although this has not been confirmed). More details about the mechanism causing the different size preference are given in the section on variation in mismatch repair between species. For the purposes of the discussion here, all that is important is that in many species the size limits of loop recognition help to explain why microsatellites with different repeat unit sizes have different mutation rates.

The size specificity of loop recognition also helps to explain variation in mutation patterns between microsatellites with different sized repeats. For example, in *S. cerevisiae*, the majority of mutations in mononucleotide repeats are additions or deletions of one repeat (i.e. plus or minus 1 bp). However, the majority of mutations at microsatellites with 5 bp repeats are additions or deletions of two or more repeats (Sia *et al.* 1997*b*). To understand this phenomenon, it is important to recognize that the mutation rate and pattern for a microsatellite is determined by a combination of the rate and type of SSM errors and how well these errors are repaired. Thus a particular mutation may occur at a high rate either because it is a common SSM error or because it is repaired poorly. For the mononucleotide repeat described above, most SSM errors are repaired about equally well (errors involving even five repeats at a time can be repaired by mismatch repair). Thus the most common mutations are those that are the most common SSM errors. In contrast, for the microsatellite with the 5 bp repeat, mismatch repair will only repair single repeat changes. Thus, although SSM errors involving two or more repeats are not very frequent, most of the mutations are changes in two or more repeats because many of the single repeat changes are repaired. The size dependence of mismatch repair also explains why 20 bp repeats are so unstable in *S. cerevisiae* (Sia *et al.* 1997*b*); mismatch repair will not recognize any SSM error involving such a large repeat. Since both the number of repeats and the size of the repeat influence microsatellite stability, it is important to compare repeats of the same unit size when studying copy-number effects, and repeats with the same number of copies when studying unit size effects.

One aspect of loop repair that has been poorly studied is the role of the type of microsatellite (e.g. GT vs. GA repeats). Since base : base mismatch repair is not uniform for all mismatches (e.g., C : C mismatches are not repaired well in many species), it is likely that loop repair will also not be uniform. Since most of the studies of microsatellite mutation mechanisms have been done on limited types of microsatellites, it will be important to determine if the results of these studies are universal to all types of repeats.

## 4.5.2 Mismatch repair variation: effects of external factors

As with SSM and proofreading, many factors in addition to the nature of the microsatellite itself can influence the effectiveness of mismatch repair. For example, the location of the mismatch within the genome is important. In *S. cerevisiae*, loop recognition appears to be biased between loops on the template versus nascent strand of replication. For loops including a single repeat, mismatch repair appears to repair preferentially those that are on the template strand, resulting in a bias towards single repeat additions. The exact mechanism of this strand bias is not known although some of the genes involved have been identified (Sia *et al.* 1997*b*). Another effect of location is whether the mismatch is in nuclear or organellar DNA. Although organellar mismatch repair has not been characterized in detail, it is likely to be quite different from nuclear mismatch repair. The surrounding DNA also influences mismatch repair. For example, studies of base : base mismatches have shown that mismatch recognition is affected by sequence context (Cheng *et al.* 1992), and by GC content (Jones *et al.* 1987). It is likely that the recognition of loops will also be affected by these factors. Finally, mismatch repair can also be influenced by conditional factors including the presence of strand-recognition signals, methylation state, and level of transcription.

### 4.5.3 Mismatch repair variation: differences within a species

Differences in mismatch repair among individuals of a particular species have been well documented. For example, many strains of *E. coli* in the 'wild' are defective in mismatch repair (LeClerc *et al.* 1996; Matic *et al.* 1997). Since there are adaptive benefits to having modest increases in mutation rates in certain circumstances (Taddei *et al.* 1997*a,b*), and since one way to alter mutation rates is by altering mismatch repair, many strains may be found to have defects in mismatch repair. Also, since mismatch recognition is involved in other cellular processes such as the regulation of interspecies recombination, there may be other selective pressures that lead to variation in mismatch repair capabilities within a species. Finally, since organisms appear to be able to turn mismatch repair on and off in certain situations (Harris *et al.* 1997; Macintyre *et al.* 1997; Torkelson *et al.* 1997), environmental conditions may play a major role in determining mismatch repair capabilities.

### 4.5.4 Mismatch repair variation: differences between species

Although mismatch repair is a highly conserved process, there are many ways in which it varies between species. For example, the mismatch recognition process is not completely conserved between bacteria and eukaryotes. The best characterized eukaryotic mismatch repair system is that of *S. cerevisiae*. As suggested above, the general mechanism of *S. cerevisiae* mismatch repair is very similar to that of *E. coli* (see Kolodner 1996 for a review). In particular, the role of the MutS and MutL proteins is highly conserved—*S. cerevisiae* uses homologues of these proteins in essentially the same way that they are used in *E. coli*. Even the use of the proteins as dimers is conserved. However, unlike *E. coli*, *S. cerevisiae* uses multiple homologues of both MutS and MutL for mismatch repair. These multiple homologues are used to make separate mismatch repair complexes with unique and distinct functions. The specificity of each of these complexes is determined almost entirely by its particular combination of MutS homologues (which are referred to as MSH proteins for *MutS H*omologue). For mismatch repair of nuclear DNA there are two recognition complexes: an MSH2–MSH6 heterodimer for recognizing and repairing base : base mismatches and loops of 1–2 bases, and an MSH2–MSH3 heterodimer for recognizing and repairing loops of 1–6 bases (and possibly even up to 14 bases—see Sia *et al.* 1997*b*). Genetic studies suggest that there may also be a mitochondrial-specific mismatch repair complex. Defects in another MutS homologue, MSH1, cause increases in the mutation rates in mitochondrial DNA. However, the details of mitochondrial mismatch repair are not well understood. In particular, it is not known what role mismatch repair plays in microsatellite stability in mitochondrial DNA. Interestingly, *S. cerevisiae* encodes two additional MutS homologues (MSH4 and MSH5) that do not function in mismatch repair, but instead appear to use mismatch recognition to regulate meiotic crossing-over and chromosome segregation. The mismatch recognition process of other eukaryotes is highly similar to that of *S. cerevisiae* (Fishel and

Wilson 1997). One of the results of the differences in mismatch repair between eukaryotes and *E. coli* is that eukaryotes can repair loops of larger sizes than *E. coli*. This explains why microsatellites with these larger sized repeats are more stable in eukaryotes than in *E. coli*.

Another major difference in mismatch repair between species is in the mechanism used to determine which strand is the recently replicated strand (and thus is the strand that contains the error). In *E. coli* the 'incorrect' strand is determined by its methylation state—the newly replicated strand is unmethylated and thus can be distinguished from the template strand. In some other species, strand recognition is thought to be based on the presence of nicks, which are more likely to occur on the newly replicated strand. In such species, there may be differences in mismatch repair efficiency between the leading and lagging strands, since nicks are more common on lagging strand.

Although the process of mismatch repair is highly conserved, some species may not have the process at all. For example, analysis of complete genome sequences shows that some bacterial and Archaeal species do not encode any likely MutS or MutL homologues (Eisen *et al.* 1997; Eisen 1998). It is likely that these species do not have any mismatch repair, since functional MutS and MutL homologues are absolutely essential to the mismatch repair process. Any species without mismatch repair should have significantly elevated levels of microsatellite instability. In addition, differences between species could arise from the number and types of MutS and MutL homologues that are present.

## 4.6 Additional factors that influence microsatellite stability

Although the studies of microsatellite mutation mechanisms have been extensive, there are still many factors that have been found to influence microsatellite stability but for which the mechanism of the effect is unknown. For example, Wierdl *et al.* (1996), following up previous studies (Datta and Jinks-Robertson 1995), showed that transcription leads to a 4–9-fold destabilization of a poly(GT) repeat. One explanation for this is that transcription will increase the likelihood of repair by the process of transcription-coupled repair and this process is mutagenic in some conditions (Wang *et al.* 1996). Alternatively, transcription could interfere with either mismatch repair or replication. Another unexplained observation is that microsatellites in the chromosome are usually more stable than identical microsatellites on a plasmid (Henderson and Petes 1992). Finally, many studies have shown that microsatellite stability is dependent on the orientation of the microsatellite within the DNA (Kang *et al.* 1995; Maurer *et al.* 1996; Freudenreich *et al.* 1997). For example, Freudenreich *et al.* showed that a microsatellite with 130 CTG repeats was more unstable when the CTG was on the lagging strand (Freudenreich *et al.* 1997). They suggested that this could be due to differences in the likelihood of slippage on the leading vs. lagging strand of replication. However, they could not rule out differences in mismatch repair, transcription, proofreading or other factors between the strands as the explanation. An alternative explanation for the orientation effect is that loops may be better recognized on the nascent strand than on the template strand (Sia *et al.* 1997*b*). More detailed studies of microsatellite mutation mechanisms will likely sort out how these and factors influence microsatellite stability.

## 4.7 Conclusions and summary

The mutation process at microsatellites can be considered to be a balance between the generation of replication errors by slip-strand mispairing and the correction of some of these errors by exonucleolytic proofreading and mismatch repair. The mutation rate and pattern for a particular microsatellite will be determined by the rate and type of SSM errors, as well as by how well these errors are recognized and repaired by exonucleases and mismatch repair. The details of the mutation mechanism explain why microsatellites are so unstable. First, SSM occurs much more frequently in microsatellites than in normal DNA. In addition, exonucleolytic proofreading, which prevents a large proportion of base misincorporation errors from becoming mutations, has only a limited role in preventing SSM errors from becoming mutations. The details of the mutation mechanism also help to understand why microsatellite stability varies within and between species. For example, the high mutation rate of microsatellites with repeats of large unit size can be explained by the inability of mismatch repair to recognize SSM errors in such large repeats. In addition, the positive correlation between number of copies of a repeat and stability can be explained by an increased likelihood of SSM errors in microsatellites with more repeats. The details not only help to understand the mutation process causing microsatellite instability, but they can be used to improve models of microsatellite evolution. Just as better models of nucleotide substitution processes have improved the analysis of DNA sequence variation, better models of microsatellite instability should improve the analysis of copy number variation at microsatellite loci.

# 5 Microsatellite evolution: inferences from population data

Arnaud Estoup and Jean-Marie Cornuet

## Chapter contents

## Abstract

A thorough understanding of the mutational events that shape microsatellite evolution is necessary to optimize the information provided by these markers. The application of statistical tests for theoretical mutation models (IAM, SMM, TPM, and KAM) to actual population data has arrived at rather contradictory or inconclusive results, reflecting the low power of these tests and the fact that basic assumptions such as mutation–drift equilibrium are not always met. However, they suggest that large differences in the mutation process exist among loci, and the involvement of more complex mutation processes than those assumed for the theoretical mutation models tested. Pedigree, sequencing, and phylogenetic studies have shown that mutations at microsatellite loci generally involve addition or loss of one repeat, and much less frequently of several repeat units. They have also indicated that the mutation of repeat arrays has a complicated dependence on allele size and purity, the mutation process is upwardly biased, and some constraints on allele length exist. The sequencing of imperfect, interrupted and/or compound microsatellite electromorphs revealed a substantial amount of size homoplasy, a result expected when considering the mutation processes of repeat arrays. Knowledge of the mutation processes at microsatellite loci is still insufficient to parametrize accurately the factors that have been found to be relevant to the evolution of these repeated sequences and hence to allow optimal representation in theoretical mutation models.

## 5.1 Introduction

Microsatellite loci are increasingly replacing or complementing other markers for numerous applications in evolutionary genetics (e.g. Bowcock *et al.* 1994; Gottelli *et al.* 1994; Taylor *et al.* 1994; Estoup *et al.* 1995*a*, 1996; Estoup and Angers 1998; Paetkau *et al.* 1995; Viard *et al.* 1996). A thorough understanding of the mutational events that shape microsatellite evolution is necessary to optimize the information provided by these markers. The evolutionary dynamics of repeat arrays have been examined using various empirical and theoretical methods (see the reviews of Freimer and Slatkin 1996, and of Jarne and Lagoda 1996). These studies have shown that the mutation processes of microsatellites may be more complex than previously believed. Mutation at microsatellite loci generally involves a change in size of one repeat, but sometimes mutation involves several repeat units (Weber and Wong 1993; DiRienzo *et al.* 1994; Primmer *et al.* 1996). Several complex evolutionary factors, including a dependence on allele size and purity (Weber 1990; Chung *et al.* 1993; Richards and Sutherland 1994; Pépin *et al.* 1995), an upwardly biased process (Amos *et al.* 1996; Primmer *et al.* 1996), and constraints on allele length (Bowcock *et al.* 1994; Garza *et al.* 1995) were found to have an effect on the evolution of microsatellites. However, insufficient data are currently available to allow adequate representations in theoretical mutation models, though this is crucial for a sound and accurate estimation of population parameters.

## 5.2 Theoretical mutation models: description and evaluation from population data

The question of which theoretical mutation model should be applied to microsatellite markers is essential for at least two reasons. Estimation of numerous population parameters (e.g. genetic differentiation, number of migrants per generation, etc.) is dependent upon the mutation model assumed for the markers.

This dependence may be especially strong in the case of microsatellites, since sensitivity to the mutation model increases with the mutation rate, which is generally very high for microsatellites. For instance, the relationships between the classical parameter $M = 4N_e\mu$, with $N_e$ the effective population size and $\mu$ the mutation rate of the locus, and the number of alleles in a sample of genes taken from a population at mutation–drift equilibrium diverge with the mutation model only for sufficiently large values of $M$ (see Fig. 5.1).

## 5.2.1 Theoretical mutation models: SMM, IAM, TPM, and KAM

Classically, two extreme models of mutation have been considered for microsatellite loci (Deka et al. 1991): the *infinite allele model* (IAM, Kimura and Crow 1964) and the *stepwise mutation model* (SMM, Kimura and Ohta 1978). The SMM describes mutation of microsatellite alleles by the loss or gain of a single tandem repeat, and hence alleles may possibly mutate towards allele states already present in the population. In contrast, under the IAM, a mutation involves any number of tandem repeats and always results in an allele state not previously encountered in the population. More recently, DiRienzo et al. (1994) introduced a *two phase model* (TPM), where mutations introduce a gain or loss of X repeats. With probability $p$, $X$ is equal to one (this corresponds to the SMM), and with probability $1 - p$, $X$ follows a geometric distribution ($g_k$) defined as $\text{Proba}(X = k) = C\alpha^k$, in which a specified variance $\sigma^2 = \alpha/(1 - \alpha)^2$

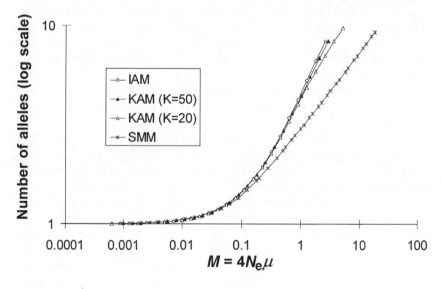

**Fig. 5.1** Relationship between the parameter $M = 4N_e\mu$ and the number of alleles at a locus in a sample of 50 genes taken from a population at mutation–drift equilibrium. This relationship is given for three different theoretical mutation models: the infinite allele model (IAM), the stepwise mutation model (SMM), and the *K*-allele model (KAM) with the number of possible allelic state (*K*) equal to 20 or 50 (see text for details about these theoretical mutation models). $N_e$ is the effective population size and $\mu$ the mutation rate of the locus.

determines the value of $\alpha$, and $C$, the normalization constant, is chosen so that $\sum_{k=1}^{\infty} g_k = 1/2$. The SMM, the TPM and *a fortiori* the IAM place no upper or lower limit on the number of repeat units in an allele.

Although rarely cited in the microsatellite literature, another classical model, called the *K-allele model* (KAM), could also be considered for microsatellite loci. This model has been used as an intermediate step in various computations concerning the IAM (Crow and Kimura 1970). Under this model, there are exactly $K$ possible allelic states, and any allele has a constant probability $[\mu/(K-1)]$ of mutating towards any of the other $K-1$ allelic states. The IAM corresponds to the KAM in which $K$ is infinite. Due to size constraints which seem to act on microsatellite loci, the KAM, in which the allelic states are consecutive positive integers (corresponding to numbers of tandem repeats), seems to be more realistic than the IAM.

## 5.2.2 Direct information from population data

Direct observations of the distribution of microsatellite alleles in population samples can be deceptive regarding inferences about the mutation model of these markers. It has been suggested that such distributions reflect an SSM because microsatellite alleles in large scale population studies often differ by multiples of the repeat unit, with most multiples represented. This conclusion is inappropriate, because a TPM with a limited proportion of mutations involving several repeats, or a KAM with a small value of $K$, could give a similar allelic pattern. In contrast, the irregular multimodal distributions observed at some loci were initially regarded as inconsistent with the SMM, and were thought to arise via a model which generated larger mutations. However, coalescence theory has established that, in a population of constant size, internal branches of a gene sample phylogeny are, on average, longer than external branches, and hence can receive more mutations. Even under the SMM, this can produce multiple-mode length distributions (Valdes *et al.* 1993; Shriver *et al.* 1993).

The best direct information about the mutational modalities of microsatellites is obtained by the analysis of alleles differing by a single mutational event. The distributions of mutation sizes observed in pedigree studies were found to comply with the SMM or the TPM (e.g. Weber and Wong 1993). However, the number of mutation events observed per locus is generally very low, and consequently most of the conclusions are based on averages across a large collection of loci, which may mask significant variation among them in mutation properties. Also, most studies characterising microsatellite mutations in human pedigrees have employed transformed lymphoblastoid cell lines, which make the distinction between somatic and germline mutations difficult (Banchs *et al.* 1994; Jeffreys and Neumann 1997; but see Weber and Wong 1993). A large set of spontaneous germline mutations was recently described at a hypervariable tetranucleotide repeat locus in the barn swallow *Hirundo rustica* (Primmer *et al.* 1996). Among the 34 mutation events observed, six (18%) involved changes of more than one repeat unit (range of changes = 2–5 repeats, mean change = 2.67 repeats). This result is incompatible with a strict SMM, but complies with a TPM. It is worth noting that this data set allowed the first direct estimate of two important parameters of the TPM model: the proportion and the distribution of multistep events. Similar analyses at additional loci are necessary to evaluate whether this result could be extended to other loci.

When mutations are not directly observable, information can be gained from the analysis of populations which have experienced a severe bottleneck during historic times. Microsatellite loci are not at mutation–drift equilibrium in these populations, but rather undergo a process of regenerating their genetic variability. In such populations, most alleles (i.e. all alleles but one or two) should be derived by a single mutational step from the most frequent allele which was supposedly conserved throughout the bottleneck. It is indeed unlikely that rare alleles constitute obligatory intermediate mutational stages. This reasoning was applied to an imperfect dinucleotide microsatellite scored in West European honey-bee populations (Estoup et al. 1995a). In these populations the rare alleles around the most frequent allele exhibited large variance in repeat number, suggesting the occurrence of large-size mutation events at this locus. It is worth mentioning that at least some of the rare alleles around the most frequent allele may have been carried through the bottleneck. Hence, this approach is probably biased towards overestimation of the proportion of large-size mutation events. In an analogous way, the microsatellite genotypes observed in parthenogenetic populations of Sitobion aphids were assumed to share a recent ancestor, because of the recent introduction of a single strain of this species in Australia (Sunnucks et al. 1996). Based on this assumption, it was suggested that five of the seven inferred mutational changes were by a single repeat unit, one by three units, and one by ten units.

The theoretical mutation models discussed above have assumed that allelic differences are entirely due to changes in the number of the basic repeat unit. The simple observation of allele size distributions indicates that microsatellite allele sizes do not always represent an increment of an entire number of repeats, suggesting that other forms of mutational change occur. In agreement with this, the sequencing of microsatellite alleles has shown that insertions and deletions also occur in the flanking sequences, especially when different species are compared (FitzSimmons et al. 1995; Garza et al. 1995; Rico et al. 1996; Angers and Bernatchez 1997; Van Treuren et al. 1997), and more rarely within species (Grimaldi and Crouau-Roy 1997). These mutation events make it difficult to compare microsatellite alleles strictly as differences in repeat number. This brings into question, for at least some loci, the validity of statistical tests of theoretical mutation models based on the comparison of expected and observed variances in repeat numbers, and the adequacy of genetic distances taking into account allele-size differences.

### 5.2.3 Statistical tests of theoretical mutation models

Two different approaches have been used to evaluate the adequacy of theoretical mutation models for microsatellite allelic distributions observed in population samples.

The first approach is based on the comparison of observed and expected values of the number of alleles and/or the heterozygosity under each mutation model (Deka *et al.* 1991; Edwards *et al.* 1992; Shriver *et al.* 1993; Estoup *et al.* 1995a). In a closed population at mutation–drift equilibrium, the heterozygosity ($H$) is a known function of the parameter $M = 4N_e\mu$ under the IAM $\{H = M/(1+M)\}$ and the SMM $\{H = 1 - (1 + 2M)^{-0.5}\}$. Under the IAM, the number of alleles ($k$) in a sample of $n$ genes has a known expectation, also depending on $M$ $\{E(k) = \sum_{j=0}^{n-1} M/(j + M)$, Ewens 1972$\}$. Under the SMM, we have only an approximate formula (Estoup *et al.* 1995a). However, from the number of alleles (the heterozygosity) in a sample of genes, it is possible to compute $M$ for each mutation model and use this estimated value of $M$ to compute the expected heterozygosity (the number of alleles). The adequacy of each mutation model is evaluated by comparing the observed and expected values of either or both parameters ($H$ and $k$). This comparison was first made in a qualitative way (Deka *et al.* 1991). Subsequently, using Ewens' (1972) sampling theory (based on the IAM), Edwards *et al.* (1992) introduced a first quantitative aspect by computing the standard error of the expected number of alleles. Additional improvements were then performed by Shriver *et al.* (1993), who assessed five and 95 percentiles under the SMM, and Estoup *et al.* (1995a), who computed $P$-values for the number of alleles under the IAM and SMM respectively. Using the simulation of the coalescent process, Cornuet and Luikart (1996) developed a method for detecting recent bottleneck events based on the comparison of observed and expected heterozygosities under the IAM and the SMM. This method also provides a way of comparing and testing the two mutation models, if loci are assumed to be at mutation–drift equilibrium.

In the second approach, the population × locus parameter considered is the variance of the number of repeats. This variance ($V$), which has no meaning under the IAM, has been shown to be a simple function of the parameter $M$ ($V = M/2$) under the strict SMM and at mutation–drift equilibrium (Moran 1975). The distribution of this variance in a set of actual microsatellite loci and gene samples is compared to its distribution in 1000 gene samples of the same size generated under the SMM (Valdes *et al.* 1993). The test used by DiRienzo *et al.* (1994) is original in two ways: (i) they considered the human Sardinian population to have expanded in the past (and consequently away from mutation–drift equilibrium), and (ii) they compared the strict SMM with the TPM. Taking parameters (mutation rates, proportion of strict SMM, and variance of the geometric distribution for the TPM) compatible with the observed variance in repeat numbers, they established confidence intervals for the homozygosity and the frequency of the most common allele through computer simulations based on coalescence theory.

The power for rejecting a given model still needs to be assessed for both approaches. The first category of tests, based on the relationship between the number of alleles and heterozygosity, has a low power for distinguishing the extreme models (SMM vs. IAM) (Jean-Marie Cornuet unpublished results). The power diminishes further when attempting to discriminate intermediate models such as the KAM or the TPM. However, the power of such tests strongly increases when considering populations that are not at mutation–drift equilibrium, in particular for populations undergoing demographic expansion (Jean-Marie Cornuet unpublished results). The second category of tests, based on the comparison of expected and observed variances in repeat numbers, appears to be more powerful than the first category of tests for discriminating between the SMM and the KAM (Jean-Marie Cornuet unpublished results). However, their power remains low when the SMM is compared with a TPM with usual probabilities of a single repeat addition or deletion ranging from 0.80 to 0.95. Similarly to the first category of tests, an increase in power is also expected when analysing non-equilibrium conditions.

## 5.2.4 Application of statistical tests to actual population data

In one of the first population studies on variation at human dinucleotide microsatellites, Edwards *et al.* (1992) concluded that the predictions of the SMM were somewhat closer to the observed allelic distributions than predictions made using the IAM. Subsequently, Valdes *et al.* (1993) found that SMM simulation results globally fit allele frequency distribution at 108 human dinucleotide microsatellites. However, Shriver *et al.* (1993) showed that if the evolution of human microsatellites with 3–5 bp repeat units was described well by the SMM, this was not the case for microsatellites with 1–2 bp repeat units. These authors also showed that minisatellite markers, that is, repeated sequences with larger repeat units (15–70 bp), strongly deviate in the direction of the IAM. Consistent with Shriver *et al.*'s (1993) conclusions, Deka *et al.* (1995) concluded that an SMM model may not be appropriate for all the dinucleotide repeats of their study. Taking into account the demographic history of the populations, DiRienzo *et al.* (1994) found that a TPM, in which most mutations are one-step and a small proportion are multistep (0.05–0.20), provided a better fit than did an SMM to the data for most of the ten human dinucleotide microsatellites considered. Similar conclusion were reached by Estoup *et al.* (1995*b*) for an imperfect dinucleotide honey-bee microsatellite, either under the hypothesis of a stable or growing population. Estoup *et al.* (1995*a*) also found that the IAM could not be ruled out for any of seven honey-bee microsatellite loci. Interestingly, four out of the seven loci had core sequences composed of repeats of two or even three different repeat lengths, which is likely to preclude a unique stepwise mutation process as postulated under the SMM.

In summary, the application of the above statistical tests to actual microsatellite population data has arrived at rather contradictory or inconclusive results.

This presumably reflects the involvement of more complex mutation processes than those assumed for the theoretical mutation models tested (see following sections, this chapter). It also reflects the low power of these tests, and the fact that basic assumptions such as mutation–drift equilibrium are not always met, due to population-size fluctuation. These statistical tests also suggest that large differences in the mutation process exist among loci. Specifically, a potential influence of molecular features of microsatellite markers such as the length and composition of the repeated motif is suggested (Jin 1994; Estoup *et al.* 1995*b*). Further analyses, based on more powerful tests and a higher number of loci, are needed to confirm and investigate the variation in the mutation model of microsatellites in relation to their structural features.

## 5.3 Estimates for size homoplasy from natural populations

### 5.3.1 Definition

Originally, homoplasy was a concept used by evolutionists to characterize the fact that a given character present in two species is not derived from the same character in a common ancestral species, but that the similarity is due to factors such as convergence, parallelism or reversion. Homoplasy is therefore considered as the 'noise' in opposition to its corollary, homology, which is the evolutionary 'signal' (Scotland 1992). Transposed to the gene level, we can say that two alleles are homoplasic when they are identical in state, though not identical by descent.

Microsatellite variation is essentially revealed through electrophoresis of PCR products, and allelic classes differ by the length (in bp) of the amplified fragments. Two PCR products of the same length may not be copies without mutation of the same ancestral sequence, introducing the possibility of size homoplasy. The occurrence of size homoplasy is tightly linked to the way mutations produce new alleles, and hence to the mutation model. If the mutation model is assumed to be the IAM, there should not be any homoplasy, because any new allele created by a mutation is distinct from the existing alleles in the population. All other mutation models (e.g. the SMM, TPM and KAM) can generate size homoplasy. Size homoplasy is also dependent on the mutation rate of the locus and of the time of divergence between populations. Intuitively, the occurrence of homoplasy will increase with the mutation rate and the time of divergence.

A substantial amount of size homoplasy is expected at most microsatellite loci for at least three reasons: (i) there is both direct and indirect evidence that microsatellites evolve in a stepwise fashion, and fit either an SMM, a TPM or a KAM, even if an IAM cannot always be rejected (see Section 5.2.4); (ii) microsatellite loci are often characterized by high mutation rates, between $10^{-2}$ and $10^{-5}$ mutations per locus per generation in humans (Weber and Wong 1993); and (iii) selective constraints act on the range of allele size, reducing the number of possible allelic states and hence favouring size homoplasy (Nauta and

Weissing 1996; Feldman *et al.* 1997). When a locus strictly follows an SMM, size homoplasy is taken into account by several distance measures (Goldstein *et al.* 1995*a,b*; Shriver *et al.* 1995; Slatkin 1995; Rousset 1996; Feldman *et al.* 1997). This is not the case, however, when a locus follows a TPM or a KAM with unknown parameters (proportion of multistep mutations and variance of the geometric distribution for the TPM, and number of possible alleles for the KAM) and/or when allele size constraints exist.

## 5.3.2 Experimental analysis of size homoplasy

### Detection through electromorph sequencing

A length-class of microsatellite alleles potentially consists of a mixture of alleles identical by descent (homology) and of alleles identical by state (size homoplasy). For *perfect microsatellites*, sequencing the core region cannot provide further information than simply determining the allele size. Two different approaches have been proposed for detecting size homoplasy within species, both based on the fact that microsatellite electromorphs are not always pure stretches of the same motif. Estoup *et al.* (1995*b*), Angers and Bernatchez (1997) and Viard *et al.* (1998) used tandem repeats interrupted by one or more additional or substituted nucleotide(s) (*interrupted* or *imperfect microsatellites*). These interruptions introduce another level of polymorphism, and alleles with the same size may differ by the number and/or the location of interruptions. Garza and Freimer (1996) and Viard *et al.* (1998) determined the core sequence of *compound repeat loci* (composed of stretches of different repeat motifs) to detect differences between identity in size and identity in sequence. Homoplasy at microsatellite loci may also be detected by looking for variation in the flanking regions, such variation being commonly observed among species and occasionally within species (Blanquer-Maumont and Crouau-Roy 1995; Garza and Freimer 1996; Grimaldi and Crouau-Roy 1997). Hence, electromorph sequencing can reveal some cases of size homoplasy. However, one should distinguish between the occurrence of size homoplasy and its detectability. Some microsatellite alleles, depending on where the mutations are likely to occur (e.g. on only one side of interruptions or on both for interrupted loci) will allow easier detection of size homoplasy. It must be also stressed that identical sequences may be identical in state but not identical by descent.

In the following, 'electromorphs' will refer to PCR products of a given size, and 'alleles' to electromorphs exhibiting the same sequence.

### Application to actual population data

Estoup *et al.* (1995*b*) established the number and position of interruptions at one imperfect locus for electromorphs from closely and distantly related populations of the honey-bee *Apis mellifera* (Hymenoptera). No sequence difference was evident when electromorphs originated from the same population, from populations belonging to the same subspecies, or from closely related subspecies. In contrast, sequence differences were often detected in distantly related subspecies, revealing that size homoplasy frequently occurs at this level of population differentiation.

Similar analyses were recently extended to additional loci and electromorphs in the honey-bee, the bumble bee *Bombus terrestris* (Hymenoptera), and the freshwater snail *Bulinus truncatus* (Gastropoda) (Viard *et al*. 1998). In this study, size homoplasy was frequently detected among populations and more rarely within populations. Variation in the amount of size homoplasy detected among loci was observed. This could be a reflection of variation in the mutation process (i.e. different mutation models and/or mutation rates) and/or in the evolutionary history of populations and species. It may also reflect differences in the detectability of size homoplasy due to differences of the molecular structure of the core sequence among loci, in particular in the distribution and the length of the interruptions. Variation in the amount of size homoplasy detected among electromorphs of the same locus was also observed. The number of alleles per electromorph was globally correlated with the size (in bp) of the variable region, suggesting that large electromorphs contain more hidden alleles than shorter ones. This supports the idea that larger electromorphs are less stable than shorter ones (Primmer *et al*. 1996). However, the large difference in the amount of size homoplasy between some electromorphs is unlikely to result only from their difference in size and in detectability. It may reflect different coalescence times between the copies of the electromorphs. Unspecified allele-specific factors could be also be involved.

Viard *et al*. (1998) found that the detection of size homoplasy through electromorph sequencing had a substantial effect on the resolution of population structure. More single locus pairwise tests of population differentiation were significant when alleles rather than electromorphs were considered, and non-stepwise estimators of genetic differentiation (see Feldman *et al*., this volume) increased in the three species studied. The sequencing of more copies per electromorph would refine this result, but this would be extremely labour-intensive. Electromorph sequencing is not adapted to routine analyses of populations, and should be reserved for particular cases in which identity of alleles is crucial, for example, in some studies of association and linkage disequilibrium (Hastbacka *et al*. 1992; Freimer and Slatkin 1996; Grimaldi and Crouau-Roy 1997), and of genetic introgression between divergent forms (Arnaud Estoup, unpublished results). Alternatively, less time-consuming techniques, such as SSCP (single-strand conformation polymorphism; Orita *et al*. 1989; Hayashi 1991), could be considered to reach this goal.

## 5.4 Correlation between structural features and level of polymorphism

The level of polymorphism varies widely among microsatellite loci, indicating that the mutation rate also varies substantially. Several studies have attempted to determine the structural factors that would predict the level of variability at a microsatellite locus.

## 5.4.1 Variability in relation to size and composition of the repeat unit

Chakraborty *et al.* (1997) tested the influence of the size of the repeat unit on the mutation rate by estimating theoretically the relative mutation rates of different categories of microsatellite loci. The method applied ANOVA to the within-population distributions of repeat counts at microsatellite loci, grouped according to the size of their repeat unit. All microsatellite loci were assumed to mutate following a generalized stepwise mutation model (see Chakraborty and Kimmel, this volume, for a more detailed treatment). Allelic data were obtained from different sets of di-, tri-, and tetranucleotide loci typed in several human populations. Results indicated that although substantial interlocus variation of mutation rates within each group of loci existed, on average microsatellite loci had mutation rates inversely related to the size of their repeat unit. Dinucleotide loci appeared to be evolving at a rate 1.5–2 times greater than the tetranucleotide loci, and the non-disease-related trinucleotide loci had a mutation rate in between di- and tetranucleotides. Interestingly, the disease-related trinucleotide loci appeared to have a mutation rate 3.7–6.9 times higher than that of tetranucleotide loci, and even higher than the dinucleotide loci. The last result was obtained using repeat-count frequency from unaffected healthy individuals, so that at each of the disease-causing trinucleotide loci analysed all alleles were within the normal size range.

The conclusions of Chakraborty *et al.* (1997) contradict the studies based on direct observations of mutations, which suggest a higher mutation rate for tetranucleotide loci (Hastbacka *et al.* 1992; Weber and Wong 1993; Zahn and Kwiatkowski 1995). However, this may be the result of a limited and nonrandom sampling of tetranucleotide loci in direct mutation assays. The statistical tests mentioned in Section 5.2.4 of this chapter suggested a potential influence of microsatellite structural features, in particular the size of the repeat unit, on the mutation process. The variances in number of repeats estimated by Chakraborty *et al.* (1997) are proportional to the product of the mutation rate and the variance of the repeat number changed by mutation. Therefore, the differences in the relative mutation rate between di-, tri-, and tetranucleotides could reflect, at least partly, differences in the modalities of mutation for these three groups of loci.

For a given size of repeat unit, the level of microsatellite polymorphism also seems to depend upon the composition of the repeat unit. This has been suggested by results of analyses of polymorphism on a large variety of tri- (Gastier *et al.* 1995), and anonymous tetranucleotide loci (Sheffield *et al.* 1995). The biological basis for differences in stability of tracts of repeat units with similar size and different nucleotide composition remains unclear, although GC content appears to be at least one factor, since the most polymorphic tri- and tetranucleotide motifs are AT-rich. However, this interpretation may be the result of a systematic selection against clones containing plasmids with long GC-rich sequences (Gastier *et al.* 1995).

## 5.4.2 Variability in relation to repeat count and length purity

The first parameter that influences microsatellite stability independently of the repeat type is the repeat count. Weber (1990) found that the polymorphism of perfect $(CA)_n$ markers increased with increasing average number of repeats. Human $(AAT)_n$ microsatellites also showed a trend toward more polymorphic loci as the repeat count in the cloned allele increased (Gastier *et al.* 1995). In agreement with this, a weak but positive correlation between some measures of diversity and repeat count was found in several population studies (Beckmann and Weber 1992; Ostrander *et al.* 1993). In *Drosophila*, Goldstein and Clark (1995) found a stronger correlation between the maximum repeat count and variance than between the mean and variance, suggesting that the mutation rate does indeed increase with repeat count, but not necessarily smoothly. Hence, microsatellite loci with higher repeat counts appear to be associated with higher mutation rates, presumably because the opportunity for a stable misaligned configuration is greater for longer repeat tracts.

The second parameter that substantially influences microsatellite stability independently of the repeat type is the 'purity' of the tract. That is, whether the tract contains different repeat types (compound repeat) and/or an inserted base(s) separates motifs of the microsatellite (interrupted repeat), and/or one of the motifs has been mutated (imperfect microsatellite) (see glossary, this volume). The longest run of perfect repeats was found to be the best predictor of polymorphism for compound and interrupted $(CA)_n$ microsatellites (Weber 1990). Interrupting bases appear to stabilize the tract of repeats, reducing the possibility of misalignment and resulting in lower levels of polymorphism than uninterrupted microsatellites with a similar number of repeats (Chung *et al.* 1993; Richards and Sutherland 1994; Pépin *et al.* 1995). This phenomenon partly explains differences in polymorphism among loci as well as differences in polymorphism at the same locus among species and/or among populations within species. To assess the effect of interruptions on population diversity, Goldstein and Clark (1995) compared the variance of dinucleotide repeats with interruptions to those without using an analysis of covariance. Interestingly, their results suggested that, when an interruption split a microsatellite sequence, the mutation rate in the larger uninterrupted tract of repeats is somewhat reduced by the presence of the smaller uninterrupted tract on the other side of the interruption. The data set is, however, too restricted to assess the generality of this unexpected result.

## 5.4.3 Differences in mutability among alleles at the same locus

The above relationships between the variability, repeat count, and tract purity should also hold for the mutability of alleles at the level of an individual microsatellite locus.

Congruent with this, the distribution of germline mutations at a single hypervariable barn swallow tetranucleotide repeat locus indicated that mutation events are biased towards alleles with large repeat counts (Primmer *et al.* 1996). This case can be compared to the expansion of trinucleotide microsatellites associated with human diseases (Caskey *et al.* 1992), and of some human tetranucleotide loci (Talbot *et al.* 1995). Studies of the trinucleotide microsatellites associated with two human diseases, the fragile X (Eichler *et al.* 1994) and the spinocerebellar ataxia type 1 (Chung *et al.* 1993), have further suggested that interruptions within the repeat array reduce the rate of tract instability. Studies in bacterial systems of CTG repeats from human disease genes have also shown that alleles with large repeat counts are more prone to mutation events than shorter ones (Kang *et al.* 1995). In a yeast system, large mutation events were found to delete or duplicate the interruption(s) within the repeated region, and consequently to change the stability of a dinucleotide repeat array (Petes *et al.* 1997). Finally, a phylogenetic approach was developed in humans (Jin *et al.* 1996) to study the pattern and rate of mutations at a single dinucleotide microsatellite locus tightly linked to an HLA locus. Results indicate that mutation rate varies drastically among alleles within this locus, and correlates positively with the repeat count, rather than with population factors such as the age of the HLA allele lineages, the effective population sizes, or a recent population expansion. In particular, some microsatellite alleles seemed never to have mutated over long periods of evolutionary time. Sequencing of these alleles suggested that they had lost their ability to mutate as a result of nucleotide substitutions that reduce the repeat count of uninterrupted repeat arrays.

## 5.5 Mutational biases, allele size constraints and evidence for selection

In addition to the dependence of the mutation process on allele size and purity, the theoretical mutation models outlined in Section 5.2 do not consider several factors which have recently been identified and which are thought to be relevant to the evolution of microsatellites.

### 5.5.1 Asymmetry in the distribution of mutations

Studies characterizing microsatellite mutations in human pedigree collections suggest that mutations at microsatellite loci involve more gains than losses of repeats (Weber and Wong 1993; Banchs *et al.* 1994; Talbot *et al.* 1995). However, only one-third of the scored mutations were definitely germline in origin, and corrected mutation data, while still showing an excess of gains, were insufficient to achieve statistical significance. By combining Weber and Wong's (1993) data with new human germline mutation data, Amos *et al.* (1996) confirmed a significant bias in favour of expansion (21 expansions vs. nine contractions; unilateral binomial test, $p = 0.021$). Additional empirical evidence was provided by Primmer *et al.* (1996), who observed significantly more gains than losses of repeat units (26 gains vs. seven losses; unilateral binomial test, $p = 7 \times 10^{-4}$) in a large set of germline mutations at a single hypervariable barn swallow tetranucleotide locus.

It remains to be determined whether asymmetry in the distribution of mutations occurs in all types of microsatellites, or specifically in the evolution of hypervariable and long microsatellites. Interestingly, such a mutational bias would reduce the probability of fixation of microsatellite loci by preventing their evolution to a low mutating state corresponding to very short repeat arrays ($n < 3$) (Tachida and Iizuka 1992; Garza *et al.* 1995; Samadi *et al.* 1998). It may also facilitate the emergence of polymorphic microsatellite loci from very short repeat arrays (Tachida and Iizuka 1992). The molecular mechanism resulting in this upwardly biased mutation process remains unclear.

## 5.5.2 Constraints on allele size

A mutation process dependent on the repeat count, combined with a propensity for gaining rather than losing repeats, would theoretically promote an expansion of microsatellite arrays towards a large (virtually infinite) number of repeats. Although a few large repeat arrays have been found in telomeric regions (Wilkie and Higgs 1992), most microsatellite loci have a finite size generally shorter than a few tens of repeat units. This strongly suggests that there must be some size constraints restricting the expansion of repeat arrays. To date, there is no direct evidence for selective constraint acting on allele length at microsatellite loci; however several pieces of evidence have recently been suggested. Bowcock *et al.* (1994) found that the allele size variance ratio between non-human primate and human populations was significantly lower than expected when considering the time of divergence between human populations and between primate species. Congruent with this, Garza *et al.* (1995) found that the difference in average repeat count at several loci between humans and chimpanzees were sufficiently small that there might be a constraint on the evolution of average repeat count. In order to account for the observed repeat count similarities, a biased mutation model was developed by Garza *et al.* (1995), with small alleles tending to increase in size while large alleles tend to decrease in size. This model showed that for some loci a weak bias can account for the observed repeat count similarities. However, loci with small differences in average repeat counts would require a substantial bias, sometimes much larger than the mutation rate itself.

Several alternative mechanisms were proposed for counteracting the elongation of microsatellite arrays. The predisposition of small gains over losses may be balanced by rare large deletions (Weber and Wong 1993; Garza *et al.* 1995). However, this in itself would invoke a mutational bias which, given the rarity of large jumps observed in germline mutations, would have to be extremely strong (Amos *et al.* 1996). Because large alleles are strongly counter-selected at loci associated with genetic diseases (Sutherland and Richards 1995), it was also argued that selection may act as a upper truncating mechanism, imposing a ceiling on alleles with large repeat counts (Feldman *et al.* 1997; Samadi *et al.* 1998). Alternatively,

Samadi *et al.* (1998) suggested that selection may act on the difference in repeat count between the two alleles borne by homologous chromosomes in a diploid individual. This last hypothesis derives from the idea that microsatellites may ensure a correct alignment of homologous chromosomes at meiosis (Pardue *et al.* 1987) and that large differences in repeat count would destabilize chromosome pairing, provoking abnormal recombination in the regions flanking microsatellites.

If dependent on the absolute size (in base pairs) of repeat arrays, constraints on the repeat count may be stricter when the size of the repeated motif increases. This hypothesis has the advantage of reconciling the apparently high mutation rates in tetranucleotides observed in direct studies (Hastbacka *et al.* 1992; Weber and Wong 1993; Zahn and Kwiatkowski 1995), with the low variance in repeat count calculated by Chakraborty *et al.* (1997). It may also explain the observation of lower $F_{st}$ values for tetranucleotide repeats (Jorde *et al.* 1995).

### 5.5.3 Other mutational biases and selective factors

The analysis of the parental genotypes of human families for which germline mutations have been detected suggested that mutation is more likely to occur in heterozygous individuals whose alleles differ greatly in repeat count (Amos *et al.* 1996; and Amos, this volume). In contrast to this observation, no significant difference in allele length was observed between mutant and non-mutant parents at a single hypervariable locus in barn swallows, when alleles of the same repeat count classes were compared (Primmer and Ellegren, personal communication). If the difference in repeat count between alleles in a heterozygote were to increase instability relative to a homozygote, then mutation rate could correlate with heterozygosity, and loci in larger populations would evolve faster than those in smaller ones (Amos *et al.* 1996). This has currently not been observed at a population level, though Rubinsztein *et al.* (1995*a,b*) suggested that microsatellites cloned from humans were, on average, longer than their chimpanzee homologues, and that this could be a result of population size. However, it has been argued that microsatellite loci will tend to be longer and more polymorphic in the species they were cloned from, as a result of biases for longer repeat arrays during the isolation procedure (Ellegren *et al.* 1995; but see Amos, this volume). Results of a reciprocal comparison of the repeat counts of microsatellites loci in cattle and sheep using markers derived from bovine genome as well as the ovine genome are in agreement with Ellegren *et al.*'s (1995) claim (Ellegren *et al.* 1997). Hence, comparisons of microsatellite evolution between species may be flawed unless they are based on reciprocal analyses or on genuinely random selection of loci with respect to repeat count. In addition to repeat count control, it is also recommended to control the structure of the repeated region when making comparison among species. Substantial discrepancies indeed potentially occur at this taxonomic level (Garza *et al.* 1995; Primmer and Ellegren, 1998; Arnaud Estoup, unpublished results).

A higher mutation rate in paternal meioses has been reported from multi-locus mutation data (Weber and Wong 1993) as well as at the level of a single microsatellite locus (Primmer and Ellegren, personal communication). A similar trend was observed at trinucleotide microsatellite loci associated with two human diseases, Huntington's disease (Duyao *et al.* 1993) and the spinobulbar muscular atrophy (La Spada *et al.* 1992; Rubinsztein, this volume). The rationale behind a male-biased mutation rate is the large difference in the number of cell divisions between spermatogenesis and oogenesis (Vogel and Motulsky 1997). However, the influence of sex on the mutation rate seems to vary among loci. Uniform mutation rates with respect to sex were found at other non-disease-related loci (Talbot *et al.* 1995) and a mutational bias in favour of maternal meiosis was found at the trinucleotide locus associated with the fragile X syndrome (Fu *et al.* 1991).

Intuitively, di- or tetranucleotide repeat arrays have generally been considered as non-coding neutral DNA, because these sequences, and the polymorphism they display, would disrupt any open reading frame. At present, there is no direct evidence for a general role of microsatellite loci in eukaryotic genomes. However, several types of functional significance associated with simple sequence repeats have been reported in the literature (reviewed in Garza *et al.* 1995; Rico *et al.* 1996; for details see chapters by Hancock, and Kashi and Soller, this volume) suggesting that selective constraints may directly act on the evolution of at least particular microsatellite loci.

## 5.6 Conclusion and perspective

The study of the evolutionary processes affecting microsatellite loci is still in its infancy. However, the increasing number of experimental and theoretical studies on this subject have identified several factors relevant to the evolution of microsatellites. These studies have shown that mutations at microsatellite loci generally involve addition or loss of one repeat, and much less frequently of several repeat units. They have also indicated that (i) the mutation of repeat arrays has a complicated dependence on allele size and purity; (ii) the mutation process is upwardly biased; and (iii) some constraints on allele length exist either due to selection or to the mutation process. It is likely that these allele-dependent processes, biases, and size constraints vary from one microsatellite locus to another, in particular depending on the size and composition of the repeat unit. The sequencing of imperfect, interrupted and/or compound microsatellite electromorphs revealed a substantial amount of size homoplasy, a result expected when considering the mutation processes of repeat arrays.

Theoretical mutation models which represent more accurately the evolutionary processes of microsatellites are needed, to obtain better estimates of population differentiation measures and demographic parameters inferred from within population variation. Regarding measures of population differentiation, the widely used SMM may provide adequate measures where closely related populations are considered (Takezaki and Nei 1996), but this simplistic model becomes clearly inadequate beyond a critical level of divergence and/or with the use of particular types of repeat arrays, such as compound microsatellites. The factors relevant to the evolution of repeat arrays are progressively incorporated into theoretical mutation models. Garza *et al.* (1995) incorporated allele size constraints by developing a model in which small and large alleles tend to mutate upward and downward respectively. Nauta and Weissing (1996) and Feldman *et al.* (1997) considered explicit allele size limits, and described the expected dynamics of a locus mutating under the SMM with a restricted set of alleles. Kimmel *et al.* (1996) and Kimmel and Chakraborty (1996) included in their mutation model the possibility of multistep and directionally biased changes in allele size. The dependence of the mutation rate on the repeat count and on the purity of alleles has still not been incorporated in theoretical mutation models.

However that may be, the general level of knowledge of the mutation processes at microsatellite loci is currently insufficient to allow adequate representation in theoretical mutation models. In addition, we often do not have the data to parametrize accurately the factors that have been found to be relevant to the evolution of these repeated sequences (proportion and range of multistep mutation, asymmetry in the distributions of mutation, dependence of the mutation rate on the repeat count and purity of alleles, constraints on allele size). Finally, it is worth stressing that the large variance in the mutation parameters among loci will be a persistent problem. In that respect, the selection of loci with similar mutation parameters appears to be crucial for accurate estimation of population parameters. The combination of empirical population and molecular studies, and the statistical testing of theoretical models as described in Section 5.2, should contribute to the realization of this objective.

# Acknowledgements

We thank Hans Ellegren and Craig Primmer for sharing unpublished results with us, David Goldstein, Hans Ellegren, Philippe Jarne and Michel Solignac, and one anonymous referee for their helpful comments, and Amanda Brooker and Jacqui Shykoff for their invaluable help in correcting our English. This work was partly supported by a grant from the Bureau des Ressources Génétiques, appel d'offre 'Recherches méthodologiques pour l'amélioration des processus de gestion et de conservation des ressources génétiques animales, végétales et microbiennes'.

# 6 A comparative approach to the study of microsatellite evolution

William Amos

## Chapter contents

## Abstract

This chapter examines microsatellite evolution in the light of comparative and empirical data. Sources include limited numbers of germline mutations from humans and birds, large numbers of human allele frequency distributions, studies where loci cloned from one species are amplified in related species and data from studies on a wide range of taxa which yield mean microsatellite lengths and frequency per kilobase of genome. These data are used to examine a range of microsatellite characteristics including: the bite size of the mutation process, whether mutations are biased in favour of expansion, whether heterozygotes are more likely to mutate than homozyogtes, whether maximum repeat number is constrained by an absorbing or reflecting boundary and the extent to which the length and frequency of microsatellites in the genome can be predicted from a species' characteristics.

## 6.1 Introduction

DNA sequences evolve through mutations, which create novelty, followed by the actions of neutral genetic drift and natural selection, which determine whether and how fast new forms spread within and between populations. Of these three fundamental processes, drift and selection are now reasonably understood through the works of pioneering figures such as Fisher and Wright. By contrast, the nature of the mutation process, as it operates in natural populations, remains largely unknown, because individual events occur too infrequently to be studied directly. Consequently, much of our interpretation of the patterns of genetic variability in nature is unavoidably based on blind assumptions about the forces which spawn genetic novelty.

Fortunately, several recent developments in molecular biology offer opportunities to learn more about how mutations come about. First, many short tandem repeat markers have mutation rates high enough that significant numbers of germline mutations can be identified in pedigrees (Wong *et al.* 1986; Jeffreys *et al.* 1988; Weber and Wong 1993). Second, the advent of PCR, with its ability to amplify DNA from as little as a single target molecule, has allowed the development of methods for identifying *de novo* mutant molecules in DNA isolated from sperm. This approach has proved particularly powerful in elucidating the mutation processes affecting minisatellite loci (Jeffreys *et al.* 1994; Monckton *et al.* 1994). Third, the combination of high throughput genotyping systems and the ability of many microsatellite markers to amplify across species has resulted in large amounts of data, both in terms of information from hundreds or even thousands of loci within key species such as humans, mice and domestic animals, and in terms of smaller numbers of loci being typed across related species. Opportunities for comparative analyses improve almost daily.

Microsatellite mutations provide an interesting testing ground. It would be easy to argue that a simple null model consisting of a symmetric stepwise mutation process is unlikely to be far wrong. Indeed, computer simulations show a reasonable fit between this model and empirically derived human allele frequency distributions (Valdes *et al.* 1993; Di Rienzo *et al.* 1994). Such simplicity has also allowed the development of rather precise formulations which relate microsatellite divergence to evolutionary time, with Goldstein and colleagues showing elegantly that their measure delta mu scales linearly with time since separation (Goldstein *et al.* 1995*a*). However, significant discrepancies between 'known' divergence times and microsatellite genetic distances (Deka *et al.* 1994; Garza *et al.* 1995; Valsecchi *et al.* 1997) imply that one or more of the basic assumptions of the null model may be flawed, and it has been suggested that significant numbers of loci do not obey the simplest stepwise models (Shriver *et al.* 1993).

In this chapter I summarize current empirical evidence concerning how microsatellites evolve, combining limited direct mutational data gleaned from pedigrees with comparative studies. In turn I consider the bite size of the mutation process, its symmetry, evidence for mutation rate variation and the nature of any length boundaries which might restrict allele length. I then examine the key question of whether homologous loci are longer in some species than in others.

## 6.2 Aspects of microsatellite evolution

### 6.2.1 How many repeats are gained and lost in each mutation?

The weight of evidence from both mutation studies (Banchs *et al.* 1994; Amos *et al.* 1996; Cooper *et al.* 1996; Primmer *et al.* 1996; Weber and Wong 1993) and theoretical treatments (Shriver *et al.* 1993; Valdes *et al.* 1993; Di Rienzo *et al.* 1994) looking for fit to the stepwise mutation model of Moran (1975) agree that the vast majority of mutations involve the gain or loss of one or, less frequently, two repeat units. Multi-repeat jumps probably occur, but are rare, and their true frequency is difficult to infer directly because germline mutations are so scarce. Rather more mutations have been recorded from immortal cell lines (Weber and Wong 1993) and carcinomas, but it is probably unwise to merge germline data with these, since the somatic events may show elevated rates of jump mutations (Weber and Wong 1993). Allele length distributions which carry gaps are suggestive of jumps, and will tend to give a better fit to models which invoke jumps (Di Rienzo *et al.* 1994). However gaps can also result from other processes such as the structure of the genealogical tree (see Donnelly, this volume), and when internal point mutations create allele lineages with different mutation rates, which then drift apart in length (Jin *et al.* 1996).

### 6.2.2 Symmetry and the mutation process

A key assumption of the null model of microsatellite evolution is that gain-in-length mutations occur with the same probability as loss-of-length mutations. However, empirical data do not support a symmetrical process. In both data sets containing enough germline mutations to perform statistically meaningful tests, a significant excess of gain mutations was found. Mutations at diverse human loci reveal 21 gains to nine losses ($p = 0.021$) (Amos *et al.* 1996), and at a hypervariable swallow microsatellite there were 26 gains and only seven losses ($p < 0.01$) (Primmer *et al.* 1996). Combined, these data allow rejection of a symmetric mutation process at $p = 0.00006$. The same trend is found among smaller numbers of mutations from sheep (Crawford and Cuthbertson 1996) and from human immortal cell lines (some of which are germline and others of which occurred during cell culture) (Weber and Wong 1993; Banchs *et al.* 1994).

Although it is dangerous to extrapolate directly between different classes of DNA sequence, it is interesting to note that other classes of short tandem repeat show a similar pattern. Both minisatellites and tracts of triplet repeats associated with diseases such as Huntington's disease show clear evidence of expansion (Duyao *et al.* 1993; Jeffreys *et al.* 1994; Monckton *et al.* 1994; Rubinsztein *et al.* 1994*a*). Furthermore, Gordenin and colleagues recently proposed a molecular model for how short tandem repeat mutations occur during DNA replication that includes an asymmetry that is predicted to favour expansion among trinucleotide repeats and, plausibly, among other short tandem repeats (Gordenin *et al.* 1997).

Ideally, mutation patterns should be determined by counting verifiable germline events, but the experimental effort required to achieve acceptable sample sizes is prodigious. A possible criticism of the swallow data of Ellegren and colleagues is that all the mutations come from a single, hyper-variable locus, which might be exceptional. Therefore, it is useful to support direct observations with indirect lines of evidence. In the case of microsatellite expansion, it is interesting to extend concepts developed by Jin and colleagues (1996). Alleles which become stabilized by internal point mutations should evolve more slowly, allowing all other alleles to evolve away from them in length. Such alleles will appear as single length-class outliers. If microsatellites are indeed expanding, these outliers should be mostly shorter than the remaining alleles.

Jin *et al.* (1996) present one example where alleles with low mutability are shorter than those with higher mutation rates. To see whether the pattern is general, 337 published human allele frequency distributions were examined for outlying alleles, using the following definitions which are arbitrary but *a priori* (Amos and Crawford, in prep.): (1) contiguous: a single unbroken cluster of alleles; (2) single outlier: a single allele separated from an otherwise contiguous group; (3) double outlier: two alleles, either contiguous or non-contiguous, which are displaced from, and lie on the same side of, an otherwise contiguous group of three or more alleles; (4) broken: all other distributions. A total of 122 contiguous distributions, 80 broken distributions, 101 single outliers, and 34 double outliers were identified, with short outliers being significantly more frequent than long outliers in both cases (63 : 38, $\chi^2 = 6.12$, 1 d.f., $p < 0.05$; 25 : 9, $\chi^2 = 7.53$, 1 d.f., $p < 0.01$). Furthermore, whereas every long single outlier can be explained by stochastic sampling from the edge of a distribution, being either rare ($<10\%$ frequency) or displaced by only one repeat unit from the main group, 16 of the short outliers were both frequent ($>10\%$) and displaced by more than 1 repeat unit, the difference in frequency being significant (47 : 16 vs. 38 : 0, $\chi^2 = 16.8$, 1 d.f., $p \ll 0.001$). A similar but non-significant pattern is found among the smaller number of double outliers (whose lengths were scored as that of the allele closest to the main group) (16 : 9 vs. 8 : 1, $\chi^2 = 1.98$, 1 d.f.).

Such an asymmetry, with outlying alleles tending to lie below rather than above the main group, is consistent with mutational bias favouring expansion, but there is an important proviso. Several studies have shown a correlation between heterozygosity and repeat number (Weber 1990), interpreted as indicating that mutation rate increases with allele length. If so, then a deep cleft in an allele phylogeny could by chance create two lineages, one characterized by low repeat number and low variability (plausibly a single outlier), the other by high repeat number and high variability. Only detailed simulation studies and sequencing of representative outlying alleles will show which model explains the data best. However, given that interrupted tracts are both common and show lower mutability than equivalent pure tracts, at least in humans (Weber 1990), it would be remarkable if the nearly 350 loci considered here did not include many with stranded alleles.

It is important to note that a correlation between length and heterozygosity does not demonstrate a length-dependent mutation rate *per se*. If mutation rate and heterozygosity are correlated (see below), and biased mutation favours expansion, length and heterozygosity would both tend to increase over time. This model yields virtually identical predictions to that of a length-dependent mutation rate, and it is unclear that any data set yet published gives unambiguous support to a length-dependent mutation rate. Indeed, if anything, limited circumstantial evidence favours a length-*independent* mutation rate. Thus, the rates of *in vitro* microsatellite slippage (Schlötterer and Tautz 1992) and germline minisatellite mutations (Monckton *et al.* 1994) both appear to be length-independent.

The concept of stranded alleles can be extended to between species comparisons involving monomorphic and polymorphic homologues. According to the biased mutation model, monomorphic loci will be expanding more slowly relative to polymorphic homologues in other species, and therefore would be expected to be shorter. A large sample of almost 400 polymorphic cattle microsatellites tested in sheep provide data which are consistent with this prediction. Loci which are monomorphic in sheep are ~6.8 times more likely to be shorter than their bovine homologues compared with loci which are polymorphic in both species (111 : 29 vs. 110 : 198) (Crawford *et al.* 1998). Just as with stranded alleles within a locus, a length-dependent mutation rate could provide an alternative explanation. However, as before, the fact that interrupted repeats are both common and relatively stable makes it likely that a significant proportion of monomorphic loci appear so because they have acquired internal point mutations.

## 6.2.3 Variation in mutation rate and heterozygote instability

Human microsatellite markers typed in chimpanzees are very much more likely to be longer in humans, even when both homologues are polymorphic and have overlapping allele frequency distributions (Rubinsztein *et al.* 1995*a,b*). This observation has been used to argue for both biased mutation and variation in the overall mean microsatellite mutation rate between closely related lineages (see below, and Rubinsztein *et al.* 1995*a*). One hypothesis which could explain the apparently greater mutation rate of human microsatellites involves greater mutability in the heterozygous state (Amos *et al.* 1996). Every base pair in the genome has a finite opportunity to mutate during DNA replication. However, during meiosis, extensive regions of heteroduplex DNA are formed between paired homologous chromosomes. In this state, heterozygous sites are often 'repaired' by gene conversion-like events, which would offer an extra opportunity for microsatellite slippage. If the length difference at heterozygous microsatellite loci is sufficient to initiate 'repair', then heterozygous sites would tend to mutate more frequently than homozygous sites.

Empirical support for this model can be divided into limited direct evidence and more extensive indirect evidence. Direct evidence is based on anecdotal data from sheep (Crawford and Cuthbertson 1996) and a small human data set comprising 19 informative mutations. In the human data set, significantly more mutations could be traced to the parent with the greater difference in length between his/her alleles (Amos *et al.* 1996). A second study based on mutation accumulation in homozygous inbred lines of the fruit fly, *Drosophila melanogaster*, appears to demonstrate the converse of heterozygote instability, homozygote stability (Schug *et al.* 1997*a*). Here, the average microsatellite mutation rate was estimated to be $6.3 \times 10^{-6}$, far lower than estimates from other species, which range from $10^{-5}$ to $10^{-3}$ (Edwards *et al.* 1992; Weber and Wong 1993; Banchs *et al.* 1994). The relative stability of the fly loci was interpreted by the authors in terms of the short microsatellites they used being less mutable than the generally longer loci found in mammals. However, a reduction in mutation rate through homozygosity provides an equally parsimonious explanation.

The concept of heterozygote instability is supported indirectly by a large body of molecular data, which shows that heterozygous sites may be both recognized and 'repaired' by gene conversion-like events during meiosis. Work on yeast shows that large regions of heteroduplex DNA are formed during synapsis (Nag *et al.* 1995; Collins and Newlon 1996), and that within these regions, heterozygous sites are repaired by gene conversion-like events (Borts and Haber 1989; Borts *et al.* 1990; Szostak *et al.* 1983), a proportion of which result in genetic crossovers (Chambers *et al.* 1996; Manivasakam *et al.* 1996). Similar processes have been invoked as a general feature of eukaryotes which allows homologous regions to pair and undergo crossing over (Carpenter 1994*a,b*). The ubiquity of these processes is suggested by the fact that many of the enzymes involved in the yeast mismatch repair system have direct homologues in mammals (Baker *et al.* 1995, 1996).

A test of the heterozygote instability model is provided by hybrid populations, where the meeting of dissimilar chromosomes would be predicted to increase the mutation rate. Microsatellites cloned from grey (*Halichoerus grypus*) and harbour (*Phoca vitulina*) seals tested in South American fur seals (*Arctocephalus australis*) and Antarctic fur seals (*A. gazella*) which have been seen to hybridize, show increases in both mean length (ten of 15 loci longer in the fur seals) and variability (Gemmell *et al.* 1997). Although levels of variability are difficult to compare, given the profoundly different sample sets used for each species, at three loci the number of alleles increased more than twofold, and in one case a locus with eight alleles in the grey seal had 25 alleles in *A. gazella*. An increase in both length and variability is unexpected on the basis of population size and allele mixing alone, since fur seals were hunted to the verge of extinction during the nineteenth century, and therefore should, if anything, be genetically impoverished (Bonner 1968).

Although originally proposed to account for observations at microsatellite loci, there seems no clear reason why heterozygote instability should not affect other loci. Indeed, in protein isozyme studies of hybrid zones, rare alleles not present in either pure population appear so frequently that they have earned their own name, 'hybrizymes' (Barton *et al.* 1983; Woodruff 1989). Hybrizymes could arise either by recombination between two pre-existing alleles or as *de novo* mutations reflecting an increased mutation rate in hybrids. Limited DNA sequence analysis support the latter, suggesting that point mutation rates are perhaps elevated in hybrid zones (Hoffman and Brown 1995).

Finally, heterozygote instability provides one possible explanation for the otherwise puzzling observation that humans have less mitochondrial DNA diversity but greater nuclear diversity, as assessed by protein isozyme and microsatellite analysis, than chimpanzees (Wilson *et al.* 1985; Wise *et al.* 1997). Human population size has expanded dramatically over the last few millennia. Such growth would tend to increase heterozygosity and hence (according to the hypothesis) mutation rate at diploid nuclear loci (Rubinsztein *et al.* 1995a), but would not affect the haploid mitochondrial genome. Interestingly, Wise and colleagues (1997) argue that the observed pattern is due to human nuclear genes evolving unusually fast, rather than due to a change in the rate of evolution of either human mitochondrial or chimpanzee sequences.

### 6.2.4 The meaning of consistent microsatellite length differences between species

Several studies have reported that microsatellites in one species are consistently longer than their homologues in other, related species (Ellegren *et al.* 1995; Rubinsztein *et al.* 1995a; van Treuren *et al.* 1997; Crawford *et al.* 1998). By definition, the lineage in which the microsatellites are longer must have experienced a greater average number of expansion mutations since the most recent common ancestor. This could come about either by a shift in the equilibrium length distribution in one or both genomes, or if a biased mutation process were accompanied by a difference in the average mutation rate between the lineages. The former explanation might involve, for example, a shift in the size of the upper length boundary, but is difficult to examine further without more data. The latter explanation has relevance to the previous section and will be discussed further.

Heterozygote instability provides one possible mechanism by which the genome-wide microsatellite mutation rate could change rapidly over short periods of evolutionary time. Assuming an expansion-prone mutation process (see above), the prediction would be that expanded populations will carry longer microsatellites than their homologues in smaller populations (Rubinsztein *et al.* 1995a). Several relevant data sets allow this prediction to be examined. Human populations have expanded dramatically, and most human microsatellites (Bowcock *et al.* 1994; Meyer *et al.* 1995; Rubinsztein *et al.* 1995a) (and, incidentally, minisatellites, Gray and Jeffreys 1991) are longer than their homologues in chimpanzees (*Pan troglodytes*). Microsatellites in the highly abundant barn swallow (*Hirundo rustica*) are consistently longer than their homologues in related species (Ellegren *et al.* 1995), and rats, whose populations have tended to grow in parallel with humans, are described as having unusually long microsatellites

(Beckmann and Weber 1992). Microsatellites in sheep (*Ovis aries*) are consistently longer than their homologues in cattle (*Bos taurus*) (Crawford *et al.* 1998), and cattle probably have the smaller long-term effective population size. Prior to domestication, the larger body size of cattle would imply smaller absolute population sizes and in modern times the practice of using a single 'prize' bull to father very large numbers of offspring would reduce severely the effective population size of cattle. Smaller studies on other species appear to follow the same trend. In a comparison between the northern and southern hairy nosed wombat, ten of 15 microsatellites were longer in the larger population (Taylor *et al.* 1994), and four humpback whale (*Megaptera novaeangliae*) microsatellites were all longer in the historically larger Antarctic populations (Valsecchi *et al.* 1997).

Unfortunately, most microsatellites are cloned from abundant species, and it has recently been suggested that it is the cloning process, not population size differences, which cause the observed length differences (Ellegren *et al.* 1995). The proposition is that, since long microsatellites are selected as markers, marker loci are likely to be longer than their homologues in other lineages, an effect referred to as an ascertainment bias. Clearly, there is a critical need to determine the size of this ascertainment bias, since locus length comparisons between populations and species need to allow for this artefactual component. Any effect over and above that ascribable to ascertainment bias would imply either a change in average mutation rate or a shift in the equilibrium state.

Assuming unbiased mutation, that all correlation between the lengths of homologous loci due to shared ancestry has been lost, uniform mutation rate over all lengths, and reflecting upper and lower length boundaries, the size of the ascertainment bias can be estimated as half of $C$, the minimum length acceptable as a marker in the focal species (Goldstein and Pollock 1997). Although the validity of most of these assumptions is questionable, this formula does provide a useful starting point for thinking about how to estimate ascertainment bias. For example, since the size of the ascertainment bias must increase with time until all correlation between the lengths of homologous loci has been lost, any observed length correlation can be taken as evidence that the ascertainment bias is less than its maximal value, estimated as $C/2$. To illustrate: the variance of the difference between two variables, $Y_1$ and $Y_2$, is given by:

$$s^2_{Y1-Y2} = s^2_{Y1} + s^2_{Y2} - r_{12}s_{Y1}s_{Y2} \quad \text{(Sokal and Rohlf 1995, p. 847)}.$$

Assume $Y_1$ = length in species A, $Y_2$ = length in species B and that there is no correlation between the lengths of homologues in species A and B, i.e. $r_{12} = 0$. Then, the variance of the length difference between the species should equal the sum of the individual variances in length. In the human–chimpanzee data set (Amos and Rubinsztein 1996) $s^2_{Y1-Y2} = 27.5$, and is significantly smaller than $s^2_{Y1} + s^2_{Y2} = 48.4 + 36.9 = 85.3$ ($F = 3.1$, $p < 0.001$), the relatively small value of $s^2_{Y1-Y2}$ indicating that the lengths of homologous loci are highly correlated ($r_{12} = 0.684$, $p < 0.001$). Thus, for this species pair, the ascertainment bias is (probably considerably) smaller than $C/2$. Goldstein (personal communication) points out that a correlation between the lengths of homologous loci does not have to be entirely due to shared ancestry, but may also arise if loci vary significantly in the range of lengths they are able to assume.

Empirical estimation of the strength of the ascertainment bias requires reciprocal tests using markers cloned from both species (Ellegren *et al.* 1995). Such experiments have now been completed for cattle and sheep. In a comprehensive study involving 472 markers which could be amplified in both cattle and sheep, loci which were polymorphic in both species are twice as likely to be longer in sheep than in cattle, regardless of which species is focal: sixteen of 20 bovine markers longer in sheep, $p = 0.012$; 198 of 308 bovine markers longer in sheep, $p < 0.0001$) (Crawford *et al.* 1998). In this comparison the direction and strength of the trend does not differ significantly between the two sets of markers ($\chi^2 = 1.74$, 1 d.f.). Hence, the ascertainment bias is probably rather weak, and the dominant effect is a genome-wide shift in mean allele length between the two species.

Great care must be exercised in interpreting these reciprocal comparisons. First, it is vital to exclude loci which are monomorphic in the non-focal species, since these are likely to be shorter according to both the ascertainment bias model (by definition) and the differential rate of expansion model (monomorphic loci are evolving slowly and would get left behind). The effect is shown dramatically by Crawford *et al.*'s study, where polymorphic bovine loci which are monomorphic in sheep are approximately seven times as likely to be longer in cattle than equivalent markers which are polymorphic in both species (111 : 29 compared with 110 : 198) (Crawford *et al.* 1998). The consequences of failure to exclude monomorphic non-focal loci is shown by data from 13 bovine markers (Ellegren *et al.* 1997), which appear directly to contradict Crawford *et al.*'s findings. However, with almost 25 times fewer markers, half of which are monomorphic in sheep, the strong ascertainment bias reported by Ellegren *et al.* is itself probably artefactual.

One further study, involving microsatellites cloned from the plant *Arabidopsis thaliana*, is of particular interest (van Treuren *et al.* 1997). The authors set out to test the heterozygote instability hypothesis, arguing that because *A. thaliana* is self-fertilising, microsatellites cloned from this species and amplified in an outcrossing relative, *Arabis petraea*, should be longer. That *A. thaliana* markers are almost invariably shorter in *A. petraea* suggests a strong ascertainment bias. However, there are two problems. First, all but one locus is either mono- or dimorphic in the non-focal species. Second, the products in the non-focal species are not just a little shorter, but most have lost all but six or fewer repeats (van Treuren *et al.* 1997). Reversing the equation of Goldstein and Pollock (1997) and substituting the observed mean length difference between the two species of 13.3 repeats yields a minimum length criterion of 26.6 repeats, long enough to exclude all but one of the loci actually used! Combined, the lack of variability (despite being in an outcrossing species) and extreme brevity of the *A. petraea* homologues suggest a radically different evolutionary state. There appear to be two possible explanations: either the short homologues never expanded, or they expanded but have since become deleted. The latter possibility is intriguing because it implies an absorbing upper boundary (see the next section).

## 6.2.5 The nature of any length constraints

One of the least understood aspects of microsatellite evolution is the factors which constrain allele length. Very long microsatellites are rare, and although the precise definition of 'long' seems to vary between taxa, there is still no clear idea of what happens to long microsatellites. There appear to be three classes of explanation: some force such as natural selection prevents expansion (a reflecting boundary); long loci become unstable and either delete or degenerate into random sequence (an absorbing boundary); or centrally directed mutation bias in which short- and medium-length alleles expand but long alleles contract (Garza *et al.* 1995; Zhivotovsky *et al.* 1997).

A reflecting boundary predicts negatively skewed allele frequency distributions among long microsatellites and a relatively sharp cut-off length, above which no microsatellites are found. Also, if microsatellites do tend to expand with time, a reflecting boundary should create a length trap near the upper limit of the permitted length distribution. What few data there are, if anything, go against these predictions: exceptionally long loci tend to exhibit strongly positively skewed allele frequency distributions (for example, García de León *et al.* 1997) and the genomic distribution of locus lengths shows no evidence either of a sharp cut-off or of a frequency peak near its upper limit (Beckmann and Weber 1992). Finally, one of the more plausible mechanisms for creating a reflecting boundary, natural selection acting against long alleles, seems unlikely because of the number of loci involved and the high levels of variability seen at most loci (Di Rienzo *et al.* 1994).

The concept of an absorbing boundary has yet to receive much attention, despite the lack of clear evidence favouring a reflecting boundary. One direct test for an absorbing boundary is suggested by the *Arabidopsis* study (van Treuren *et al.* 1997) mentioned above. Microsatellites amplified in closely related species will yield 'adjacent' evolutionary states which are of similar length. However, near an absorbing upper boundary, homologous loci may be very long and variable in one species but short and monomorphic in another. Long–short pairs should be particularly prevalent when crossing to a species whose microsatellites show a tendency to be longer, for example amplifying very long chimpanzee-derived microsatellites in humans. Small numbers of loci from cattle amplified in sheep (Ellegren *et al.* 1997), and from chimpanzees amplified in humans (G. Cooper, personal communication), together with the *Arabidopsis* data, lend anecdotal support, but more work is needed.

A centrally directed mutation bias is an interesting concept, which deserves further work. However, the predictions of this model do not appear to be supported by what few observation exist. First, the distribution of microsatellite lengths in the human genome does not exhibit the sort of peak that centrally directed bias would create. Second, allele frequency distributions tend to be skewed in the direction of any mutational bias, and published data for human loci show positive skew over all marker lengths (Swinton and Amos, submitted). This finding is difficult to reconcile with a system in which contraction and expansion phases occur at the same rate, but does not preclude one in which contraction is relatively rapid.

## 6.2.6 A comparison of locus length distributions between taxa

In my opinion, the evidence presented so far tends to favour a model of microsatellite evolution which includes progressive expansion operating alongside heterozygote instability and an absorbing length boundary. Consequently, I would like to take the opportunity in this final section to speculate how these characteristics have the potential to explain broader taxonomic trends. Though the fit between the predictions and the data is intriguing, my primary aim is to draw attention to an area of comparative analysis which has thus far been largely overlooked. The key point to bear in mind is that the frequency and length distribution of microsatellites in each species' genome represents a pseudo-equilibrium which provides information about the rates at which loci are gained and lost, and the proportion of time spent by an average microsatellite at any particular length during its 'life'.

Consider how a typical microsatellite life history might run. Individual loci are probably 'born' out of regions of high cryptic simplicity (Tautz et al. 1986; Messier et al. 1996), and expand under the influence of biased mutation. At first the mutation rate is likely to be extremely low (Weber 1990; Schug et al. 1997a) but, as heterozygosity increases, so will the mutation rate, creating a feedback loop in which both allele length and heterozygosity increase ever more rapidly towards some upper stability threshold. The locus then either deletes or stabilizes and degenerates into the background sequence by acquisition of point mutations.

Large populations are expected to carry greater levels of neutral variability than equivalent smaller populations. Consequently, the proposed microsatellite life history will begin earlier and accelerate more rapidly in larger populations, implying a shorter average 'lifespan' and less time spent at longer lengths. If, as seems reasonable, the rate of origin of new microsatellites is independent of population size, decreased microsatellite longevity in larger populations predicts that larger populations will carry fewer microsatellites per kilobase, and that the mean length of these microsatellites will be shorter.

A second prediction concerns body temperature. Microsatellites probably mutate by molecular slippage (Schlötterer and Tautz 1992), a process which might be expected to increase in rate with increasing body temperature, thereby decreasing the length of the microsatellite life cycle. Species with higher body temperatures might also have a lower length-stability threshold. Combining the population size and body temperature effects, the prediction is that birds and mammals should have fewer, shorter microsatellites than fish, and that invertebrates such as insects, which have much larger population sizes, would have very few, very short microsatellites. These predictions appear to be born out by published studies (see Fig. 6.1). Insects have large population sizes and few, short microsatellites. Among the vertebrates, both frequency per kilobase and repeat copy number decrease with increasing body temperature, with birds having the fewest (Primmer *et al.* 1997), shortest microsatellites, and fish the most and longest loci. Indeed, one study reports that fish microsatellites on average carry approximately twice as many repeats as mammalian loci (Brooker *et al.* 1994).

Anecdotal reports also follow this trend. Thus, although the vast majority of experimental effort has been directed towards mammals, the current records for microsatellite longevity across taxa are both held by cold-blooded vertebrates, turtles and fish (FitzSimmons *et al.* 1995; Rico *et al.* 1996). Also, most successful insect studies involve species such as bees (Estoup *et al.* 1993), ants (Peters 1997), wasps (Hughes and Queller 1993), and aphids (Sunnocks *et al.* 1996), all of which would be expected to have rather low heterozygosity, due variously to predominantly asexual reproduction, eusociality or haplo–diploid life cycles. Studies on several other species of insect have either failed, or yielded few useful loci, as with, for example, the damselfly *Ishnura elegans* (Cooper 1995).

Although these trends are interesting, it is clear that more systematic testing is required. Data gleaned from the literature are inevitably complicated by a range of important publication biases, most of which will tend to minimize differences between published studies. First, negative results are difficult to publish, therefore species with abundant, long, variable microsatellites will be over-represented. Second, a general preference for long microsatellites will tend to compress the range of reported lengths. Third, failure to find suitable dinucleotide repeats may cause people to turn their attentions to tri- or tetranucleotide motifs. Thus, among

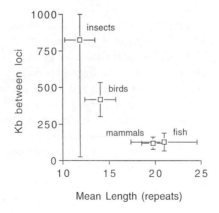

**Fig. 6.1**

the 56 studies used for Fig. 6.1, the relative numbers of dinucleotide : longer repeats motif loci decreases from fish and reptiles (60 : 0, 100% dinucleotide), through mammals (316 : 10, 97%) and birds (41 : 4, 91%) to insects (74 : 54, 58%). The same trend emerges if these data are expressed in terms of which markers each study used, with the proportion of studies based only on dinucleotides being greatest in fish and reptiles (9 : 0 : 0, dinucleotide repeats only : mixed length repeat types : no dinucleotides), and decreasing through mammals (20 : 3 : 0) and birds (7 : 3 : 0) to insects (9 : 2 : 3).

## 6.3 Conclusions

The most widely used model of microsatellite evolution is essentially a null model, formulated at a time when there was little good evidence relating to the actual process of mutation, and based on simple assumptions such as a symmetric, stepwise mutation process and a constant mutation rate. However, data are now emerging which allow these assumptions to be tested, and the preliminary analyses suggest a more complicated model is required. Two observations in particular stand out as being difficult to reconcile with the null model and perhaps pointing the way to more realistic models.

First, many, if not most, microsatellite markers appear subject to a biased mutation process favouring expansion. Such asymmetry raises important questions about what forces prevent the genome from accumulating ever more, ever longer microsatellites. A variety of models have been proposed to account for why such an accumulation does not occur. These include centrally directed mutations (short and medium length loci expand but long loci contract), a reflecting upper length constraint, and a stability threshold above which microsatellites either decay by incorporation of internal point mutations, or suffer deletion of most or all of the repeat array.

A second pattern concerns the observed variation in locus lengths between species and populations. Over brief evolutionary timescales, homologous microsatellites in related species appear to show genome-wide shifts in relative length. Over greater periods, both the frequency and lengths of microsatellites in the genome can vary markedly between major taxonomic clades, with, for example, cold-blooded vertebrates carrying larger numbers of longer and longer-lived microsatellites than birds, which are mostly carry few, rather short microsatellites. From the speculative analysis presented in the preceding section, such patterns can, in principle, be explained by a single, relatively simple model based on microsatellite mutability increasing with both heterozygosity and body temperature. However, it is important not to dismiss more complex models in which, for example, whatever factors create the upper length boundary vary from species to species.

Direct germline mutation data are unlikely ever to be available in large numbers or from more than a handful of species. Consequently, the construction of more realistic models of microsatellite evolution will depend on making best use of an increasing wealth of indirect evidence available from comparative studies. In particular, while there are now many studies comparing locus lengths in closely related populations and species, little attention has yet been paid to the directionality of any trends. Similarly, even though the frequency distribution of microsatellite lengths in the genome effectively translates into a summary of how much time an average microsatellite spends at each length during its 'life', rather profound differences between taxa have yet to receive more than passing comments. If nothing else, it seems clear that the field of microsatellite evolution will remain open and fascinating for many years to come.

# 7 Trinucleotide expansion mutations cause diseases which do not conform to classical Mendelian expectations

David C. Rubinsztein

## Chapter contents

## Abstract

Trinucleotide repeat diseases are caused by expansions of normally polymorphic trinucleotide repeat sequences in various genes. These diseases are associated with the clinical phenomenon of anticipation, where the age-at-onset tends to decrease in successive generations. This phenomenon can be explained by a tendency for mutant alleles to increase repeat number when transmitted through the germline and by the inverse correlation between age-at-onset and repeat number. This review will introduce the clinical and molecular characteristics of these diseases and consider the mutation processes of both normal and mutant alleles at these loci.

## 7.1 Introduction

A number of monogenic diseases show features which do not to conform to the classical Mendelian expectations of a constant phenotype and similar ages of onset within affected families. First, the severity of these diseases increases and/or the age at onset of symptoms decreases in successive generations, a phenomenon called anticipation. Second, in X-linked diseases, obligate male carriers are expected to manifest symptoms, but in Fragile X syndrome, some obligate male carriers of the disease-associated chromosome show no phenotype. Third, in autosomal dominant diseases, one expects that children inheriting a maternally derived mutation will show the same degree of severity as those inheriting a paternally derived mutation. Yet juvenile-onset Huntington's disease and congenital myotonic dystrophy, which have the most severe phenotypes, are associated with paternally inherited and maternally inherited disease chromosomes respectively.

The dynamic nature of the trinucleotide-repeat mutations which cause these diseases provides a molecular explanation for these phenomena. To date, five classes of trinucleotide-repeat diseases have been described. This review will first discuss the clinical and molecular features of these diseases, and then focus on the mutational processes of the repeats.

## 7.2 Clinical and molecular features of trinucleotide repeat diseases

### 7.2.1 Fragile sites

Fragile X mental retardation syndrome is the most common form of inherited mental retardation, with a prevalence of about 1/4000–1/5000 males (Murray *et al.* 1996). The disease is inherited in an X-linked dominant fashion with 80 per cent penetrance in males and 30 per cent penetrance in females (Sherman *et al.* 1984, 1985).

Fragile X syndrome shows unusual inheritance, since the likelihood of an individual being affected is related to his or her position within a pedigree, and a trend towards increasing penetrance is observed in successive generations (Fig. 7.1) (Sherman *et al.* 1984, 1985). This phenomenon is called the 'Sherman paradox'. Fragile X is associated with a cytogenetically detectable folate-sensitive fragile site at Xq27.3 called FRAXA (Ashley and Warren 1995). This disease was shown to be caused in most cases by a CGG repeat expansion in the 5′ untranslated portion of an RNA-binding protein coded for by the *FMR1* gene (Fu *et al.* 1991). The CGG repeats on normal chromosomes (<50 repeats) are

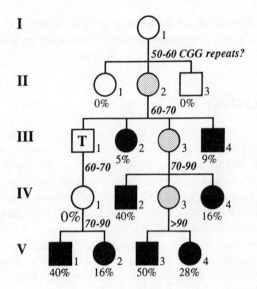

**Fig. 7.1** Empirical risk of mental retardation varies with pedigree position. This pedigree shows the per cent risk of mental retardation based on pedigree position from studies of fragile X families. Suggested CCG repeat numbers are given in bold below transmitting parents. The variability in risk of expansion to full mutation dependent on the size of the premutation accounts for the variation in risk, based on pedigree position. The male marked T is a normal transmitting male. Black-filled figures indicate mentally-retarded individuals, while grey-filled figures are unaffected carriers. From Fu *et al.* (1991), with permission from Cell Press.

generally stably transmitted from one generation to the next. Premutation chromosomes have 50–230 repeats, and show meiotic instability with a tendency to increase in size in meiosis, but are not associated with disease. Full mutations (>230 repeats) are associated with methylation of the CpG dinucleotides within the CGG repeats and the nearby CpG island in the gene, and with transcriptional repression of *FMR1* (Pierietti *et al.* 1991, Sutcliffe *et al.* 1992; Hornstra *et al.* 1993). The phenomenon of normal-transmitting males was resolved by the discovery of premutations, since these were found in males who had inherited the same chromosomal haplotype as other affected family members with full mutations (Fu *et al.* 1991) (Fig. 7.1).

Full mutations show meiotic instability but also exhibit mitotic (i.e. post-zygotic) instability, since a number of different-sized repeat lengths are frequently observed in affected males. Recent studies (Malter *et al.* 1997; Moutou *et al.* 1997) have shown that germline instability accounts for the change in repeat size from one generation to the next, disproving the alternative hypothesis that the germline was spared from the mutational process and that the intergenerational instability was the result of mitotic instability in early embryogenesis.

While females with full mutations and premutations have children with full mutations and premutations, males with these mutations only pass on premutations. This sex bias in disease transmission is thought to be the result of selection against sperm with full mutations (Malter *et al.* 1997).

FRAXE is another fragile site on the long arm of the X chromosome. It is associated with expansion of a GCC repeat stretch (Knight *et al.* 1993) and is also thought to be a rare cause of mental retardation.

## 7.2.2 Myotonic dystrophy (DM)

Myotonic dystrophy or dystrophia myotonica (DM) is the commonest adult-onset muscular dystrophy (prevalence of about 1/8000 in Europe). This autosomal dominant disease is associated with a wide range of severity and ages at onset, from congenital disease, which is often lethal and associated with mental retardation, to disease which only presents in late adulthood with ocular cataracts as the only manifestation (Harper 1989).

DM is associated with a CTG repeat expansion in the 3' untranslated part of a protein kinase gene (Aslandis *et al.* 1992; Brook *et al.* 1992; Buxton *et al.* 1992; Fu *et al.* 1992; Harley *et al.* 1992; Mahadevan *et al.* 1992). Normal chromosomes have up to 50 repeats, and are generally stably transmitted from one generation to the next, while disease chromosomes tend to expand in successive generations. Since the severity of disease and earlier age at onset broadly correlate with increasing expansion size (Fig. 7.2), this leads to the phenomenon of anticipation.

The degree of expansion of small DM mutations of <100 repeats is greater through the male line (Brunner *et al.* 1993). However, congenital DM is almost exclusively associated with maternal inheritance. This is thought to result from

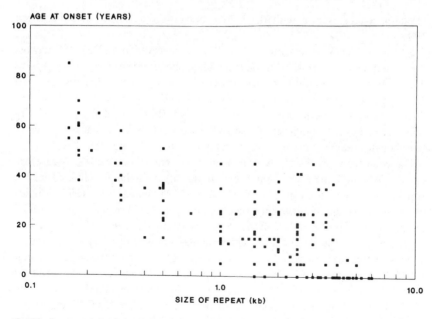

**Fig. 7.2** Age at onset for 218 myotonic dystrophy patients, plotted against CTG repeat length (logarithmic scale). From Harley *et al.* (1993), with permission from the University of Chicago Press.

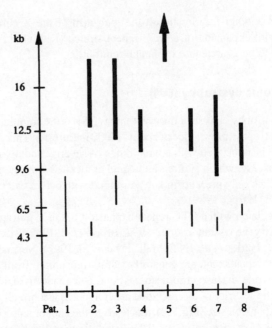

**Fig. 7.3** Schematic illustration of the DNA fragment expansion sizes in lymphocyte DNA (thin lines) and muscle tissue DNA (thick lines) in patients with myotonic dystrophy. From Anvret *et al.* (1993), with permission from Oxford University Press.

selection against sperm with the largest expansions (Paulson and Fischbeck 1996).

Like the FRAXA repeats, mitotic instability is observed in the DM CTG repeats associated with disease. In addition to a wide range of repeats sizes being present in a given tissue, variability is also observed between tissues. For instance, the size range of repeat sizes seen in muscle from a DM patient is generally longer than that seen in lymphocytes (Anvret *et al.* 1993) (Fig. 7.3). Somatic instability seems to continue throughout life, since the size of the CTG expansions seen in lymphocytes seems to increase appreciably over time (Wong *et al.* 1995).

The molecular processes by which the CTG expansion causes disease are unclear. Although the expression of mature mRNA from the protein kinase gene appears to be decreased in DM (Krahe *et al.* 1995; Wang *et al.* 1995), studies of heterozygote and homozygote knockout mice do not show all of the cardinal features of the disease, although they have myopathy (Jansen *et al.* 1996; Reddy *et al.* 1996). This suggests that this disease may not be caused by a simple gain or loss of function of the DM protein kinase gene. Some have suggested that the CTG expansion may cause disease by affecting neighbouring genes (Klesert *et al.* 1997; Thornton *et al.* 1997), although this is controversial (Hamshere *et al.* 1997). Recently, another mechanism has been proposed for this CTG expansion mutation, where the repeat bind CUG-binding proteins, which are important regulators of mRNA splicing of certain transcripts (Phillips *et al.* 1998). These authors showed that the DM CTG expansion modified splicing of the troponin T gene by a gain-of-function mechanism and it is possible that other gene transcripts are also affected.

### 7.2.3 Diseases associated with CAG expansions coding for polyglutamine tracts

This class of diseases includes spinobulbar muscular atrophy (SBMA, Kennedy's disease) (La Spada *et al.* 1991), Huntington's disease (Huntington's Disease Collaborative Research Group 1993), dentatorubral-pallidoluysian atrophy (Koide *et al.* 1994; Nagafuchi *et al.* 1994), and the spinocerebellar ataxias SCA1 (Orr *et al.* 1993), SCA2 (Imbert *et al.* 1996; Pulst *et al.* 1996; Sanpei *et al.* 1996), SCA3 (Kawaguchi *et al.* 1994), SCA6 (Zhuchenko *et al.* 1997), and SCA7 (David *et al.* 1997). All of these diseases are associated with specific and distinct patterns of neurodegeneration (Ross 1995), and all are inherited in an autosomal-dominant fashion, except for SBMA, which is X-linked. In each case, the disease is caused by an abnormal expansion of a transcribed CAG trinucleotide tract, which is translated into a polyglutamine stretch (Table 7.1).

Although the largest disease chromosomes in these diseases have smaller numbers of repeats than those seen typically in FRAXA and DM (Table 7.1), they share a number of features. In general, disease chromosomes show meiotic instability, compared with normal chromosomes (Fig. 7.4). Mitotic instability is also observed (Fig. 7.5), although this is more subtle than in DM or FRAXA (Telenius *et al.* 1994; Chong *et al.* 1995). In most of these diseases the repeats tend to expand in successive generations, and this tends to be more pronounced in male transmissions (Kremer *et al.* 1995). This results in anticipation, since the age at onset of symptoms tends to correlate inversely with the repeat number (Fig. 7.6). Thus, juvenile onset HD is more often paternally-inherited.

A number of lines of evidence suggest that these mutations are associated with a deleterious gain of function by the mutant proteins, which are known to be expressed. First, SBMA is associated with CAG expansions in the androgen receptor gene. Loss of function mutations elsewhere in this gene result in androgen insensitivity and testicular feminization. Patients with these loss of function mutations do not have neurodegeneration, which is characteristic of SBMA, while SBMA patients show mild features of androgen insensitivity. Second, the *HD* gene maps to the region of the short arm of chromosome 4 which is deleted in Wolf–Hirschhorn syndrome. While patients with Wolf–Hirschhorn syndrome have haplo-insufficiency for the *HD* gene, they show none of the cardinal features of HD. This observation, coupled with the apparently normal phenotypes of mice with h aplo-insufficiency for the HD gene, argue for a gain-of-function mutation (Duyao *et al.* 1995; Zeitlin *et al.* 1995). An intriguing feature of Huntington's disease, which may apply to other CAG-repeat diseases, is that the rare individuals who inherit two mutant HD chromosomes appear to have similar phenotypes to the more usual heterozygotes (Wexler *et al.* 1987). This apparent

**Table 7.1**
Summary of different classes of trinucleotide repeat disease comparing repeat type, size ranges of normal and disease chromosomes, site of repeat and role of repeat in disease

| Disease | Repeat | Normal range | Disease range | Site of repeat | Gain/loss of function |
|---|---|---|---|---|---|
| FRAXA | CGG | 6–52 | 60–200 (premutation) 200 to > 200 (full mutation) | 5′ untranslated | loss |
| FRAXE | GCC | 6–25 | 43–200 (premutation) > 200 (full mutation) | ? | ? |
| Myotonic dystrophy | CTG | 5–37 | > 50 | 3′ untranslated | ? |
| Huntington's disease | CAG | 6–35 | 36–120 | coding | gain |
| Dentatorubral-pallidoluysian atrophy (Haw River syndrome) | CAG | 7–25 | 49–88 | coding | gain |
| Spinobulbar muscular atrophy (Kennedy's) | CAG | 11–34 | 40–52 | coding | gain |
| Spinocerebellar ataxia 1 | CAG | 6–39 | 41–81 | coding | gain |
| Spinocerebellar ataxia 2 | CAG | 15–25 | 35–59 | coding | gain |
| Spinocerebellar ataxia 3 (Machado-Joseph disease) | CAG | 13–36 | 68–79 | coding | gain |
| Spinocerebellar ataxia 6 | CAG | 4–16 | 21–27 | coding | gain |
| Spinocerebellar ataxia 7 | CAG | 7–17 | 38–130 | coding | gain |
| Friedreich's ataxia | GAA | 7–22 | 200 to > 900 | intron | loss |
| Synpolydactyly | GCG/A/T | 15 | 22–29 | coding | ?gain |

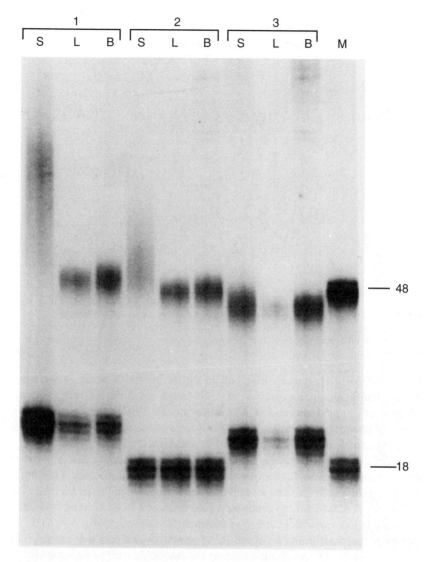

**Fig. 7.4** Comparison of Huntington's disease repeat lengths in matched sperm, lymphoblast, and blood DNA. The Huntington's disease CAG repeat was amplified by PCR from matched DNA samples of subjects 1 to 3. S = sperm DNA; L = lymphoblast DNA; B = blood DNA. Lane M contains PCR products from two cosmids known to contain 18 and 48 repeats, respectively. From MacDonald *et al.* (1993), with permission from the BMJ Publishing Group.

absolute dominance is very rare in genetic diseases, and may be an important clue to the underlying pathogenic mechanisms. Third, since the repeats are translated, and mice with expanded repeats in an HD gene which produces mRNA but no detectable protein (due to the introduction of a stop codon downstream of the repeats) show no disease, it is thought that these diseases result from

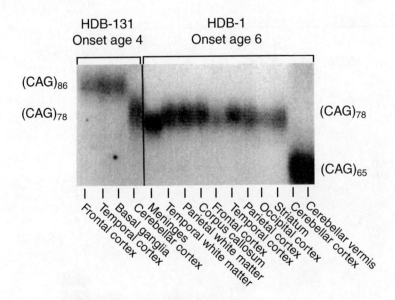

**Fig. 7.5** Mosaicism of the Huntington's disease (HD) gene's CAG repeats in brain tissues from two juvenile-onset HD patients. The upper HD alleles are shown, displaying a shift in size of the major repeat between the cerebellum and other regions of the brain. From Telenius *et al.* (1994), with permission from *Nature Genetics*.

the expression of abnormally long polyglutamine tracts (Goldberg *et al.* 1996). Indeed, transgenic mice expressing the full-length mutant HD (Hodgson *et al.* 1997; Reddy *et al.* 1997) and SCA1 (Burright *et al.* 1997) gene products show abnormal phenotypes compatible with the human diseases, while no abnormality was seen in mice overexpressing the wild-type genes.

Recent data suggest that the expanded polyglutamines can cause cell death when isolated from most of the rest of the gene product. For instance, a construct expressing 79 glutamines and a small amount of surrounding sequence from the *SCA3* gene was sufficient to cause increased cell death in COS cells, compared to a homologous construct with 35 glutamines (Ikeda *et al.* 1996). A similar 79-glutamine construct from part of the SCA3 protein caused ataxia in a transgenic mouse, while the 35-repeat construct was not associated with any disease (Ikeda *et al.* 1996).

Recent abstracts suggest that this process is a feature of other polyglutamine diseases, since toxicity is induced in transient transfection experiments with amino-terminal fragments of huntingtin with expanded repeats (Cooper *et al.* 1997; Hackam *et al.* 1997), and with fragments of the DRPLA with expanded repeats (Koide *et al.* 1997). Mangiarini and colleagues (1996) have created a mouse model which has features like those of human HD, which contains a transgene expressing only exon 1 of the *HD* gene with about 130 glutamine repeats. The control mouse with 18 repeats shows no symptoms.

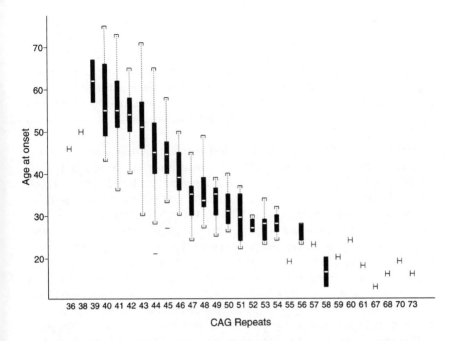

**Fig. 7.6** Relationship between CAG repeat number on the Huntington's disease chromosome and age of onset of disease. For each repeat size, the median value is indicated as an open bar, the 95 per cent confidence interval is indicated as the solid box, the range is indicated by brackets, and the outlying points by solid lines. From Rubinsztein *et al.* (1997), ©1997 National Academy of Sciences, USA.

A central issue in these diseases is to understand how the polyglutamine expansion causes disease. There has been exciting recent progress which suggests that cells expressing expanded HD repeats develop neuronal intranuclear inclusions, which have a ribbon-like fibrillar morphology reminiscent of scrapie prions and the β-amyloid fibrils in Alzheimer's disease (Davies *et al.* 1997; DiFiglia *et al.* 1997; Scherzinger *et al.* 1997). This phenomenon was observed in Bates' mouse model which only expresses exon 1 of the *HD* gene (Davies *et al.* 1997), and similar inclusions have been observed in HD brains (DiFiglia *et al.* 1997), and in *in vitro* systems expressing the amino-terminus of huntingtin with expanded repeats (Cooper *et al.* 1997; Hackam *et al.* 1997). These changes are not confined to HD and also appear in SCA3 (Paulson *et al.* 1997), SCA1 (Skinner *et al.* 1997) and DRPLA (Koide *et al.* 1997). These inclusions stain with anti-ubiquitin antibodies (Davies *et al.* 1997; Paulson *et al.* 1997). Ubiquitination of pathological structures is thought to reflect abnormal folding, aggregation, or degradation processes. The distribution of the neurons containing these inclusions roughly correlates with the regions which are affected in HD or SCA3 (Davies *et al.* 1997; Paulson *et al.* 1997).

There are a number of theories which could explain how the expanded polyglutamines result in protein aggregates, including 'polar zipper' formation between proteins with glutamine repeats (Stott *et al.* 1995), or that proteins with expanded polyglutamines may become substrates for transglutaminase activity, which could result in cross-linked products involving an $\epsilon$-$\gamma$ isopeptide (Green 1993). The cross-linked proteins may be degraded, but the isopeptide may be resistant to proteolysis and result in toxicity. It is possible that SCA6 results from a different pathological process, since the repeat expansions are much smaller (<30 repeats) than those seen in the other polyglutamine diseases (>36–40 repeats) (see Table 7.1).

It is not clear how these intranuclear inclusions cause neurodegeneration in HD. It is possible that they alter transcription of various genes, or perturb intranuclear proteolysis affecting the regulation of the levels of nuclear factors (Paulson *et al.* 1997). Another possibility is that the inclusions alter nuclear matrix-associated structures, as seems to be the case in SCA1 (Skinner *et al.* 1997).

All of the proteins which are implicated in these diseases are widely expressed, and the expression patterns alone cannot account for the specific patterns of neuronal death, which are distinct for each of these diseases (Ross 1995). Accordingly, there has been much interest in identifying proteins which interact with the disease gene products. The HD, DRPLA (Burke *et al.* 1996), SCA1, and SBMA (Koshy *et al.* 1996) gene products interact with the glycolytic enzyme glyceraldehyde-3-phosphate dehydrogenase (GAPDH). A novel polypeptide, HAP-1, interacts with the HD gene products in a CAG repeat length-dependent fashion (Li *et al.* 1995), and interactions have been shown to occur with a ubiquitin-conjugating enzyme (Kalchman *et al.* 1996). Perhaps the most interesting of these HD-interacting proteins is the human homologue of the yeast Sla2p gene (Kalchman *et al.* 1997; Wanker *et al.* 1997), which is thought to play a role in cytoskeletal integrity, and shows impaired binding to the HD protein with expanded repeats (Kalchman *et al.* 1997).

Some workers have argued that there are familial factors distinct from the CAG repeat size which influence the age of onset of HD (Kremer *et al.* 1993) and SCA1 (Ranum *et al.* 1994). While the smallest HD mutations have 36 repeats, the HD mutation can be non-penetrant in some cases with 36–39 repeats, who fail to show clinical and neuropathological features of disease in their ninth and tenth decades (Rubinsztein *et al.* 1996). Furthermore, while a number of studies have shown that the CAG repeat number on HD disease chromosomes correlates inversely with the age of onset of symptoms, this appears to account for only 50–70 per cent of the variance of the age of onset (reviewed by Ashley and Warren 1995; Paulson and Fischbeck 1996).

Genes coding for participants in excitotoxic pathways are logical candidates to consider as modifiers in HD, since rodents and primates injected intrastriatally with excitatory amino acids, such as kainate, have similar striatal pathology and neurochemistry to that seen in HD. HD brains show a loss of binding sites for excitatory amino acids, suggesting that cells expressing these receptors are made vulnerable by the mutation. The role for excitotoxicity is strengthened by the finding of an association between certain genotypes at the GluR6 kainate receptor locus and earlier age of onset of symptoms of HD (Rubinsztein *et al.* 1997). The identification and confirmation of such modifying genes provides another approach to characterising the pathways leading to neurodegeneration in these diseases.

### 7.2.4 Friedreich's ataxia

Friedreich's ataxia is the most common early-onset autosomal recessive ataxia, with a carrier frequency of about 1/100. In addition to progressive ataxia, which usually (but not always) presents below 20 years, other features of the disease include diabetes and cardiomyopathy. The causative gene, called frataxin, has recently been identified. More than 95 per cent of disease chromosomes are associated with a GAA repeat expansion in intron 1 (Campuzano *et al.* 1996). Normal chromosomes have 7–22 repeats, while disease chromosomes have 200–900 repeats. In a small proportion of cases the disease chromosome is associated with a point mutation in the gene.

The disease is thought to be associated with decreased levels of frataxin mRNA. Frataxin is a homologue of a yeast mitochondrial protein which is involved in iron homeostasis and respiratory function (Babcock *et al.* 1997; Koutnikova *et al.* 1997; Wilson and Roof 1997), and human frataxin localizes to mitochondria. These studies suggest that Friedreich's ataxia may be a mitochondrial disease resulting from a nuclear gene mutation.

GAA repeat length, particularly on the smaller of the two disease chromosomes, correlates with the severity of the disease, lower age at onset, and presence of diabetes and cardiomyopathy (Filla *et al.* 1996). Recent work has confirmed that this GAA expansion behaves similarly to the other trinucleotide repeat sequences, since it shows both meiotic and mitotic instability (Campuzano *et al.* 1996) and the presence of premutation alleles which are not pathogenic, but seem to be more mutable than typical normal alleles. These premutation alleles are thought to represent a reservoir for the creation of new mutations (Cossee *et al.* 1997; Montermini *et al.* 1997).

### 7.2.5 Synpolydactyly

The mutation causing this variably penetrant disease associated with extra digits and fused digits on the hands and feet has been recently shown to be an expansion of a polyalanine tract coded for by GCG, GCA or GCT codons in the HOXD13 gene. I believe that this can be categorized as a trinucleotide repeat disease, particularly since there seems to be a relationship between the severity of the phenotype and the expansion size (Goodman *et al.* 1997). However, in contrast to many of the other trinucleotide repeat diseases, the expanded alleles do not show obvious hypermutability (Goodman *et al.* 1997). This may be because the repeats are interrupted (i.e. not pure repeated sequence, see below) or just because they are short, since the comparatively small expansion at the SCA6 locus is also not associated with obvious instability (Reiss *et al.* 1997).

# 7.3 Mutational processes of trinucleotide repeats

## 7.3.1 General principles

There has been much interest in the mutational processes of trinucleotide repeats associated with disease genes, as these can give clues about the evolution and population genetics of the diseases and inform our understanding of microsatellites in general.

It is comparatively easy to characterize the behaviour of disease chromosomes, since they are highly mutable and therefore can be examined in pedigrees. By contrast, the behaviour of the more stable non-disease chromosomes needs to be inferred from population-genetic studies. The unimodal and bimodal allele-length distributions, and the limited number of mutations that have been observed, are consistent with normal trinucleotide alleles mutating in a stepwise manner (i.e. changes of one repeat unit occur most frequently), as is thought to be the case with microsatellites in general. However, the highly mutable disease alleles can show large expansions and occasionally contractions. In HD, where the transmission of many mutant chromosomes has been observed, large mutations seem to be more common in male meioses, while disease chromosomes transmitted by females are generally confined to single-repeat or two-repeat size changes (Kremer *et al.* 1995) (Fig. 7.7). In many of the CAG trinucleotide repeat diseases (HD, DRPLA, SCA1, and SCA3), the mutation rate of mutant chromosomes appears to be greater in the male line, leading to a tendency for juvenile-onset disease to be paternally inherited (Paulson and Fischbeck 1996). This trend has also been suggested for normal microsatellites (Weber and Wong 1993). These observations can be explained by the approximately tenfold increased number of cell divisions between zygote and sperm, as compared to zygote and ovum.

Trinucleotide repeats show length-dependent mutation rates, since normal alleles are generally meiotically stable, while disease alleles show marked instability (Fig 7.4). It is important to note that there is no absolute correlation between mutability and pathogenicity, since chromosomes at the upper limit of the normal (non-pathogenic) length distribution show a degree of instability (which is probably intermediate between that of normal and disease chromosomes). These are thought to represent the pool of normal chromosomes from which the pathogenic expansions are derived. This concept is supported by observations in HD (Rubinsztein *et al.* 1994*a*; Almqvist *et al.* 1995), DRPLA (Burke *et al.* 1994), and DM (Goldman *et al.* 1994), showing a correlation between disease prevalence in different populations and the frequency of long normal alleles at the relevant loci, and because certain haplotypes are over-represented on both disease chromosomes and long normal HD, FA, or FRAXA chromosomes, compared to

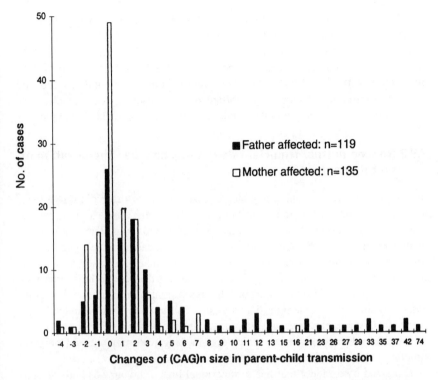

**Fig. 7.7** Distribution of intergenerational changes of CAG size in 254 affected parent–child transmission in Huntington's disease. From Kremer *et al.* (1995), with permission from the University of Chicago Press.

average-length normal chromosomes (Almqvist *et al.* 1995; Ashley and Warren 1995; Rubinsztein *et al.* 1995c; Cossee *et al.* 1997). The evolution of disease chromosomes from a small pool of large normal chromosomes can account for the haplotype data supporting apparent 'founder' mutations in HD, FRAXA, and other diseases like FA (Cossee *et al.* 1997). Unlike most other diseases, the 'founder' chromosomes for trinucleotide repeat disorders are likely to be long normal alleles, that is, non-pathogenic ones. However, these alleles are more likely to evolve into disease chromosomes due to their slightly greater mutability and their proximity to the disease size range.

The observation that disease chromosomes in HD and myotonic dystrophy (DM) both show strong linkage disequilibrium with certain flanking markers compared to normal chromosomes has led to the suggestion that these haplotypes mark mutation-prone chromosomes, which were more likely to have expanded into the disease-size range than similar-sized repeats associated with other haplotypes (Imbert *et al.* 1993; MacDonald *et al.* 1993; Neville *et al.* 1994). Subsequent studies (Rubinsztein *et al.* 1994b, 1995c; Tishkoff *et al.* 1995; Zerylnick *et al.* 1995) provide more parsimonious explanations, dispensing with the need to invoke mutation-prone haplotypes. However, a study using single-sperm analysis of HD chromosomes from the Hayden laboratory presents preliminary data which are consistent with this intriguing possibility (Chong *et al.* 1997), although data from more individuals are required to show a statistically significant effect. In addition, this phenomenon seems to occur at VNTR loci (Monckton *et al.* 1994), although it is likely that they behave differently from trinucleotide repeats and other microsatellites. This is an important issue to resolve, as the results have implications for our general understanding of trinucleotide repeat behaviour.

### 7.3.2 How do normal trinucleotide repeats behave? The question of mutational bias

In order to try to understand the processes responsible for the evolution of these diseases, the CAG trinucleotides in the Huntington's disease gene were studied in a wide range of human and non-human primate populations. The CAG repeats in all normal human populations that were examined had positively skewed distributions, with more chromosomes having larger allele lengths than the mode (Rubinsztein *et al.* 1994*a*). The ancestral length was deduced to be shorter than present-day lengths in humans, as non-human primate chromosomes had comparatively shorter repeat lengths than humans. This gradual increase in the length of the HD CAG repeats over time in humans, and the positive skews, could not be accounted for by meiotic drive, which favours longer normal alleles. Computer simulation experiments were used to show that these empirical findings could be explained by a model that had a mutational bias in favour of longer repeats as a critical component (Rubinsztein *et al.* 1994*a*). Thus, it was argued that the mutational bias that is a feature of abnormally expanded disease chromosomes may also be part of the mutation process in the normal size range. If this model is correct, it predicts a gradual increase in the CAG allele length in the *HD* gene in all human populations. Since the prevalence of HD in a population is related to the proportion of long normal alleles, this mutational bias would lead to an accompanying increase in the prevalence of this disease over an evolutionary time scale.

Subsequent work suggests that a mutational bias in favour of longer repeats may be a feature of microsatellites in general (Rubinsztein *et al.* 1995*a*; Amos and Rubinsztein 1996; Amos *et al.* 1996; Primmer *et al.* 1996). This issue is discussed in detail in the chapter by Amos in this book.

### 7.3.3 Imperfect repeats

Trinucleotide-repeat diseases have provided good examples supporting the idea that interruptions of perfect repeats (e.g. $(CAG)_n$ CTG $(CAG)_y$) are associated with lower mutation rates. SCA1 (Chung *et al.* 1993), SCA2 (Imbert *et al.* 1996; Pulst *et al.* 1996; Sanpei *et al.* 1996), and FRAXA (Eichler *et al.* 1994; Hirst *et al.* 1994; Kunst and Warren 1994) disease chromosomes have long stretches of perfect repeats, while most normal chromosomes are associated with interruptions. A possible mechanism which could give rise to hypermutable normal chromosomes would be a point mutation within an imperfect repeat which led to a much longer perfect repeat stretch (e.g. $(CAG)_4$ CTG $(CAG)_5$ could mutate to $(CAG)_{10}$).

### 7.3.4 Meiotic drive/segregation distortion in trinucleotide repeat diseases?

Another process which may challenge classical Mendelian principles in some trinucleotide-repeat diseases is segregation distortion, or a failure of random segregation of alleles from one or other parent. This was first suggested by studies at the DM locus by Carey et al. (1994) and subsequently by Chakraborty et al. (1996), which suggested that larger normal alleles were preferentially transmitted compared to smaller alleles in heterozygotes. Reanalysis of these data by Hurst et al. (1995) and Morton (1997), respectively, has cast doubt upon the validity of the statistical bases for these conclusions. The HD repeats do not show any obvious segregation distortion (Rubinsztein et al. 1994a).

Recent data from Ikeuchi and co-workers (1996) suggest that there is preferential transmission of the mutant alleles at the SCA3 and DRPLA loci from affected males. This was not observed in affected females, and these data were consistent with previous examples in other species where meiotic drive occurs in males or females at different loci, but not in both for a given locus (Hurst et al. 1995). These segregation data have been confirmed in single sperm experiments at the SCA3 locus (Takiyama et al. 1997). Curiously, when this locus was examined in normal individuals, non-Mendelian transmission was confined to females, with preferential transmission of alleles with smaller CAG repeats (Rubinsztein and Leggo 1997). Although these findings remained significant after correction for multiple tests, this may represent a chance phenomenon, and therefore these findings need to be replicated in an independent sample.

### 7.3.5 Mouse models of trinucleotide repeat behaviour

A number of transgenic mice have been created which contain human tri-nucleotide-repeat genes, or fragments of these genes, with expanded repeat tracts. While some of the initial experiments suggested that repeats which were unstable in man may be stable in mouse (Bingham et al. 1995; Burright et al. 1995; Goldberg et al. 1996), this is not the case with some of the more recent models which show clear instability (Gourdon et al. 1997; Mangiarini et al. 1997; Monckton et al. 1997). However, the differing degrees of instability of similar-sized repeat expansions in the same construct in different lines suggest that the degree of instability may be related to the site of integration of the transgene. This could account for the stability which has been observed in some mice. An alternative possibility is that the degree of expansion may be related to expression, since the HD mouse line of Mangiarini et al. (1997), which failed to show instability, was associated with no expression, while the transgenes in the unstable mouse lines were transcribed.

The mouse lines which show instability seem to show less instability than one would expect from the human mutation in man. For instance, the offspring of male mice derived from founders with 142 and 116 HD CAG repeats show a smaller overall expansion of repeats and a smaller variance of repeat lengths (Mangiarini *et al.* 1997) than one would have expected from such an expansion in man (Kremer *et al.* 1995). This may be because the number of germ-cell divisions from male sexual maturity to reproduction is less in mice than in man. This model is supported by the finding in the HD mice of increasing repeat size in the progeny of older mice (Mangiarini *et al.* 1997). Indeed, this is compatible with previous observations of an inverse relationship between paternal age and age at onset of HD in offspring.

The mice lines which showed instability seem to have different patterns of transmission in males and females. The mice expressing exon 1 of the HD with expanded repeats seemed to show a bias favouring expansion through the male line, and a bias favouring contraction through the female line (Mangiarini *et al.* 1997). A similar pattern was observed in the mice with the DM repeats from Monckton *et al.* (1997).

The mice which had repeat instability also showed somatic mosaicism (Gourdon *et al.* 1997; Mangiarini *et al.* 1997; Monckton *et al.* 1997), which is a feature of human trinucleotide repeat disease alleles. Another parallel between humans and mice was shown by the tendency for somatic mosaicism of the HD CAG repeats to become more pronounced with age (Mangiarini *et al.* 1997), a phenomenon which has been observed in DRPLA (Takano *et al.* 1996).

A fascinating finding described by Monckton *et al.* (1997) was segregation distortion favouring transmission of the transgene (encoding expanded DM repeats). This was reminiscent of observations in DM, DRPLA, and SCA3 in man (see above). However, it is interesting to note that all of the accepted segregation-distortion models only occur in one sex for a given locus (Hurst *et al.* 1995), while Monckton *et al.* (1997) showed this phenomenon in both males and females.

The creation of transgenic mice with unstable expanded repeats demonstrates that instability can be mediated by intrachromosomal events. This is not surprising, given the existence of a number of polymorphic microsatellites on the non-recombining part of the Y chromosome which show mutation rates within the ranges previously estimated for autosomal loci (Cooper *et al.* 1996). However, these data do not exclude the possibility of interchromosomal events in humans.

## 7.4 Conclusions

The mutation patterns of trinucleotide repeat mutations have provided a molecular explanation for many of the atypical inheritance patterns of the above diseases. Thus, this mutational mechanism has been invoked for a number of other diseases which appear to show anticipation, including schizophrenia, bipolar affective disorder, and hereditary spastic paraplegia (McInnis 1996).

Although the behaviour of the mutant expanded chromosomes has been carefully assessed in many cases, a number of controversies need to be resolved, particularly the possibility of mutation-prone chromosomal lineages and meiotic drive. The dynamics of the less mutable normal chromosomes are more difficult to assess, and the issue of biased mutation in favour of longer repeats needs to be considered.

Our understanding of the molecular processes governing repeat instability and mismatch repair has progressed rapidly, probably as a result of the interest in mutations in mismatch repair genes involved in non-polyposis colon cancer. Nevertheless, the molecular events which lead to mutations in trinucleotide repeats are still largely hypothetical.

The different classes of trinucleotide repeat diseases are associated with different pathogenic processes. A more comprehensive understanding of these molecular pathologies will advance our knowledge of disease processes and hopefully provide bases for therapeutic intervention strategies.

## Acknowledgements

The work from my laboratory which has been referred to in this review has been supported by the Huntington's Disease Association (UK), the Rehabilitation and Medical Research Trust and The Leverhulme Trust. I am grateful to Glaxo Wellcome for a Research Fellowship.

# 8 Mutation and migration in models of microsatellite evolution

Marcus W. Feldman, Jochen Kumm and Jonathan Pritchard

## Chapter contents

## Abstract

Dynamic and statistical properties of generalized stepwise mutation models are described, and used to compare data on human di-, tri-, and tetranucleotide polymorphisms. The time-dependent behaviour of an island model with stepwise mutation is analysed, and its equilibrium properties used to estimate the product $Nm$ of the population size and the migration rate. Population statistics are derived from the complete equilibrium analysis of the model, and these are combined to give an estimate of $Nm$ that is related to Slatkin's 1995 estimate which used $R_{ST}$. When this new statistic is applied to a set of 85 human microsatellite polymorphisms, the resulting population clusters match those in the original tree of Bowcock et al. (1994).

## 8.1 Introduction

Because of their high level of polymorphism and ubiquity in eukaryotic genomes, microsatellites are widely preferred markers in evoluutionary and ecological studies. Databases of microsatellites isolated for population-level analyses are under development. Statistical evaluation and evolutionary interpretation of microsatellite polymorphisms demand models for the origin and maintenance of this variability, and the models used so far are variants of the stepwise muta-tion model (SMM), which was originally introduced by Ohta and Kimura (1973) to model electrophoretically detectable enzyme variation in finite populations.

The first part of this paper describes these models and illustrates how stepwise mutation, in combination with genetic drift, affects important measures of within- and between-population variation. Using empirical estimates of overall mutation rates, these measures may then provide estimates of average population size or divergence times between populations. In the second part of the paper, the dynamics of a set of populations subject to migration according to Wright's (1943) island model are developed. From the equilibrium structure of this model, estimators of the extent of migration based on population statistics are suggested.

We shall assume throughout that the microsatellites are perfect: that is, each allele at a locus is completely specified by the number of times a motif is repeated. Generations are non-overlapping and each is produced from the previous by multinomial sampling from the parental generation's array of alleles. Our anal-ysis follows that of Moran (1975), who developed recursions for the population central moments under the simplest stepwise mutation model permitting alle-les of arbitrary (positive and negative) repeat number. The modifications to this model that we discuss below include asymmetric mutations of arbitrary size, a simple model of linear bias in which the rate of mutation depends on the number of repeats in the allele, and range constraints on the permitted repeat score.

## 8.2 Moran's formulation: a generalized SMM

Consider a population of $N$ diploid individuals and a locus at which each allele is characterized by its repeat score, which may take any positive or negative integer values. Let $\mu_c$ be the probability that mutation changes an allele by $c$ repeat units, irrespective of its original count. The total mutation rate is then $\mu = \sum_{c \neq 0} \mu_c$ and the expected change in repeat score for any allele is $\bar{c} = \sum c \mu_c$. The variance in repeat score change is $\sigma_m^2 = \sum c^2 (\mu_c/\mu) - \bar{c}^2$. We write $w = \mu \sigma_m^2$. The standard one-step symmetrical model has $\mu_{+1} = \mu_{-1} = \mu/2$, say, and $\mu_i = 0$ otherwise, in which case $\bar{c} = 0$ and $w = \mu$. We shall refer to this special case as the one-step symmetrical SMM.

Let $p_i$ be the frequency of the allele carrying $i$ repeats in the parental generation. Then, following mutation, the frequency of this allele is:

$$\tilde{p}_i = \sum_c \mu_{(i-c)} p_c. \tag{8.1}$$

The $2N$ gametes undergo multinomial sampling with probabilities $\{\tilde{p}_i\}$ to pro-duce the $2N$ copies in the offspring generation. The process of mutation changes the population average repeat score $r = \sum_i i p_i$ to $\tilde{r} = \sum_i i \tilde{p}_i = r + \bar{c}$, while the variance changes from:

$$V = \sum_i p_i (i - r)^2 \qquad (8.2)$$

to:

$$V = \sum_i \tilde{p}_i (i - \tilde{r})^2 = V + w. \qquad (8.3)$$

Multinomial sampling has no effect on $\tilde{r}$ in the sense that $E_m(r') = r + \bar{c}$, where $E_m(\cdot)$ is the expectation with respect to sampling, and the prime refers to the offspring generation. As shown by Moran, however, sampling reduces the variance by a proportion $1/(2N)$:

$$E_m(V') = \left(1 - \frac{1}{2N}\right)(V + w). \qquad (8.4)$$

These results are minor modifications of Zhivotovsky and Feldman (1995), who also developed analogous recursions for the skew and kurtosis in the case $\bar{c} = 0$. From eqn (8.4), taking expectations with respect to initial conditions, we obtain the equilibrium variance

$$\hat{V} = (2N - 1)w. \qquad (8.5)$$

Goldstein *et al.* (1995b) used a similar recursion in the one-step symmetrical model for the quantity $D_0$, the average squared difference in repeat score between two alleles chosen randomly from a population of $N$ diploids. Note that $D_0 = 2V$.

In the symmetrical case, the variance across loci of the equilibrium random variable $V$ in eqn (8.2) was given by Zhivotovsky and Feldman (1995), who showed that this variance will usually be quite large, of the order of $\hat{V}^2$. Pritchard and Feldman (1996) obtained the small sample analogue of this variance (see also Roe 1992) and showed that as the sample size $n$ increases, the two variances coincide. Goldstein *et al.* (1996) used properties of this variance to derive a confidence interval for the population variance. Consider $\bar{V}$, the mean over $L$ loci of $\hat{V}$. We have recently obtained an expression for the variance $\text{Var}_u(\bar{V})$ over evolutionary realizations of $\bar{V}$ for samples of size $n$ when the loci are unlinked:

$$\text{Var}_u(\bar{V}) = \frac{N\mu\gamma(n^2 + n) + 4N^2\mu^2\left(\sigma_m^2\right)^2(2n^2 + 3n + 1)}{6L(n^2 - n)},$$

where $\gamma = \sum c^4(\mu_c/\mu)$. The corresponding result when the $L$ loci are completely linked is

$$\text{Var}(\bar{V}) = \text{Var}_u(\bar{V}) + \frac{2N^2\mu^2\left(\sigma_m^2\right)^2(L - 1)(n^2 + n + 3)}{9L(n^2 - n)}.$$

## 8.3 An application of $\hat{V}$

The expectation in eqn (8.5) may be used to estimate the effective population size $N$ when the mutation parameter $w$ is known. Goldstein *et al.* (1996) used the variance obtained by averaging the variance in the pooled worldwide sample of human genotypes across the 30 dinucleotide loci studied by Bowcock *et al.* (1994). This variance was 10.1, and with Weber and Wong's (1993) estimate of $5.6 \times 10^{-4}$ for $\mu$, an estimate of about 9000 for the effective human population size was obtained, under the assumption that $\sigma_m^2 = 1$ and $\bar{c} = 0$, the simplest symmetric SMM. Among the 30 loci in the study by Bowcock *et al.*, one locus exhibited a variance some tenfold greater than the others. When this apparently aberrant gene is deleted, the variance drops to 6.827, giving a corresponding estimate for $N$ of about 6100. In a situation such as this, it may pay to examine the molecular properties of the outlying locus in more detail.

Recent work in Cavalli-Sforza's laboratory has increased the number of dinucleotide loci studied to 85, with a corresponding average across loci of the pooled worldwide variance (using the same human sample as in the study by Bowcock *et al.*) of 8.652. Under the same assumptions as above, this corresponds to an effective population size of about 7700.

The assumption in these calculations that $\sigma_m^2 = 1$ is critical; it is clear that for a given observed $\hat{V}$, the estimate of $N$ decreases by the factor $1/\sigma_m^2$. Dib *et al.* (1996) reported on 5264 dinucleotide polymorphisms in CEPH families. The mean heterozygosity for these polymorphisms was 70% and the overall mutation rate was $6.2 \times 10^{-4}$, while the average mutation resulted in an increase of about 0.39 repeat units. Although some 90 per cent of the mutations produced repeat number changes of one or two, the mutation distribution had long tails on both positive and negative sides that resulted in an observed variance of about 4.5. A symmetrized geometric distributed with its parameter based on the average rate of one-step increases and decreases would produce a variance of about 2.5, considerably less than that observed. (The observed variance would, of course, be strongly influenced by a few atypical loci of large size that contribute many mutations.)

For the purposes of illustration, assume that the overall mutation rate is $6.0 \times 10^{-4}$ and the variance in mutational changes is 2.5. Then, with a worldwide pooled average variance of 8.652, as observed for the 85 dinucleotide polymorphisms mentioned above, the estimate of $N$ would be 2900. This would drop to about 1800 with $\sigma_m^2 = 4$. The reduction in estimated average effective human population size due to $\sigma_m^2$ is, therefore, substantial. It must be remembered, however, that in the model leading to these estimates of $N$, the probability of mutation is independent of the size of the allele in which the mutation occurs. We have yet to investigate the effect of this 'stationarity' assumption on these estimates.

## 8.4 Tri- and tetranucleotides

Data on a scale comparable to that discussed above for dinucleotide mutation and variation are not available for tri- and tetranucleotides. Weber and Wong (1993) directly observed mutations at 28 loci on chromosome 19, and reported that the average mutation rate for tetranucleotides was nearly four times that for dinucleotides. The small number of loci and mutations suggests that it would be risky to take this factor very seriously. Heyer *et al.* (1997) directly estimated the mutation rate across seven Y chromosome tetranucleotides as 0.002, in reasonable agreement with Weber and Wong (1993).

Chakraborty *et al.* (1997) used a relation for the population variance, $\hat{V}$, similar to that of Zhivotovsky and Feldman (1995; see eqn (8.5) above) to obtain indirect estimates of the relative mutation rates of di-, tri-, and tetranucleotides from polymorphisms reported in a number of different data sets. Dinucleotides had mutation rates 1.5–2.0 times higher than tetranucleotides, with non-disease-causing trinucleotides intermediate between the di- and tetranucleotides. In their analysis, they assumed that all three motifs had the same mutational variance, $\sigma_m^2$. (In fact, as mentioned above, Chakraborty *et al.* make the same approximation as Kimmel *et al.* (1996), namely $\bar{c}^2 = 0$; their assumption therefore amounts to all second moments about zero being the same.)

Using the same individuals sampled in the original study of dinucleotide polymorphisms by Bowcock *et al.* (1994), Bennett *et al.* (1998) have examined 22 trinucleotide and 21 tetranucleotide polymorphisms. For the trinucleotides, the pooled worldwide average variance was 3.994, and for the tetranucleotides it was 4.148. Recalling the estimate of 8.652 for the 85 dinucleotides, and since the same populations are involved, we may regard the ratios $8.652/3.994 = 2.166$ and $8.652/4.148 = 2.086$ as estimates of the ratios $w_{di}/w_{tri}$ and $w_{di}/w_{tetra}$, respectively. Of course, these ratios are products of the ratios of overall mutation rates times the ratios of mutational variances. If the latter are unity (i.e. the mutational variances of di-, tri-, and tetranucleotides are all the same), then the above numbers are in good agreement with the estimates of Chakraborty *et al.* Until we have reliable direct estimates of the mutation rates and variances, interpretation of the observed variances will be difficult. If, for example, Weber and Wong (1993) and Heyer *et al.* (1997) are correct, and the mutation rate for tetranucleotides in four times that of dinucleotides, then the values of the variance ratios above entail that the mutational variance of tri- and tetranucleotides should be about one eighth that of dinucleotides, but we estimated that $\sigma_m^2$ should be in the range 2–4 for dinucleotides, and $\sigma_m^2$ must be no less than 1. This contradiction suggests either that the factor 4 is incorrect or that the assumption of equilibrium under the SMM is incorrect. Also, we must remember that these calculations were predicated on the assumption that the size of mutational jumps does not depend on the size of the original allele, an assumption which has yet to be given a rigorous empirical test. Conceivably, the constraints on the ranges of tri- and tetranucleotides could be much more stringent than on dinucleotides, and this would have an important effect on the relative variances (Feldman *et al.* 1997).

## 8.5 Genetic distances for stepwise mutation models

In order to estimate times of separation between populations (of the same or different species), it is desirable that the measure of separation, in our case the genetic distance, have an explicit relationship with time. Thus, for the infinite alleles model (IAM), Nei's (1972) standard genetic distance, $D_s$, between populations that diverged $t$ generations ago is expected to be close to $2\mu t$, where $\mu$ is the mutation rate per locus per generation. When mutation occurs according to the SMM, no explicit form for the dependence of $D_s$ on time has been found.

In considering properties of the variances in repeat numbers within and between populations under the SMM, we were led to an expression for the genetic distance between two applications which we called $(\delta\mu)^2$:

$$(\delta\mu)^2 = (m_x - m_y)^2, \tag{8.6}$$

where $m_x$ and $m_y$ are the mean repeat numbers at a locus in populations $x$ and $y$ (Goldstein *et al.* 1995a,b). For samples of size $n_x$ and $n_y$ from the two populations, the appropriate unbiased estimate of the population value $(\delta\mu)^2$ is (Goldstein and Pollock, 1997)

$$(\delta\mu)^2 = \sum_i \sum_j (i - j)^2 x_i y_j - \tilde{V}_x - \tilde{V}_y = D_1 - \tilde{V}_x - \tilde{V}_y, \tag{8.7}$$

where $D_1 = \sum_i \sum_j (i - j)^2 x_i y_j$, with $x_i$ and $y_j$ the frequencies of repeat lengths $i$ and $j$ in the samples from $x$ and $y$, respectively, and $\tilde{V}_x$ and $\tilde{V}_y$ are the usual unbiased estimates of repeat length variance in populations $x$ and $y$. Goldstein *et al.* (1995a) showed that if populations $x$ and $y$ diverged from a common ancestral population $t$ generations ago, then

$$E(\hat{\delta}\mu)^2 = 2\mu t \tag{8.8a}$$

under the simple symmetrical SMM, while Zhivotovsky and Feldman (1995) showed that in the symmetrical SMM with mutational variance $\sigma_m^2$, eqn (8.8a) is replaced by

$$E(\hat{\delta}\mu)^2 = 2\mu\sigma_m^2 t = 2wt. \tag{8.8b}$$

It is not difficult to see that eqn (8.8b) also applies in the asymmetrical case. In applications, the sample values of this distance are calculated for each microsatellite locus, and then the average is taken over loci, under the assumption that the correlations between pairs of loci are weak.

It is worth comparing our result, eqn (8.8b) with the corresponding expression in eqn (19) of Slatkin (1995). He estimates the time of divergence as

$$T_R = \frac{4R_{ST}}{(1 - R_{ST})},$$

where $R_{ST}$ (with equilibrium assumed in each population) can be expressed in terms of our eqns (8.5) and (8.7) above as

$$R_{ST} = \frac{D_1 - V_x - V_y}{D_1 + V_x + V_y},$$

when sample sizes are large. Hence we may write

$$T_R = \frac{2(D_1 - V_x - V_y)}{V_x + V_y}.$$

The corresponding ratio of expectations (assuming each population is at equilibrium) is

$$T_R \cong \frac{E(\delta\mu)^2}{E(V)} = \frac{2wt}{2Nw} = t/N$$

as in Slatkin's equations (17)–(19). Again, in the coalescent argument he uses, the factor $2N - 1$ is replaced by $2N$. An argument similar to this relating $R_{ST}$ to our $(\delta\mu)^2$ through the use of the variances was made by Kimmel *et al.* (1996).

Shriver *et al.* (1995) suggested a distance that replaces the squared differences in eqn (8.7) by the absolute value of differences. This measure, $D_{sw}$, is difficult to relate analytically to the time of divergence between the groups under comparison, although simulation studies indicate that it is almost linear with time (Takezaki and Nei 1996).

Although the distances $(\delta\mu)^2$ and $D_{sw}$ were developed specifically for the SMM, they have high sampling variances (Goldstein *et al.* 1995a,b; Takezaki and Nei 1996). Thus, when $(\delta\mu)^2$ was applied to the data of Bowcock *et al.* (1994), the resulting population tree was poorly supported, with bootstrap values much lower than those obtained using an allele-sharing distance or the chord distance of Cavalli-Sforza and Edwards (1967). On the other hand, $(\delta\mu)^2$ was able to separate African from non-African populations, indicating that for groups at this level of divergence it can be used for reconstruction of population relationships. In the simulation study of Takezaki and Nei (1996), the coefficients of variation for $(\delta\mu)^2$ (and $D_{sw}$) were consistently greater than for the traditional allele-frequency-based distances, although $(\delta\mu)^2$ was relatively insensitive to changes in population size.

Zhivotovsky and Feldman (1995) showed that with complete sampling, the variance of $(\delta\mu)^2$ between two populations $t$ generations after their divergence from a common ancestral population, with each in mutation–drift equilibrium according to a symmetric SMM, is

$$\text{Var}(\delta\mu)^2 = 8w^2t^2. \tag{8.9}$$

Thus, if $(\delta\mu)^2$ is computed as the average over $L$ loci of the population values at each locus, all satisfying eqn (8.9), and if all loci have the same mutation rate and mutational variance, then the coefficient of variation,

$$CV_{(\delta\mu)^2} = \sqrt{2/L}. \tag{8.10}$$

If the mutation parameter $w$ (recall $w = \mu\sigma_m^2$) varies among the loci, then, according to Zhivotovsky and Feldman (1995, eqn 25),

$$CV_{(\delta\mu)^2} = \{2(1 + CV_w)/L\}^{1/2}, \tag{8.11}$$

which suggests an empirical method of evaluating $CV_w$, the coefficient of variation among the $w$ values for each locus, from an observed $CV_{(\delta\mu)^2}$ (Reich and Goldstein 1998).

We have compared the variance and mean of the corrected $(\hat{\delta}\mu)^2$ (i.e. eqn 8.7) as a function of time since separation and the mutation rate, using a coalescent simulation of a simple symmetrical SMM at 30 loci that were absolutely linked or completely unlinked, but all with the same mutation rate. Following separation of an initial population into two populations (identical to the progenitor population) that subsequently evolve independently, the value of $CV_{(\delta\mu)^2}$ is tracked over time. Table 8.1 reports the approximate time $\tau$ in units of $2N$ generations at which $CV_{(\delta\mu)^2}$ settles around the value $\sqrt{2/30} = 0.258$ predicted for unlinked loci. Linkage delays this convergence by retarding the convergence of the variance. As suggested by Takezaki and Nei (1996), smaller values of $\Theta = N\mu$ produce faster approach of $CV_{(\delta\mu)^2}$ to the asymptote $\sqrt{2/L}$ with $L$ loci, although in all cases convergence is slow.

**Table 8.1**
Times to convergence of $CV_{(\delta\mu)^2}$ to $\sqrt{2/L}$

|  | Linked | Unlinked |
|---|---|---|
| $\theta = 10$ | $\tau \simeq 70$ | $\tau \simeq 40$ |
| $\theta = 0.1$ | $\tau \simeq 20$ | $\tau \simeq 10$ |

## 8.6 Application of $(\delta\mu)^2$

Removing the aberrant locus from the data set of Bowcock *et al.* (1994), the average variance across the remaining 29 dinucleotide loci (pooling all individuals from all 14 populations) was 6.827. The fourteen populations were grouped by Goldstein *et al.* (1995a) into seven clusters: I, Australian, New Guinean; II, Central African Republic Pygmies, Lisongo; III, Chinese, Japanese, Cambodian; IV, Northern Italian, Northern European; V, Karitiana, Surui, Mayan; VI, Melanesian; VII, Zaire Pygmies. In their analysis, Goldstein *et al.* included all 30 loci, giving an average variance of 10.1, and in their calculations of $(\hat{\delta}\mu)^2$ they did not correct for small sample size. Table 8.2 reports the 21 pairwise values of $(\delta\mu)^2$ corrected for small sample size and including 29 loci. Table 8.3 gives the same matrix for 85 dinucleotide loci.

**Table 8.2**
Corrected $(\delta\mu)^2$ distances for 29 dinucleotide loci*

|     | I     | II    | III   | IV    | V     | VI    |
|-----|-------|-------|-------|-------|-------|-------|
| I   |       |       |       |       |       |       |
| II  | 2.732 |       |       |       |       |       |
| III | 0.702 | 2.292 |       |       |       |       |
| IV  | 1.491 | 3.102 | 1.343 |       |       |       |
| V   | 1.101 | 2.618 | 0.919 | 1.745 |       |       |
| VI  | 0.983 | 2.244 | 0.751 | 1.364 | 0.483 |       |
| VII | 5.803 | 2.447 | 4.466 | 6.410 | 5.571 | 6.459 |

* Population groups are: I, Australian, New Guinean; II, Central African Republic Pygmies, Lisongo; III, Chinese, Japanese, Cambodian; IV, Northern Italian, Northern European; V, Karitiana, Surui, Mayan; VI, Melanesian; VII, Zaire Pygmies.

**Table 8.3**
Corrected $(\delta\mu)^2$ distances for 85 dinucleotide loci

|     | I     | II    | III   | IV    | V     | VI    |
|-----|-------|-------|-------|-------|-------|-------|
| I   |       |       |       |       |       |       |
| II  | 3.833 |       |       |       |       |       |
| III | 1.602 | 2.507 |       |       |       |       |
| IV  | 2.033 | 2.721 | 1.471 |       |       |       |
| V   | 2.835 | 4.132 | 1.731 | 2.517 |       |       |
| VI  | 1.285 | 5.132 | 2.081 | 2.457 | 2.499 |       |
| VII | 5.082 | 2.058 | 3.664 | 4.099 | 5.940 | 6.145 |

The average in Table 8.2 of the distances between African and non-African populations is 4.270, while from Table 8.3 it is 4.326. These values are remarkably similar, and differ substantially from the value 6.47 used originally by Goldstein *et al.*, primarily because loci with extremely high variances have been omitted here. From eqn (8.8), with knowledge of the mutation rate and the mutational variance, the time of divergence of the populations can be estimated from the observed $(\delta\mu)^2$ value. Using the value 4.326, a generation time of 27 years (Weiss 1973), a mutation rate of $6.0 \times 10^{-4}$ and mutational variance of 2.5, the estimated time of divergence of African and non-African populations is about 38 900 years.

Table 8.4 reports the pairwise $(\hat{\delta}\mu)^2$ values for 22 tri- and 21 tetranucleotide polymorphisms studied by Bennett *et al.* (1998) in exactly the same individuals for which the distances for the 85 (and 29) dinucleotide loci are presented in Tables 8.2 and 8.3. Under the assumptions of our analysis, the ratio $(\delta\mu)^2_{\text{di}}/(\delta\mu)^2_{\text{tetra}}$, for example, should be $w_{\text{di}}/w_{\text{tetra}}$. For the African–non-African distance, the tetranucleotide value of $(\hat{\delta}\mu)^2$ is 2.137. The ratio $w_{\text{di}}/w_{\text{tetra}}$ is, therefore,

**Table 8.4**
Corrected $(\delta\mu)^2$ distances for 22 tri- and 21 tetranucleotides

|     | I tri | I tetra | II tri | II tetra | III tri | III tetra | IV tri | IV tetra | V tri | V tetra | VI tri | VI tetra |
|-----|-------|---------|--------|----------|---------|-----------|--------|----------|-------|---------|--------|----------|
| I   |       |         |        |          |         |           |        |          |       |         |        |          |
| II  | 1.214 | 1.446   |        |          |         |           |        |          |       |         |        |          |
| III | 0.414 | 0.493   | 0.939  | 1.109    |         |           |        |          |       |         |        |          |
| IV  | 1.147 | 0.858   | 0.442  | 0.608    | 0.532   | 0.919     |        |          |       |         |        |          |
| V   | 0.580 | 0.977   | 1.131  | 1.761    | 0.472   | 0.400     | 0.678  | 1.047    |       |         |        |          |
| VI  | 0.669 | 0.830   | 1.081  | 2.099    | 0.498   | 1.361     | 0.967  | 1.801    | 0.606 | 2.327   |        |          |
| VII | 1.054 | 3.047   | 0.650  | 0.706    | 0.940   | 2.418     | 1.221  | 2.203    | 1.419 | 3.002   | 1.000  | 3.767    |

estimated to be 4.326/2.137=2.024. Recall that from the world (pooled) variances we estimated $w_{di}/w_{tetra}$ as 2.086. For the tetranucleotides, the variance and distance calculations thus produce consistent ratios of the compound mutation parameter. For the trinucleotides, however, the average African–non-African distance in Table 8.4 is only 1.044, giving an estimate for $w_{di}/w_{tri}$ of 4.144, much greater than the estimate 2.166 obtained from the variances.

The assumption underlying the use of $(\delta\mu)^2$ entails that the two populations being compared have evolved independently since their formation from one ancestral population. An alternative view of the distance between two groups is an inverse measure of the extent of admixture or gene flow between them. We now discuss how statistics like $(\delta\mu)^2$ and the within-population variances produce estimates of gene flow for a special migration model.

## 8.7 Dynamics of microsatellites in an island model

Consider a set of $d$ populations in each of which we follow one locus subject to a simple symmetric SMM. All populations have $N$ diploid individuals, and the mutation rate $\mu$ is the same in all locations. The migration parameter $m$ is defined as follows: after mutation, each population is constructed by taking a fraction $1 - m$ of that population and a fraction $m$ made up of the average of all $d$ populations after mutation. Thus, if $n_i^k(t)$ is the number of chromosomes with $i$ repeats in population $k$ at time $t$, the next generation in population $k$ is produced by multinomial sampling of chromosomes using the probabilities

$$\pi_i^k(t) = \frac{1-m}{2N}\left\{(1-\mu)n_i^k(t) + \frac{\mu}{2}\left[n_{i-1}^k(t) + n_{i+1}^k(t)\right]\right\}$$
$$+ \frac{m}{2Nd}\sum_{l=1}^{d}\left\{(1-\mu)n_i^l(t) + \frac{\mu}{2}\left[n_{i-1}^l(t) + n_{i+1}^l(t)\right]\right\}. \tag{8.12}$$

The new average repeat score in population $k$, $r_k(t+1)$, given those at time $t$, becomes

$$E\left[r_k(t+1|t)\right] = (1-m)r_k(t) + \frac{m}{d}\sum_{l=1}^{d}r_l(t). \tag{8.13}$$

Hence

$$E\left[r_k(t+1) - r_l(t+1)|t\right] = (1-m)\left[r_k(t) - r_l(t)\right],$$

and all populations are expected eventually to have the same average repeat score.

Following the analysis by Moran (1975, as used by Goldstein *et al.* 1995*b*), write $V_k(t)$ for the variance in repeat scores in population $k$ at time $t$. Then, after a considerable amount of algebra, we find

$$EV_k\,(t+1|t) = \left(1 - \frac{1}{2N}\right)$$

$$\times \left\{ (1-m)V_k(t) + \frac{m}{d}\sum_{l=1}^{d} V_l(t) + \mu + \frac{m}{d}\sum_{l=1}^{d}[r_k(t) - r_l(t)]^2 \right.$$

$$\left. - \frac{m^2}{d^2}\left[\sum_{l=1}^{d}(r_k(t) - r_l(t))\right]^2 \right\}. \tag{8.14}$$

Write $\tilde{V}(t) = \sum_{k=1}^{d} V_k(t)$. Then eqn (8.14) may be rewritten as

$$E\tilde{V}\,(t+1|t) = \left(1 - \frac{1}{2N}\right)\left\{ \tilde{V}_k(t) + \mu d + \frac{m}{d}\left(1 - \frac{m}{d}\right)F(t) - \frac{m^2}{d^2}G(t)\right\},$$
$$\tag{8.15}$$

where

$$F(t) = \sum_{k}\sum_{j}[r_k(t) - r_j(t)]^2$$

and

$$G(t) = \sum_{k}\sum_{j}\sum_{\substack{l \\ j \neq l}}[r_k(t) - r_j(t)]\,[r_k(t) - r_l(t)].$$

The reader is now referred back to eqn (8.7) where $D_1$ is defined as the average of the squared difference between pairs of alleles, one chosen from each of two populations. For populations $k$ and $l$ we define analogously

$$D_1^{kl}(t) = \sum_{i}\sum_{j}\left[i^{(k)} - j^{(l)}\right]^2 n_i^k(t)n_j^l(t)/(2N)^2,$$

where $n_i^k$ and $n_j^l$ are the numbers with length $i$ and $j$, respectively, in populations $k$ and $l$, respectively, and the sum

$$\tilde{D}(t) = \sum_{\substack{k \\ k \neq l}}\sum_{l} D_1^{kl}(t).$$

After a considerable amount of algebra we have the elegant simplification

$$E\tilde{D}_1(t+1|t) = D_1(t) - \frac{2m}{d}\left(1 - \frac{m}{d}\right)F(t) + \frac{2m^2}{d^2}G(t) + 2\mu d(d-1). \tag{8.16}$$

To complete the iteration of eqns (8.15) and (8.16) we need recursions for $F(t)$ and $G(t)$. These are the most algebraically demanding but lead ultimately to the following iterations

$$
EF(t+1|t) = F(t)\left[(1-m)^2 - \frac{1}{2N}\left(1 - \frac{2m}{d} + \frac{2m^2(d-1)}{d^2}\right)\right]
$$
$$
+ \frac{\tilde{D}_1(t)}{2N} - \frac{2m^2(d-1)G(t)}{2Nd^2} + \frac{2\mu d(d-1)}{N} \qquad (8.17)
$$

and

$$
EG(t+1|t) = \tilde{V}(t)\frac{(d-1)(d-2)}{2N}
$$
$$
+ F(t)\left\{\frac{(d-2)(1-m)^2}{2} + \frac{(d-2)(d-1)}{2N}\frac{m}{d}\left(1 - \frac{m}{d}\right)\right\}
$$
$$
- \frac{(d-2)(d-1)m^2}{2Nd^2}G(t) + \frac{\mu d(d-1)(d-2)}{2N}. \qquad (8.18)
$$

The system of eqns (8.15), (8.16), (8.17) and (8.18) provides a complete linear recursion for the expectations of moments $\tilde{V}(t)$, $\tilde{D}_1(t)$, $F(t)$ and $G(t)$.

On inspection of eqns (8.15) and (8.16), we see the first equilibrium result, namely

$$
E\tilde{V} = (2N-1)\mu d^2.
$$

Notice that with two populations (i.e. $d=2$), $E(\tilde{V}) = E(V_1 + V_2) = 2E(V_1) = 2E(V_2) = 4\mu(2N-1)$, so that in each population, the expected variance is $\hat{V}_A = 2\mu(2N-1)$. In other words, with two populations exchanging immigrants at any positive rate, the equlibrium repeat score variance in each is double that in the absence of gene flow (eqn (8.5) above; Moran 1975; Goldstein *et al.* 1995b). Pritchard and Feldman (1996) also observed this phenomenon.

Equilibrium values for $E\tilde{D}_1$, $EF$ and $EG$ are obtained on inversion of the complete linear system of eqns (8.15), (8.16), (8.17), (8.18). These equilibria are given below:

$$
E\tilde{V} = \mu(2N-1)d^2, \qquad (8.19a)
$$

$$
E\tilde{D}_1 = \frac{2\mu d(d-1)\{d[1 - m^2(2N-1) + 4mN] - 4m\}}{m(2-m)}, \qquad (8.19b)
$$

$$
EF = \frac{2\mu(d-1)d^2}{m(2-m)}, \qquad (8.19c)
$$

$$
EG = \frac{\mu d^2(d-1)(d-2)}{m(2-m)}. \qquad (8.19d)
$$

It is worth noting that $E(\tilde{D}_1)$ may be expressed as a linear combination of $E(\tilde{D}_0)$ and $E(F)$, namely $E(\tilde{D}_1) = \beta_1 E(\tilde{V}) + \beta_2 E(F)$, where $\beta_1 = 2(d-1)$ and

$\beta_2 = 1 + 2m(1 - 2/d)$. With two populations, $\beta_1 = 2$, $\beta_1 = 1$, we have a result equivalent to eqn (8.7) above.

In analyses of the island model, it is usual to ignore terms of order $m^2$ and smaller, to set $2N - 1$ to $2N$ and to ignore $m$ relative to $Nm$. Then, to this degree of approximation, from eqn (8.19b)

$$E\tilde{D}_1 \simeq \frac{\mu d^2(d-1)(1+4Nm)}{m},$$ (8.20)

so the average pairwise value of $D_1$ at equilibrium is

$$\bar{D}_1 = \frac{1}{d(d-1)}E\tilde{D}_1 = \frac{\mu d(1+4Nm)}{m}.$$ (8.21)

Also, from eqn (8.19a), the equilibrium average within population variance is

$$\bar{V} = E(\tilde{V}/d) \simeq 2N\mu d.$$ (8.22)

From eqns (8.21) and (8.22), using the definition in eqn (8.7), the average equilibrium value of $(\delta\mu)^2$ over all pairs of populations is

$$\overline{(\delta\mu)}^2 = \bar{D}_1 - 2\bar{V} = \frac{\mu d(1+4Nm)}{m} - 4N\mu d = \mu d/m.$$ (8.23)

Finally we have

$$L_m = \bar{V}/\overline{(\delta\mu)}^2 = 2Nm.$$ (8.24)

In other words, at equilibrium, the ratio $L_m$ of the average variance within populations to the average distance between populations provides an estimate of $2Nm$.

In order to reconcile this approximation with Slatkin's (1995) use of $R_{ST}$, we will assume that the size $n$ of the sample from each of the $d$ sampled populations is large enough that we may write $(2n-1)/(2nd-1) \simeq d^{-1}$ and $2n(d-1)/(2nd-1) \simeq (d-1)/d$. Then, as indicated by Slatkin (1995), his $S_W$ is equivalent to $2\bar{V}$ in our notation, while his $S_B$ is equivalent to our $\tilde{D}_1/d(d-1)$ and, according to his eqn (15),

$$\left[\frac{1}{R_{ST}} - 1\right] = \frac{d}{d-1}\left[\frac{S_W}{S_B - S_W}\right] = \frac{d}{d-1}\left[\frac{2\bar{V}}{\bar{D}_1 - 2\bar{V}}\right].$$

His $M_R$, used to estimate $Nm$, is defined by

$$M_R = \frac{d-1}{4d}\left(\frac{1}{R_{ST}} - 1\right) = \frac{1}{2}\left(\frac{\bar{V}}{\bar{D}_1 - 2\bar{V}}\right)$$ (8.25)

in our terminology, and referring to eqn (8.24), under the approximations we have made, $\bar{V}/(\bar{D}_1 - 2\bar{V}) = \bar{V}/\overline{(\delta\mu)}^2$ estimates $2Nm$. $L_m$, defined by $\bar{V}/\overline{(\delta\mu)}^2$, is suggested as an estimate of $2Nm$ when sample sizes are at least moderately large, but not necessarily equal. $M_R$ (in eqn (8.25)) was originally derived under the assumption of equal sample sizes, but in light of eqn (8.25), for moderately large samples, the values of $L_m$ and $2M_R$ should be fairly close.

If the samples from each population are not really small, the estimate given by eqn (8.24) of the average variance to the average pairwise genetic distance should be adequate to estimate $2Nm$ in this island model.

In the case of just two populations ($d = 2$), it is usual to express admixture in terms of a parameter $v$, say, where population I in generation $t + 1$ is produced by fractions $1 - v$ from population I and $v$ from population II at time $t$. Thus, $v$ is the extent of admixture, and in the model the admixture continues at each generation during the evolution. In terms of the parameter $m$ from the island model, we have $v = m/2$. For two populations, the dynamics under migration are simpler. Write $H$ for $(r_1 - r_2)^2$. Then we have

$$E\left[D_1(t+1)|t\right] = D_1(t) - 2v(1-v)H(t) + 2\mu, \qquad (8.26a)$$

$$E\left[H(t+1)|t\right] = D_1(t)/2N + \delta H(t) + 2\mu/2N, \qquad (8.26b)$$

where $\delta = (1 - 2v)^2 - (1 - 2v + 2v^2)/2N$. Hence there is convergence to

$$E(D_1) = 2\mu \left\{ \frac{1}{2v(1-v)} + 2(2N-1) \right\}, \qquad (8.27a)$$

$$E(H) = \frac{\mu}{v(1-v)}. \qquad (8.27b)$$

Recalling that $D_1$ in eqn (8.19b) is $2D_1$ in eqn (8.27a) and $F$ in eqn (8.19c) is $2H$, these equations (8.27) are the same as the equilibria (8.19b) and (8.19c). Of course, as in eqn (8.19a), $E(\tilde{V})$, which is the sum of the expected variances in populations I and II, comes to $4\mu(2N - 1)$. Substituting this into eqns (8.27a) and (8.27b), we obtain the two-population version of eqn (8.24): $4Nv = 2Nm$ is estimated by the average of the two populations' variances divided by the estimate of $(\delta\mu)^2$ between them. This ratio may be regarded as a measure of population affinity; the more migration (i.e. admixture) the greater is the ratio (and the smaller the standardized distance).

## 8.8 More on statistical measures of subdivision

Both $L_m$ and $2M_R$ were shown in the previous section to estimate $2Nm$ for an island model of population subdivision with migration. Although the model of mutation considered here is specifically chosen to represent microsatellites, measures similar to $L_m$ and $R_{ST}$ have been used for many years to partition variation in gene frequencies into within- and between-group components. The most widely used measures are $F$-statistics; for example the classical island model produces an equilibrium $F_{ST}$ of $(1 + 4Nm)^{-1}$.

These $F$-statistics involve analysis of variance in allele frequencies with a rationale similar to classical random effects in ANOVA models. The partition is described in some detail by Weir (1996, pp. 170–184) and has been adapted in a straightforward way by Michalakis and Excoffier (1996) to the case of microsatellites. Here, the repeat score of allele $j$ from population $i$ ($i = 1, 2, \ldots, d$) is viewed as an observation on the random variable $\Upsilon_{ij} = \mu + A_i + \epsilon_{ij}$, where $\mu$ is an overall mean, $A_i$ is a random variable respresenting the random effect of population $i$, with zero expectation and variance $\sigma_A^2$, and $\epsilon_{ij}$ is an error random variable within population $i$, with zero mean and variance $\sigma_w^2$. In this framework, the intraclass correlation coefficient $\theta = \sigma_A^2/(\sigma_A^2 + \sigma_w^2)$ tells us the relative magnitude of between-group contribution to the total variation (Weir 1996). The standard estimator $\hat{\theta}$ of $\theta$ is given by

$$\hat{\theta} = \frac{MSB - MSW}{MSB + (n_0 - 1)MSW},$$ (8.28)

where $MSB$ and $MSW$ are the mean squared deviations between and within populations, respectively,

$$n_0 = \frac{1}{d - 1}\left[2\mathcal{N} - \frac{\sum_{i=1}^{d}(2n_i)^2}{2\mathcal{N}}\right],$$ (8.28a)

with $2n_i$ the number of chromosomes sampled from population $i$, and $2\mathcal{N} = \sum_{i=1}^{d} 2n_i$ the total number of chromosomes sampled. When the samples from each population are the same size, that is $2n_i = 2n$ for $i = 1, 2, \ldots, d$, $n_0$ reduces to $2n$. Michalakis and Excoffier (1996) denote the ratio of the averages over loci of the numerators in eqn (8.28) to the average over loci of the denominators in eqn (8.28) as $\hat{\phi}_{ST}$ and claim that $\hat{\phi}_{ST} = (1 - c)R_{ST}/(1 - cR_{ST})$, where with equal $n_i$, $c = (2n - 1)/(2nd - 1)$.

Slatkin's original formulation of his $R_{ST}$ was in terms of quantities $S_B$, $S_W$, and $\bar{S}$ with

$$R_{ST} = \frac{\bar{S} - S_W}{\bar{S}} = \frac{S_B - S_W}{S_B + (2n - 1)/(2n(d - 1))S_W}.$$ (8.29)

As is remarked by Slatkin, $\bar{S}$ and $S_W$ have interpretations in terms of the variance in the whole system (without partitioning), and the average variance within groups, respectively. Under this interpretation, if we use the notation standard in analysis of variance, and follow the suggestion of Slatkin after his eqn (10), then

$$R_{ST} = \frac{MSB - MSW}{MSB + [d(2n - 1)/(d - 1)]MSW}.$$ (8.30)

Then, comparing eqns (8.28) and (8.30) we see that

$$\hat{\phi}_{ST} = \frac{R_{ST}}{1 - c + cR_{ST}}.$$ (8.31)

From eqn (8.31) it is obvious that

$$R_{ST} = \frac{(1-c)\hat{\phi}_{ST}}{1 - c\hat{\phi}_{ST}}. \tag{8.32}$$

which is the same as eqn (9) of Michalakis and Excoffier (1996) but with $R$ and $\hat{\phi}$ interchanged. Note that the correct version, eqn (8.32) is given by Rousset (1996, eqn 17). It is worth noting that when $n_i = n$, the expectation of $R_{ST}$ is $[1 + 4Nmd/(d-1)]^{-1}$ while that of $\hat{\phi}_{ST}$ is $[1 + 4Nm(2N/(2N-1))]^{-1} \approx (1 + 4Nm)^{-1}$.

Both $R_{ST}$ and $(\delta\mu)^2$ have been used as genetic distances from which to construct evolutionary trees of relationship among populations. From eqns (8.24) and (8.25), we see that in an island model (and in particular, in a two-population model of admixture, as shown with eqn (8.27)), $M_R$ or $L_m/2$ both estimate $Nm$. $Nm$ may be regarded as an index of closeness between two populations. By the same token, a matrix of pairwise values of $(M_R^{-1})$ or $(L_m/2)^{-1}$ may be treated as a set of distances from which statistical clustering of the populations may be visualized. Applying these measures to 85 dinucleotide loci from Jin *et al.* (unpublished data) produces clusters that match the earlier allele-sharing tree quite closely, with $(L_m/2)^{-1}$ apparently closer than $(M_R)^{-1}$ to the tree obtained by Bowcock *et al.* (1994) using an allele-sharing distance on 30 dinucleotide loci.

## 8.9 Models with mutational constraints

Two departures from Ohta and Kimura's infinite range SMM have been proposed. One sets a finite interval of length $R$ in which repeat numbers vary and forces alleles at the boundaries to mutate only to the interior of the range. This might be termed a *hard boundary* model. The other model takes a focal repeat score $r_m$ and assumes that repeat scores greater than $r_m$ tend to mutate to lower scores while those lower than $r_m$ mutate upwards. This introduces a bias $b_i$ in the mutation from alleles with $i$ repeats which is usually assumed to be linear: $b_i = B(i - r_m)$ with $B < 0$. This model might be described as having *soft* boundaries.

For the hard boundary model, Goldstein *et al.* conjectured that the expected value of $D_1$ (see eqn 8.7) would approach the equilibrium $(R^2 - 1)/6$. This was shown to be the case by Nauta and Weissing (1996). Because $D_1$ converges, in order to construct a distance that is linear with time for a reasonable period, a correction must be made to $(\delta\mu)^2$. Feldman *et al.* (1997) suggest a distance

$$D_L = \log\left[1 - \sum_{i=1}^{L}(\delta\mu)_i^2/LM\right] \tag{8.33}$$

with $L$ loci, where $M$ is the average value of the distance at maximal divergence. $D_L$, which was derived in the special case where $R$ and $\mu$ do not vary across loci, is linear for a reasonable period of time which increases as the number of loci, $L$, increases. Pollok $et\ al.$ (1998) have improved on $D_L$ by using a weighted least squares technique originally introduced by Goldstein and Pollock (1994). Interestingly, these corrected distances do not seem to be sensitive to moderate variation in the repeat range or the mutation rate.

The linear soft boundary model was introduced by Garza $et\ al.$ (1995), who compared average allele sizes in humans and chimpanzees and found that these were sufficiently similar that some kind of evolutionary constraint on repeat lengths was likely to have been in effect. They concluded from their analysis that the bias, measured by $B$, is likely to have been quite weak, substantially less than the mutation rate itself. A different mathematical approach to the same model was taken by Zhivotovsky $et\ al.$ (1997), who developed an estimator for $B$ of the form

$$\hat{B} = \frac{-\sigma_m^2}{2[\text{Var}(\bar{r}) + \bar{V}]}, \tag{8.34}$$

where $\bar{r}$ estimates the mean repeat number at a locus, $\text{Var}(\bar{r})$ is the variance across loci of these estimated means, $\bar{V}$ is the average over loci of the within-locus variances, and $\sigma_m^2$ is the usual variance in mutation sizes. The data of Garza $et\ al.$, when inserted into eqn (8.34), produced $\hat{B}$ between $-0.0064$ and $-0.013$, substantially higher than the mutation rate $\mu$. Their analysis also allowed Zhivotovsky $et\ al.$ to estimate the time since two populations subject to this biased mutation diverged from a common ancestral population. Assuming $\sigma_m^2 = 2.0$, for example, and $B = -0.02$, the data of Bowcock $et\ al.$ (1994) gave an estimated divergence time for African and non-African populations of 100 000 years. With $B = -0.0064$, the estimate was 84 000 years. In both cases, a 27-year generation was used. Estimated divergence times are sensitive to $\sigma_m^2$ as well as to $B$.

# 9 The coalescent and microsatellite variability

Peter Donnelly

## Chapter contents

## Abstract

The coalescent captures the genealogical relationships amongst DNA sequences at a locus within a population. We motivate and describe its probabilistic structure in a variety of demographic scenarios, and discuss models for microsatellite mutations. The paper outlines applications of coalescent methods for microsatellite data, an urn scheme for simulating samples is given, qualitative and quantitative insights are described, and statistical methods for multilocus estimation and testing of demographic scenarios, and likelihood-based inference, are highlighted.

## 9.1 Introduction

One of the important advances in theoretical population genetics over the last two decades has been the use of so-called coalescent methods. The idea of the approach is to focus attention on the genealogical tree which relates the ancestral history of segments of DNA sampled from populations within a species. At a technical level, study of the probabilistic structure of this tree is often simpler than the classical diffusion approximations, which trace the composition of the entire population through time. Further, the patterns of genetic variability observed in samples, and the dependence structure in such data, is a direct consequence of the shape of the underlying genealogical tree, so that coalescent methods have proved important in the development of statistical approaches to modern population genetics data.

Extended reviews of the coalescent approach in various genetical contexts have been given, for example, by Hudson (1991, 1992), and by Donnelly and Tavaré (1995). Our aim here is more limited. We hope to explain the idea behind coalescent methods, and describe the coalescent itself in two commonly studied demographic scenarios. The coalescent provides a useful tool at a variety of different levels, and we aim, informally, to give a flavour for its application in connection with microsatellite loci for simulation, qualitative insights, calculations, and statistical inference.

## 9.2 The coalescent

Suppose we take a sample of $n$ chromosomes from a population and focus attention on a particular locus. That is, we take as the unit of interest the segment of DNA containing this locus.

Consider first the case in which only two such segments are sampled. Going back one generation, either these two segments will be descended from the same segment in the previous generation, or they will be descended from different segments. In the latter case, we can trace the two segments back a further generation, and they will either have the same or distinct ancestral segments, and so on. (We will often speak of an 'ancestor' as a shorthand for 'ancestral segment'.) This process can be continued until the two segments first share an ancestor. The number of generations this takes will depend on the demographic history of the population at that locus. Picture this ancestral history as a very simple tree with a root and two tips. The tips represent the sampled segments and the root represents their most recent common ancestor. The depth of the tree is just the number of generations since this common ancestor.

The relationships we observe between the genetic types in the two sampled copies of the locus will depend on the depth of this ancestral tree. If there were no mutation acting at the locus, then both segments would be identical. If there is mutation, then it will act independently along the two branches leading from the root to the sampled segments. The longer these branches are (i.e. the longer the time since the common ancestor of the segments) the more likely it is for mutations to have made the two segments different.

More generally, imagine tracing back the ancestry of the sample of $n$ segments. This will give rise to a genealogical tree which describes the ancestral relationships amongst the sequences (analogous to the way in which a phylogenetic tree describes relationships among species). The tree will tell us how many generations ago specific segments first shared a common ancestor, how many sequences in the population at particular times in the past were ancestral to the sampled sequences, and which of the sampled sequences were descended from each of these ancestors, and so on. The root of the tree represents the most recent common ancestor (MRCA) of the sampled copies of the locus, and the depth of the tree the time since that ancestor. Again, the shape of the tree will depend on the demographic history of the population at that locus.

The patterns we observe in the genetic data at the locus in question will depend on the shape of the tree. In particular, they result from the interplay between the forces of mutation and the tree structure. Again, in the absence of mutation all sampled segments would be identical. Mutation acts along the branches of the tree. The shorter these are, the more likely it is that the segments will share the same genetic type, and conversely.

What sorts of tree shapes are likely in this context? In general that depends on our model for demography. If we have a stochastic model for the demography of the population, then this induces a stochastic model for the genealogical tree of a sample: if we knew all the relevant demographic events (i.e. which segments were descended from which in the ancestry of the population) then we would simply be able to draw the tree by tracing the ancestral history of the sample back in time. If, as is usual, we have a probability model for the demography, then this leads (at least implicitly) to a probability model for the genealogical tree describing ancestral history.

One simple model for the demography at a neutral locus in a panmictic population of constant size $N$ is the so-called Wright–Fisher model. (We will use $N$ to denote the number of copies of the locus in the population. For a diploid locus this will be twice the number of individuals.) The model posits non-overlapping generations, and the probability structure is *as if* each segment in the current generation independently and uniformly chose its 'parent' (the segment from which it is copied) from amongst all the copies of the locus in the previous generation.

It turns out that when the population size $N$ is large, the changes in gene frequency caused by the randomness inherent in the demography (so-called 'genetic drift') only happen over periods of time of the order of $N$ generations. In other words, to see changes in gene frequencies we need to 'speed time up' by a factor of $N$, or equivalently, to measure time in units of $N$ generations.

Now we may return to a more specific version of our earlier question: what kinds of genealogical trees should we expect to see for a neutral locus in a panmictic population of constant size $N$, when $N$ is large and the demography is described by the Wright–Fisher model. The answer is that with time measured in units of $N$ generations, the (random) genealogical tree is an object called the coalescent (or strictly, for a sample of size $n$, the $n$-coalescent). That is, the coalescent is a probability model for genealogical trees which arises naturally in the models for population demography.

The probability structure of coalescent trees is simple. Starting from the $n$ tips, they have $n$ branches for a period of time $T_n$, after which a randomly chosen pair of branches coalesce. The tree then has $n - 1$ branches for a period of time,

$T_{n-1}$, after which a randomly chosen pair of branches coalesce. The process continues, so that there are $k$ branches for a period $T_k$, $k = n, n - 1, \ldots, 2$. The number of branches always decreases by exactly one, and when it does so, two of the existing branches are chosen uniformly at random to coalesce. The periods of time $T_n, T_{n-1}, \ldots, T_2$, are independent exponential random variables with means $2/[n(n - 1)]$, $2/[(n - 1)(n - 2)], \ldots, 1$, respectively.

If we take a time-slice through the tree at a particular time in the past (more recently than the MRCA of the sample) there will be a random number of branches. Each branch corresponds to an ancestor of the sample in the population at that time. Of course there were many other copies of the locus in the population in that generation. Most will not have contributed genetic material to the sample. In looking back in the past, the coalescent focuses attention *only* on segments ancestral to those in the sample. One practical consequence of this is that coalescent-based simulations are much more efficient than traditional methods.

As we have described it, the coalescent provides a very good approximation to the random genealogical history of samples taken from a (neutral, panmictic, constant-population-size) Wright–Fisher model when the population size $N$ is large, with time measured in units of $N$ generations. In fact, it can be shown that the coalescent also arises for a wide range of other (neutral, panmictic, constant-population-size) demographic models. To use the coalescent in these other models all that is needed is to measure time in units of $N_e$ generations, where $N_e$ is the so-called variance effective population size: the actual population size divided by the variance of the number of copies in the next generation of a particular copy of the locus in the current generation. (For the Wright–Fisher model, $N_e = N$.) This robustness is reassuring. It means that coalescent methods apply to a wide range of demographic models. (It is closely related to traditional diffusion approximations in population genetics, in which particular diffusion processes provide a good approximation to gene frequency changes in large populations for a range of demographic models, provided time is measured in units of $N_e$ generations.)

Much is known about the structure of coalescent trees. We merely point to some features of the constant-population-size case here. At a particular locus within a population there will have been a specific version of the coalescent tree. We can consider properties of these trees as they vary between (conceptual) repetitions of the evolutionary process at that locus. Such properties will also pertain to coalescent trees at distinct, unlinked, loci within a single realization of the evolution of the population. The mean depth of an $n$-coalescent tree is $2(1 - 1/n)$. The variance of the depth is about 1.16, so the standard deviation of the depth is more than half its mean. The mean and variance of $T_2$, the time for which there are only two ancestors, are both 1. In other words, on average the time for which there are two ancestors will account for over half the depth of the tree, and most of the variability in this depth. Some pictures of coalescent trees are given, for example, in Donnelly and Tavaré (1995) and Donnelly (1996). The features just described, of considerable variability between trees, and much of the depth being accounted for by the final two branches, are striking.

What happens to the shape and structure of these genealogical trees if we weaken some of the demographic assumptions? Quite a lot is now understood in more general contexts, for example variation in population size or geographical population structure, and we refer the interested reader to Donnelly and Tavaré (1995) and references therein (or Donnelly 1996 for some pictures). One part of this story concerns the effect on the shape of coalescent trees of differing assumptions about the size of the population. The general position is rather complicated (Donnelly and Kurtz 1999): in a loose sense the effect depends on the realized population sizes *and* the internal dynamics of the birth rates in the population. Most authors consider a particular special case which arises if, for example, individuals in the population behave independently reproductively, or if, in a suitable sense, the variation in population size is extrinsic to the population (Kingman 1982; Griffiths and Tavaré 1994*b*). For simplicity we describe this in the context of a population which has grown in size, forward in time. Looking backwards in the coalescent, the effect is to decrease coalescence rates when the population is large, and to increase these rates when it is small, relative to coalescence rates in a population with a constant, intermediate, total size. This is equivalent to stretching the coalescent tree near the tips, and contracting it near the roots. In the usual coalescent tree, the branches near the root (i.e. near the MRCA) are typically substantially longer than those near the tips. In an extreme case of very sudden growth, the tree can be changed so that effectively all the coalescences happen at or near the root. The tree then becomes 'star-shaped' in the sense that all ancestral lineages remain separate until the MRCA, at which they all coalesce.

Two particular effects on expected genealogies are noteworthy in this setting of sudden substantial population growth. (They apply to a lesser extent for less extreme versions of population growth.) The first is that there is very little variability in the shape and depth of trees between realizations of evolution, or equivalently, between the trees which arise at different, unlinked loci within a population, which is in stark contrast to the shape of coalescent trees at unlinked loci within a constant-sized population. The second stems from the fact that all the lineages in the tree separate at the root. This means that each sampled copy of the locus has evolved independently (under the effects of mutation) since the common ancestor of the sample. There is then much less dependence between the types of sampled chromosomes at the locus than there is in a sample from a population of constant size. We will see below that these differences in tree structure *between* unlinked loci within the same population can be very helpful for testing between different demographic scenarios.

We have considered genealogy at a single locus at which there is no intra-locus recombination. Genealogical structure is also understood at linked loci, or more generally for DNA sequences, between or within which recombination acts, and at a neutral locus linked to a locus undergoing various forms of selection. (See Donnelly and Tavaré 1995 for references.) Recent progress has also been made when selection acts directly on the locus of interest (Neuhauser and Krone 1997, Donnelly and Kurtz 1999).

# 9.3 Modelling the microsatellite mutation process

One of the attractions of the coalescent approach is that (for neutral loci) it separates the randomness inherent in the demography of the population from consideration of any particular genetic system. All of the discussion of the preceding section applies, whether one is thinking of a classical bi-allelic marker, a locus at which sequence information is available, a micro- (or mini-)satellite at which only repeat-copy number is measured, an Alu insertion, or a tightly linked combination of several loci from one or more of these categories. In the context of a specific genetic system (and we concentrate attention here on microsatellites) one simply superimposes the effects of mutation on the ancestral history of the genes in a sample.

In any population genetics model of microsatellite evolution, it is necessary to specify a model for the effect of mutation. Throughout, we will measure allele length by the number of copies of the repeat unit. Most authors have focused on versions of the so-called *generalized stepwise mutation model*. This model makes two key assumptions. The first is that the rate of mutation to a particular allele is the same for all alleles at the locus. The second is that the random amount by which a mutation changes allele length does not depend on the length of the progenitor allele. As specified here, the distribution of length change can be quite general—it need not have mean zero, nor necessarily be symmetric. Many authors have imposed additional restrictions, usually to facilitate analysis of the model. The most commonly studied case assumes in addition that all mutational changes involve only a single repeat unit, and that the gain or loss of a repeat unit are equally likely. As we note below, analysis of properties of the model is possible without any restrictions on step-size distribution, both by coalescent and more classical methods.

Limited direct information on the mutation process in humans is available from family studies (e.g. Weber and Wong 1993), and although microsatellite loci tend to have much higher mutation rates than is typical for many other loci, these are still small enough that information on mutation changes must be pooled across loci. One obvious disadvantage is that heterogeneities in mutation mechanism between loci will be lost. Another is that the results will be biased by loci with higher mutation rates, so that if there were, for example, a systematic difference in mechanism for loci with differing mutation rates (perhaps because of the length of the repeat unit) the picture from combining observations of germ line mutations across loci would not necessarily be typical of particular loci. Some indirect information comes from studies in yeast (e.g. Sia *et al.* 1997*b*; Wierdl *et al.* 1997), and cross-species comparisons. A recent study (Di Rienzo *et al.* 1998) examined somatic mutations in colon cancer patients at 20 microsatellite loci, and argued, on the basis of comparisons of their estimates with the variability observed in samples from three different populations, that the observed patterns in somatic mutations reflected those in the germline. One conclusion of the study was that there is substantial heterogeneity between loci in the mutation mechanism, and further that many mutations involved changes of more than a single repeat unit.

At the very least, in view of continuing uncertainty over these mechanisms, and evidence for heterogeneity of mutation sizes within and across loci, it would seem prudent to base theoretical studies on as general a model for mutation as possible. Alternatively, results based for example on an assumption of single-step changes and no mutational bias should be checked thoroughly for sensitivity to departures from these assumptions.

## 9.4  Applications of coalescent methods

We briefly review the ways in which coalescent methods may be helpful in analysing population genetics models for, or data from, microsatellite loci. In many ways they provide tools which are complementary to, and often simpler than, those of classical population genetics theory. We consider four related types of application: simulation, qualitative insights, calculation, and statistical inference. In each case, space constraints mean that we can do little more than give a flavour of the application. In particular, many important applications of coalescent methodology to microsatellites must be omitted. Our choice should be seen as a matter of personal taste rather than systematic deliberation.

### 9.4.1 Simulations

Coalescent methods are ideally suited to simulation of samples from a particular locus in an evolving population. A 'traditional' approach to this problem would be to simulate the evolution of the whole population for long enough to ensure that the population is at stationarity, and then to take a sample of the required size from the simulated population. Much of this simulation is wasted because very few of the chromosomes present in the population in past generations contribute genetic material to the sample. As a consequence, from the point of view of the sample, there is no point in keeping track of the alleles on these chromosomes at the locus, nor in troubling to decide which are subjected to mutation and what changes result from these mutations. In a sense, the whole point of the coalescent tree is that it tells us *exactly* which of the chromosomes in past generations matter. It is only those which are ancestral to the sample. Simulation of the coalescent tree involves much less computational expense than simulating the entire demography of the population into the past. Further, one only needs to worry about possible mutation events to the chromosomes on the coalescent tree rather than to the entire population.

It turns out that in the setting of populations of constant size, there is a very simple urn-type scheme for simulating a sample of size $n$ using the coalescent. We describe it in the specific context of a microsatellite locus under the generalized stepwise model of mutation. For a description in a more general context, see Donnelly and Tavaré (1995). As Moran (1975) and subsequent authors have noted, there is no sensible stationary distribution for this model. One way round this is to measure allele length relative to the common ancestor of the sample.

We do this in what follows. If the mutation probability is $u$ per gene per generation at the locus of interest, write $\theta = 2Nu$ for the usual scaled mutation rate. Further, denote by $F$ the distribution which models the change in allele length caused by a mutation. (Thus, for example, in the simplest version of the model this distribution would assign probability $1/2$ to a gain of one repeat unit, and probability $1/2$ to a loss of one repeat unit.) We describe the following algorithm in terms of an arbitrary value $A$ for the number of repeats in the common ancestor of the sample. All lengths can be measured relative to this, so that in practice this should be set to some arbitrary value. We describe the algorithm in terms of an urn containing balls. With each of these balls is associated a type or allele length. All probabilistic choices made in the algorithm are independent.

## 9.4.2 Algorithm for generating samples from a constant-sized population with the generalized stepwise model for mutation

0. Start with two balls in an urn, each of which is of type A.
1. Choose a ball uniformly at random from those currently in the urn. Denote by $k$ the number of balls currently in the urn.
   (i) With probability $1/(k - 1 + \theta)$, return the ball to the urn together with an additional ball of the same type.
   (ii) With probability $\theta/(k - 1 + \theta)$, change the type of the ball by choosing a mutation size from the distribution $F$ and adding (or subtracting, for a loss of repeat units) this number of repeats from the type of the ball. Return the ball (and no additional balls) to the urn carrying its new type.
2. If there are fewer than $(n + 1)$ balls in the urn, return to step 1. Otherwise discard the last ball added to the urn and stop.

The types of the $n$ balls in the urn when the process stops have the same distribution as a sample of size $n$ from the model (where, as noted, only the allele lengths relative to the length of the common ancestor of the sample are important).

## 9.4.3 Qualitative insights

We have seen that the probability distributions describing patterns of variability to be found in samples result from the superposition of the effects of mutation on the ancestral tree given by the coalescent. An understanding of the structure of coalescent trees can then be extremely helpful in gaining qualitative insights into the behaviour of the models. We illustrate this with two examples.

Recall that for populations of constant size, on average half the depth of the coalescent tree consists of the time for which there are only two branches. Further, much of the variability in the depth of the tree will be associated with this time, and for many trees it will account for considerably more than half the depth. Consider the consequences for the sample of a genealogical tree in which this time is large relative to the remainder of the tree. Imagine going forward in time from the MRCA, superimposing the effects of mutation. Immediately 'after' the root, the types of the two ancestors of the sample will be the same. During the period during which there are two ancestors, mutations will be changing the types of these two ancestors. For a tree for which this time is substantial, the two ancestors may well have rather different types X and Y, say, at the end of this time period. Now each of the sampled alleles will be descended from one or other of these original two ancestors. It turns out that in the coalescent the number descended from a particular one of the ancestors is uniformly distributed on each of the possible values, $\{1, 2, \ldots, n - 1\}$ (Kingman 1982). Divide the sample into two groups depending on which of the original two ancestors they are descended from. If there were no further mutation on the tree, then all of the alleles in one group would be of type X and those in the other of type Y. In fact, there will be mutation on the remainder of the tree, but in the (typical) situations when the length of the tree is dominated by the period for which there are two ancestors, there will not be much time for mutation to act to change the types in the groups. In this case, the sample will tend to consist of two 'clumps' of alleles, one clump consisting of allele lengths close to X, with the other having alleles close in length to Y. If we were to plot a histogram of allele lengths in the sample, this would then be bimodal, with one mode for each of the clumps. In summary then, we gain the insight that it will be commonplace for samples (from populations of roughly constant size, evolving under neutrality with, say, the generalized stepwise model for mutation) to exhibit a bimodality in allele lengths, a fact easily borne out by simulations. We note that some authors, on observing exactly such a pattern in data, have assumed it to be evidence of much more complicated phenomena. In particular, a bimodal distribution of allele lengths does not indicate a complicated mutation process with a tendency to produce bimodal distributions of allele lengths.

Our second example of qualitative insights comes from comparisons of trees across loci under different demographic scenarios. Recall that for a constant population size model there is very substantial variability in the shape and depth of trees across unlinked loci from the same population. In contrast, for populations which have grown substantially in size forward in time (under the sort of dynamics described above), trees for different loci have similar depths and similar, star-like, shapes. Actual data at these loci will result from the effects of mutation on the genealogical trees. Nonetheless, it follows that, in a sense which we will make more precise below, there will be much greater variability in patterns of data *between* loci in the constant population size case than in the setting in which the population has grown rapidly.

## 9.4.4 Calculations

Coalescent methods provide an additional collection of tools for quantitative analysis of population genetics models. Exactly because many properties of such models are easily related to genealogical structure, these tools are often very convenient for analysis.

For data from a microsatellite locus, one commonly used statistic which summarizes the observed diversity is $S^2$, the variance of allele lengths in a sample from the population at that locus. (That is, $1/(n-1)$ times the sum of squared differences of the observed allele lengths from the mean allele length in the sample.) Some algebra shows that $S^2$ can be rewritten in terms of squared differences of allele lengths:

$$S^2 = \frac{1}{2n(n-1)} \sum_{i \neq j} (X_i - X_j)^2, \tag{9.1}$$

where $X_1, X_2, \ldots, X_n$, are the allele lengths in the sample. Analysis of the properties of $S^2$ is facilitated by the representation (eqn (9.1)). For example, $E(S^2) = E(X_1 - X_2)^2/2$, from which it can be shown that

$$E(S^2) = Cu\eta_2, \tag{9.2}$$

where, as above, $u$ is the mutation probability per gene per generation, and $\eta_2$ is the mean square mutation size, that is the average squared size of a mutation. In eqn (9.2) the expectation refers to averages over independent repetitions of evolution, or equivalently, the average over unlinked loci, and the constant $C$ is the expected number of generations since the common ancestor of a pair of chromosomes. The result (eqn (9.2)) depends on the mutation following the generalized stepwise model of mutation, but is otherwise rather general. In particular it holds whatever the demographic model for the population, although the value of the constant $C$ will, of course, depend on the demography. For example for a population of constant (haploid) size $N$ evolving according to the Wright–Fisher model, we have $C = N$, while for a population which expanded rapidly from an initial small size $T$ generations ago, we have $C \approx T$.

We mention in passing that eqn (9.2) has interesting consequences for the amount of diversity to be expected under various different assumptions about mutation sizes. For example, suppose we compare two models, each with the same mutation rate, in which all mutations involve a change of only one repeat unit: in the first model, gains and losses are equally likely, while in the second all mutations increase allele length. It might be argued that in the second case, since all mutations are in the same direction, there would be less diversity in samples from populations. (An alternative argument, in the other direction, is that in the first case mutations might 'cancel out', thus reducing observed diversity.) In fact, in both models for mutation, the average squared mutation size, $\eta_2$, is 1, so that if diversity is measured here by $S^2$, an application of eqn (9.2) shows that the average diversity would be the same in each case. Slightly more generally, for models in which changes of only one repeat unit are possible, the expected diversity (measured by $S^2$) is the same *regardless of the degree of bias in the mutation mechanism*. Note in general that the mean of $S^2$ only depends on the distribution of mutation sizes through $\eta_2$, the mean square mutation size. (For unbiased mutation distributions, in which the mean change is 0, $\eta_2$ equals the variance of the mutation size.)

Rather more complicated analyses lead to expressions for the variance of $S^2$, although these are currently available only for the constant population size case and the case of the star-shaped genealogy which arises under rapid expansion from a small size. In the former case the result is due initially to Roe (1992) in the case of symmetric distributions for mutation size, and extended to general distributions by Kimmel and Chakraborty (1996). For special cases see also Zhivotovsky and Feldman (1995) and Pritchard and Feldman (1996). In the star-shaped case, see Kimmel and Chakraborty (1996), (although there appears to be a misprint in their equation 23), and Di Rienzo *et al.* (1995, 1998). We will not repeat the expressions here, but instead highlight the fact that in the constant-population-size case the variance, between unlinked loci, in $S^2$, will be very large. This variance is due largely to stochastic effects in evolution ('genetic drift'), and cannot be reduced by increasing the sample size at each locus. In contrast, for the rapid expansion scenario considered, the variance in $S^2$ between unlinked loci is much smaller, and in particular it decreases like the inverse of the size of the sample.

### 9.4.5 Statistical inference

In principle, samples from microsatellite loci contain important information about the demographic and evolutionary processes which have been important in the population's history. The problem of accurately retrieving this information from data is fundamentally a statistical one. Each of the three application areas just discussed (simulation, qualitative insight, and quantitative calculations) plays a role.

We illustrate this by describing recent approaches which use multi-locus microsatellite data to infer aspects of underlying population demographic processes. Analyses based on a single locus can be extremely sensitive to particular, possibly undetected, features of the locus, such as the effect of selection either directly, or at nearby linked loci. In addition, it is well known (e.g. Donnelly and Tavaré 1995) that exactly because of evolutionary variability, inferences based on single-locus data have low precision. There are thus considerable advantages in pooling information across loci. Their ease of typing makes microsatellite loci convenient for the collection of multi-locus data.

As one example, it follows from eqn (9.2) that $\bar{S}^2$, the average of observed $S^2$ values across loci, is a natural estimator of $C\bar{u}\bar{\eta}_2$, where $\bar{u}$ is the average mutation rate across loci, and $\bar{\eta}_2$ is the average value of the mutation mean square across the loci studied. Information about the value of either $\bar{u}$ and $\bar{\eta}_2$ might be available from other sources, in which case $\bar{S}^2$ can be used to estimate

$C$. Recall that for a constant-sized population, $C$ is just the effective (haploid) population size, while for a population which underwent rapid expansion from a small value $T$ generations ago, $C \approx T$. In other words, if some information about the mutation rate and process is available, averages of $S^2$ values across loci can be used to estimate either the effective population size or the time since population expansion, depending on the demographic scenario. See Di Rienzo *et al.* (1995, 1998) for application to human demographic parameters.

Recall that the variability in $S^2$ values across loci should be very different in the constant population size and rapid expansion scenarios. Di Rienzo *et al.* (1995, 1998) have exploited this fact to test between these demographic scenarios. Their conclusion is that there is too little variability in their data (for 16 microsatellite loci) to be consistent with a roughly constant size for the human population. Subsequently, other authors, for example Goldstein *et al.* (1998) and Reich and Goldstein (1998), have exploited the same intuition to develop a variety of other tests for distinguishing between these demographic scenarios. Various genetic distance-type measures, notably Slatkin's (1995) $R_{ST}$, an analogue of the usual $F_{ST}$ measure of population differentiation, and Goldstein *et al.*'s (1995a) $(\delta\mu)^2$, have also been developed and studied specifically for microsatellite loci.

All of the methods just described base inference on various statistics which aim to summarize information in the data. These statistics are not *sufficient*, in the technical sense that they do not capture all the information in the data. Their use has the advantage of relative simplicity, in that exact or simulation studies of the statistics are typically possible. On the other hand, these methods are throwing away some of the information in the data, and it is often difficult to gauge the extent of this loss of information. One exciting recent development within population genetics in general, and in the context of microsatellite loci in particular, has been the application of modern, computationally intensive, statistical methods for the development of inference techniques which extract the maximum available information from the data. These methods use Monte Carlo techniques to approximate the full likelihood of the data. We will not go into details here, but refer the interested reader to Nielsen (1997), Wilson and Balding (1998), or the chapter by Beaumont and Bruford in this volume for applications to microsatellite loci, and their references for more general applications. At this stage, these methods are only practicable, even on very powerful computers, for relatively small data sets. The field is in its early stages, however, and substantial advances seem likely before too long.

We close with a cautionary note in connection with inference from microsatellite loci. As we have noted, little direct information is available about the underlying mutation processes. The various statistical methods and analyses published in the literature, of necessity, make assumptions about these unknown aspects of microsatellite evolution. The better of these typically either perform the analyses under very general assumptions, or carry out detailed sensitivity studies to assess the extent to which conclusions depend on assumptions. Considerable care should be exercized in interpreting the conclusions of studies which make restrictive assumptions on the mutation process, either in its simplicity at each locus (for example all steps of size one, with no bias) or its homogeneity between loci. Even for more realistic analyses, care may be warranted until more is known about the process of mutations at microsatellite loci.

## Acknowledgements

This research was supported in part by EPSRC Advanced Fellowship B/AF/1255, and NSF grant DMS 95-05129

# 10 Estimating the age of mutations using variation at linked markers

David E. Reich and David B. Goldstein

## Chapter contents

## Abstract

We present a general method for estimating the dates of mutations using variation at linked microsatellite markers. Risch *et al.* (1995) take a similar approach to estimating the age of the mutation causing idiopathic torsion dystonia among Ashkenazic Jews, but they do not describe how to produce a confidence interval for the date. Here, we not only obtain a confidence interval for the date by assessing the degree of correlation among samples, but also describe how to use a Markov transition matrix approach to take full account of the complexities of the recombination process. Finally, we show how the method has been applied to a specific example: estimation of a date for a mutation that confers resistance to HIV-1 infection (Stephens *et al.* 1998).

## 10.1 Introduction

It is possible to estimate the age of a mutation because of the non-random association of alleles (i.e. linkage disequilibrium) that is generated whenever a new mutation occurs. The immediate descendants of a mutant chromosome will be monomorphic for a set of markers linked to the locus of interest. Over time, however, as recombination and mutation undo the linkage disequilibrium, the pattern of variation among mutant chromosomes will gradually reflect the pattern of variation in the population as a whole. By making a quantitative assessment of the extent to which the disequilibrium has been undone, and using known rates of mutation and recombination, we can estimate an age for the most recent common ancestor of mutant chromosomes.

## 10.2 Estimating the age of the mutation when almost all chromosomes have the ancestral haplotype

To estimate the date of the mutation when almost all mutant chromosomes are of a single type, we employ a two-pronged strategy. First, we assume that the common haplotype is the ancestral haplotype, a questionable assumption if the genealogical tree of relationships among individuals includes only a few ancient lineages, and in particular, if an early mutation or recombination event occurred on a lineage that was ancestral to the majority of current chromosomes. To determine the ancestral haplotype unequivocally, we use markers that are relatively close to the gene locus of interest. We then use the frequency ($r$) of mutation and recombination events that have the potential to unlink some chromosomes from the ancestral haplotype to find the most likely number of generations that have passed since the ancestral mutant chromosome.

To obtain the maximum likelihood estimate for the date of the mutation, we begin by considering a particular lineage of the genealogy, the chain of ancestors linking a present-day haplotype to the haplotype at coalescence. The probability that a haplotype remains ancestral during the time tracing back to the most recent common ancestor is given by the depth of the genealogy in generations, $G$, and the frequency $r$ of mutation and recombination:

$$p = e^{-Gr}. \tag{10.1}$$

Here, $p$ is just the zero term in a Poisson series with parameter $Gr$.

To find $p$, we note that for a dramatically expanded population, one for which all lineages are essentially independent, an unbiased estimate of $p$ is the proportion of observed haplotypes that are ancestral (Stephens *et al.* 1998). A surprising fact is that this statement is true even for constant-sized populations in which many lineages are highly correlated in the sense that pairs of alleles share extensive periods of co-ancestry during the time tracing back to the most recent common ancestor of the sample. The reason why the age estimate is independent of topology is that as long as mutations at the marker loci have no selective effect, the correlations in the tree amount to a process of pseudo-replication of lineages. This process will affect the variance of our estimate of $p$ (see below); however, because the lineages that are replicated are selected independent of allelic state, the proportion of ancestral haplotypes will not be systematically affected.

Finally, to obtain $G$ in terms of the estimate of $p$, we transform eqn (10.1):

$$G = -\ln(p)/r. \tag{10.2}$$

As discussed previously, this holds true whatever the shape of the genealogical tree.

## 10.3 A comprehensive approach for estimating the age of a mutation

The previous method produces an appropriate estimate for the age of the mutation when the large majority of observed chromosomes have become unlinked from the ancestral haplotype. However, when enough mutant chromosomes have become unlinked from the ancestral haplotype, the date estimate must account not only for the rate of loss of the ancestral haplotype by mutation or recombination, but also for regeneration of the ancestral haplotype among chromosomes that currently do not have it (Risch et al. 1995). When this process is included in our analysis, the estimated date of mutation becomes systematically older than that predicted by eqn (10.2).

To provide a complete description for a system in which a single locus is typed, we use a Markov transition matrix $\mathbf{K}$. Note that Risch et al. (1995) have used an alternative approach to the same problem, involving differential equations. However, we have chosen to use the transition matrix approach instead because we find it to be very flexible, and because it allows us to easily incorporate mutation and recombination events into the same evolutionary process. Specifically, the entries in the Markov matrix give the probabilities, per generation, that any one haplotype will transform into any other. To calculate $\mathbf{K}$, we take a weighted sum of matrices corresponding to recombination ($\mathbf{R}$), mutation ($\mathbf{M}$), and no event occurring ($\mathbf{I}$):

$$\mathbf{K} = c\mathbf{R} + \mu\mathbf{M} + (1 - c - \mu)\mathbf{I}, \tag{10.3}$$

where $c$ is the frequency of recombination, $\mu$ is the frequency of mutation, and $1 - c - \mu$ is the frequency of no event occurring. We now consider a single lineage tracing its ancestry back to the original mutation, and by multiplying $\mathbf{K}$ by the state vector generation by generation, evaluate the probability that after $n$ generations, the mutation will have lost its linkage to the ancestral haplotype. This is exactly analogous to the method described in Section 10.2, except here we take into account regeneration of the ancestral haplotype as well as the rate of loss of that haplotype.

Consider the case in which only a single microsatellite marker has been typed. For this case, the state vector is represented as $(q, 1 - q)$, with the first entry the probability that the allele is of the ancestral type and the second the probability that it is not. The matrices $\mathbf{R}$ and $\mathbf{M}$, and hence the Markov transition matrix, can then be derived straightforwardly from the distribution of alleles in non-mutant chromosomes. We begin with the recombination matrix ($\mathbf{R}$). After a recombination event, the probability that the allele will end up ancestral, regardless of the initial state, can be estimated as the proportion of alleles in the population that have the ancestral haplotype ($a$). The probability that the allele will be non-ancestral type is then $1 - a$:

$$\mathbf{R} = \begin{bmatrix} a & a \\ 1 - a & 1 - a \end{bmatrix}. \tag{10.4}$$

We now calculate the mutation matrix ($\mathbf{M}$). According to the stepwise mutation model for microsatellites (Goldstein and Pollock 1997), mutations change the length of an allele by a single unit, with an equal chance of increasing or decreasing the length of the allele. Using this model, we estimate the probability that a mutation will transform a non-ancestral allele into an ancestral one as $b/2$, where $b$ is the proportion of alleles that are one mutation step away from the ancestral haplotype, and the division by 2 occurs because only half of mutations at these alleles produce the ancestral type. Note that in the case of a mutation that occurs on an ancestral allele, the outcome is even simpler: the probability that an allele will remain ancestral is 0.

$$\mathbf{R} = \begin{bmatrix} 0 & b/2 \\ 1 & 1 - b/2 \end{bmatrix}. \tag{10.5}$$

To find $b$ in any generation, we require information that is not contained in the two-dimensional state vector: specifically, the frequencies of alleles that are one mutation step away from the ancestral chromosome. Thus, to describe the frequencies of all $k$ possible alleles in the system, we require a $k$-dimensional state vector—a complicated circumstance because the $\mathbf{R}$ and $\mathbf{M}$ matrices would now have to be $k \times k$ rather than $2 \times 2$. Nevertheless, it is often possible to simplify the analysis when recombination occurs much more frequently than mutation. In this case, the distribution of non-ancestral alleles among mutant chromosomes is expected to be the same as in the control population, and $b$ can be estimated directly from the proportions of alleles in the control population.

We now use eqn (10.3), and the matrices $\mathbf{R}$ and $\mathbf{M}$, to obtain the Markov transition matrix $\mathbf{K}$. Errors in $\mathbf{K}$ could arise either from misestimation of $c$ and $\mu$ (since information about these parameters is often inaccurate), or from errors in $a$ and $b$ that might occur due to inappropriate selection of control populations or failure to type a sufficient number of chromosomes in the control population, or changes in the proportions of alleles in the population over the course of recent history. Since none of these sources of error is taken into account in our method for estimating a date of mutation, experimenters should consider a range of possible values of $c$, $\mu$, $a$, and $b$, as a way of assessing how much variability in the estimate of the age of the mutation could arise from misestimation of parameters.

Under the assumption that $\mathbf{K}$ is correct, we can now consider a particular lineage of the genealogy—the chain of ancestors linking a present-day haplotype to the haplotype at coalescence—and use $\mathbf{K}$ to determine the probability that the lineage remains ancestral at any given generation. We begin with the state vector representing the ancestral mutant chromosome, which has the form (1, 0) where the first entry is the probability that the lineage has the ancestral type. To evaluate the fate of the lineage in every subsequent generation, we multiply **K** by the state vector until we obtain a probability of observing an ancestral haplotype that is closest to the observed proportion, p, of mutant chromosomes. The number of times that **K** has been multiplied tells us the number of generations that have passed since the ancestral mutant chromosome.

## 10.4 Variance of the age estimate

The variance of the age estimate (unlike the age estimate itself) is systematically affected by a population's demographic history. The reason for this is that populations with different demographic histories have differently shaped genealogical trees. For example, in a population that has undergone a relatively recent and dramatic expansion, almost all lineages will trace their ancestry independently back to the time of the expansion, and the number of independent assessments of the age of the tree will be equal to the number of samples. For a constant-sized population, there will be high degree of shared ancestry among sampled chromosomes, as explained above, and the number of independent assessments of the age of the tree will therefore be much smaller than the number of sampled chromosomes. The relatively large number of age assessments in an expanding population means that the date estimate is more accurate.

To determine confidence intervals for the date, we use computer simulations based on a coalescent algorithm by R.R. Hudson (1990) to describe a wide variety of population histories from constant population size to fast growth (final population size and exponential growth rate are the variable parameters in our simulation). For each set of demographic parameters, the simulation generates a large number of genealogical trees and distributes mutation and recombination events along them according to a random (Poisson) process (we use the Markov transition matrix to determine which events turn an ancestral haplotype into a non-ancestral one and vice versa). Thus, the final distribution of haplotypes along a genealogical tree is affected by two sources of error: first, variability in the shapes of the genealogical tree, and second, variability in the mutation and recombination events that occur on those trees. The simulations allow us to take account of both these sources of error, generating a 95 per cent central confidence interval for the number of ancestral haplotypes that could be expected to be seen in such a sample. We can then reject certain combinations of demographic parameters if the confidence intervals do not contain the number of ancestral haplotypes that was actually observed.

To find allowed dates for the mutation, we consider each combination of demographic parameters separately, simulating many genealogical trees and considering only those simulations that result in the observed number of ancestral haplotypes (i.e. we condition the simulations on the observed results). From the subset of trees obtained in this manner, we can then produce a 95 per cent central confidence interval for the date of the mutation. To obtain an allowed range of dates that is inclusive of all possible demographic histories, we then take the union of confidence intervals for each combination of parameters. The range of allowable dates can be constricted even further if we have additional information about the demographic history—for example, if the observed distribution pattern of non-ancestral haplotypes forbids particular combinations of demographic parameters, as explained in Section 10.6, below.

## 10.5 Age of the *CCR5-Δ32* AIDS resistance allele

The *CCR5* gene encodes a protein that serves as part of the primary entry port for HIV-1 in immune cells (Deng *et al.* 1996). Individuals homozygous for a particular 32 base-pair deletion mutation in the gene, which we designate as *CCR5-Δ32*, are resistant to HIV-1 infection (Dean *et al.* 1996). Indeed, as many as 26 per cent of northern Europeans carry at least one deleted copy of the gene, while the frequency of carriers drops to zero along a north–south gradient (no copies are observed among Africans). The pattern of distribution of the gene makes it seem likely that the mutation occurred recently, and it is therefore of interest to obtain a direct estimate for the date of origin of the mutation.

The data we use consist of 46 chromosomes carrying the *CCR5-Δ32* deletion, and 146 controls that do not carry the mutation. Each chromosome was typed at two microsatellite markers on the same side of the *CCR5* gene: GAAT12D11 (GAAT) and AFMB362wb9 (AFMB), with GAAT closest to the deletion locus. The ancestral haplotype is taken to be the one in which the GAAT marker carries the 197 base-pair allele and the AFMB marker carries the 215 base-pair allele. This haplotype occurs among 85 per cent of mutant chromosomes but only 36 per cent of the control population.

To calculate the Markov transition matrix for this system, we note that two polymorphic markers were typed, and that there are therefore four possible states in the system. Specifically, the states can be classified as follows: (1) both GAAT and AFMB are ancestral; (2) only GAAT is ancestral; (3) only AFMB is ancestral; and (4) neither GAAT nor AFMB is ancestral. The state vector can be represented as $(q_1, q_2, q_3, 1 - q_1 - q_2 - q_3)$, and the transition matrices, corresponding to mutation at GAAT, mutation at AFMB, recombination at GAAT or recombination at AFMB, will be four-dimensional (4×4) as well. The overall equation for the transition matrix $\mathbf{K}$ is then:

$$\mathbf{K} = \mu_{GAAT}\mathbf{M}_{GAAT} + \mu_{AFMB}\mathbf{M}_{AFMB} + c_{GAAT}\mathbf{R}_{GAAT} + c_{AFMB}\mathbf{R}_{AFMB}$$
$$+ (1 - c_{GAAT} - c_{AFMB} - \mu_{GAAT} - \mu_{AFMB})\mathbf{I}, \tag{10.6}$$

where $\mu_{GAAT}$, $\mu_{AFMB}$, $c_{GAAT}$, and $c_{AFMB}$ are the rates of mutation and recombination for the GAAT and AFMB markers, and $\mathbf{M}_{GAAT}$, $\mathbf{M}_{AFMB}$, $\mathbf{R}_{GAAT}$, and $\mathbf{R}_{AFMB}$ are mutation and recombination matrices.

We must now estimate the parameters $\mu_{GAAT}$, $\mu_{AFMB}$, $c_{GAAT}$, and $c_{AFMB}$. To obtain the recombination rates $c_{GAAT}$ and $c_{AFMB}$, we use physical distances that were determined from radiation hybrid mapping, and convert these to recombination distances using a linear regression that applies on average across the chromosome on which the mutation was found. To estimate the mutation rates $\mu_{GAAT}$ and $\mu_{AFMB}$, we use the published value for dinucleotide microsatellites, $\mu = 0.00053$ (Weber and Wong 1993). In this analysis, error in estimation of the recombination rate was much more of a worry to us than error in the mutation rate, since the recombination rate is so much larger in absolute terms.

To obtain the mutation matrices, we use the frequencies of alleles in the control population that are one mutation step away from the ancestral GAAT ($b_1$) and ancestral AFMB ($b_2$) alleles (see eqn (10.4)). It follows that for mutation at the GAAT marker, the matrix is $\mathbf{M_{GAAT}}$, while for mutation at the AFMB marker, the matrix is $\mathbf{M_{AFMB}}$.

$$\mathbf{M_{GAAT}} = \begin{vmatrix} 0 & 0 & b_1/2 & 0 \\ 0 & 0 & 0 & b_1/2 \\ 1 & 0 & -b_1/2 & 0 \\ 0 & 1 & 0 & -b_1/2 \end{vmatrix}, \quad \mathbf{M_{AFMB}} = \begin{vmatrix} 0 & b_2/2 & 0 & 0 \\ 1 & -b_2/2 & 0 & 0 \\ 0 & 0 & 0 & b_2/2 \\ 0 & 0 & 1 & -b_2/2 \end{vmatrix}.$$

$$(10.7)$$

To obtain the recombination matrices, we follow eqn (10.5), dealing first with the case in which the recombination occurs between the gene locus of interest and GAAT, and then the case in which the recombination event occurs between GAAT and AFMB. In the first case, the situation is exactly analogous to eqn (10.4), and the frequencies of each possible outcome can be estimated as the proportion of alleles in the control population that are of each haplotypic state. We designate these frequencies, respectively, as $a_1, a_2, a_3$, and $a_4$, recalling that $a_4 = 1 - a_1 - a_2 - a_3$. The resulting matrix is designated $\mathbf{R_{GAAT}}$. In the second case, in which the recombination occurs between GAAT and AFMB, the alleles change at only a single locus (AFMB), and the only relevant parameters are the frequency of alleles for which the AFMB marker had the ancestral type ($a_1 + a_3$), and the frequency of alleles for which the AFMB marker was non-ancestral ($a_2 + a_4$). The overall $4 \times 4$ transition matrix, $\mathbf{R_{AFMB}}$, then becomes:

$$\mathbf{R_{GAAT}} = \begin{vmatrix} a_1 & a_1 & a_1 & a_1 \\ a_2 & a_2 & a_2 & a_2 \\ a_3 & a_3 & a_3 & a_3 \\ a_4 & a_4 & a_4 & a_4 \end{vmatrix}, \quad \mathbf{R_{AFMB}} = \begin{vmatrix} a_1+a_3 & a_1+a_3 & 0 & 0 \\ a_2+a_4 & a_2+a_4 & 0 & 0 \\ 0 & 0 & a_1+a_3 & a_1+a_3 \\ 0 & 0 & a_2+a_4 & a_2+a_4 \end{vmatrix}.$$

$$(10.8)$$

We now use eqn (10.6) to calculate $\mathbf{K}$. Ignoring any error in the Markov transition matrix (more likely to be due to errors in estimation of the recombination rate and recombination parameters rather than errors in the mutation rate), the most likely age for the *CCR5-Δ32* mutation is 29 generations, or 725 years assuming a generation time of 25 years. For comparison, if the calculation is done according to the method of Section 10.2, the estimate is 28 generations, slightly younger because no Markov transition matrix is used to take into account regeneration of the ancestral haplotype. Note that the estimated date of the mutation is likely to be systematically lower than the date of first appearance of the mutation, since the estimation procedure only finds information about the age of the most recent common ancestor of the sampled chromosomes. Thus, our estimate of the date must be interpreted with caution: if a dramatic expansion occurred in the population of mutant chromosomes, it is likely that the most recent common ancestor of the mutant chromosomes dates to before the expansion (although it is difficult to say how much earlier). Note that Slatkin and Rannala (1997) provide an approach for dating mutations that takes this systematic bias into account.

## 10.6 Estimating a variance for the date by reconstructing the genealogy of CCR5-$\Delta 32$

To obtain a confidence interval for the date estimate, we use simulations that take into account all possible combinations of demographic parameters and genealogical trees, as described in Section 10.4. To place further restrictions on the allowed dates of the mutation, we forbid certain genealogical trees—in the simplest case by using prior knowledge of population history. For the CCR5-$\Delta 32$ data, for example, we assume that during the past 10 000 years, northern European populations have had a certain minimum size. By specifying that the initial effective population size was at least 5000, we conclude that the date of the most recent common ancestor was between 11 and 75 generations in the past (275–1875 years, assuming 25 years per generation).

In a much more fundamental way, it is also possible to use the distribution of non-ancestral haplotypes among mutant chromosomes to put restrictions on the shape of the genealogical tree. For example, if the haplotypes all derive from separate mutation or recombination events, the lineages of the genealogical tree are uncorrelated, and consistent with a dramatically expanded population. If the non-ancestral haplotypes derive from relatively few mutation or recombination events (which have been recopied and amplified within the lower branches of the genealogical tree), then the history of the mutant chromosomes is more likely to be consistent with a constant-sized population. By focusing on the distribution of non-ancestral haplotypes among CCR5-$\Delta 32$ chromosomes, we are then able to directly assess the degree of correlation in the tree, and from there to assess the variability of the date estimate.

To implement this approach, we consider the fact that of the seven non-ancestral CCR5-$\Delta 32$ chromosomes that were observed, there were four distinct haplotypes. The number of mutation and recombination events that actually gave rise to the four haplotypes was probably larger than four, since the distribution of non-mutant CCR5 chromosomes indicates that given six or seven chances, several haplotypes would be generated more than once (and, as expected from this hypothesis, the non-ancestral haplotypes we observe are the ones that are most frequent in the control population). We surmise that the non-ancestral haplotypes derive for the most part from separate mutation and recombination events, and that in the present sample, we are observing the results of at least six and perhaps seven different events. Note that it would have been possible to determine the number of events with even more precision if more that two microsatellite markers had been typed.

To make explicit use of this information, we modify the simulation described in Section 10.4 to report not only the number of non-ancestral haplotypes but also the number of distinct mutation and recombination events that gave rise to these haplotypes. Thus, for each set of demographic parameters in the *CCR5-Δ32* data set, we simulate a large number of genealogical trees that gave rise to seven out of 46 non-ancestral haplotypes, and then determine the proportion of these replicates that were derived from seven distinct events. If we require that no fewer than 5 per cent of replicates have fewer than seven distinct haplotypes, we can restrict the date of the mutation to between nine and 214 generations in the past (225–5350 years, assuming 25 years per generation). While this restriction on the date of the mutation is less stringent than the one derived from a historical assumption about effective population sizes, it is valuable precisely because it is independent of such assumptions.

## 10.7 The analysis of new data sets

In applying the method to a new data set, it is always appropriate to begin by picking microsatellite markers that have the proper distance from the gene locus of interest. The markers should be chosen to be close enough to the locus of interest to define the ancestral haplotype, but far enough away to allow as many lineages as possible to have had a chance to become non-ancestral. A good strategy for identifying markers is to select a panel that are at varying distances from the gene locus of interest, and then to pick out ones that comply with the criteria described above.

The analysis of data from a single microsatellite locus can often extract most of the relevant information about the date of a mutation. However, the use of multiple markers (e.g. in the *CCR5-Δ32* experiment) may have a particular value in assessing the variance of the date estimate, allowing for a better assessment of the shape of a genealogical tree than would be possible with a single marker. The reason for this is that multiple markers allow us to reconstruct more accurately the history of mutation and recombination events. If even more markers are typed, it becomes possible to pinpoint the exact number of distinct mutation and recombination events that had led to the observed number of non-ancestral haplotypes, further restricting the allowed range of genealogical tree. On the other hand, multiple markers have a drawback because they can make an analysis more complicated, forcing the estimation of more matrix parameters, recombination distances and mutation rates.

Another factor to consider in designing future experiments is that some mutations will be sufficiently old that only markers close to the locus will display disequilibrium. In this case, it will be difficult to determine the recombination distances of markers from the locus, and it is appropriate to use markers that are sufficiently close to the gene that mutation serves as the main molecular clock for estimating a date for the mutation. Errors in estimating the mutation rate (and not the recombination rate) then become the main source of systematic error in determining the age of the mutation, and to reduce this error, it is appropriate to use several markers that are close to the gene locus of interest, with an average mutation rate that in general will be more predictable than that of a single marker (Goldstein and Pollock 1997). In practice, however, it may be difficult to find enough markers that are sufficiently close to the gene locus of interest to make this possible, except perhaps on the Y chromosome, where a large number of microsatellites are completely linked.

# 11 Statistics of microsatellite loci: estimation of mutation rate and pattern of population expansion

Ranajit Chakraborty and Marek Kimmel

## Chapter contents

## Abstract

Applications of microsatellite loci in evolutionary studies are based on examination of the expectations of summary statistics of genetic variation at these loci, under a model for mutation and genetic drift. Usually, the single-step stepwise mutation model with unrestricted allelic states is assumed, and the effect of genetic drift is studied by assuming that the population size is constant over time. In this chapter, we illustrate the effects of relaxing both of these assumptions. We show that an expansion or contraction of population size over time leaves a signature on the genetic variation that can be detected by estimating the composite parameter (the product of population size and the mutation rate) from heterozygosity and allele size variance. Allele size restrictions also produce an imbalance between heterozygosity and allele size variance. These two statistics are insensitive to directional (expansion/contraction) bias of mutations if allele-size constraints are absent. Further, the feature that microsatellite alleles can be ordered by their sizes allows estimation of relative mutation rates of loci grouped by their biological properties. Applications of these principles to microsatellite data from several human populations indicate that the mutation rates are inversely related to the motif size, and the African populations exhibit indications of the most ancient human bottleneck.

## 11.1 Introduction

Microsatellites, also known as short tandem repeats, are currently the most efficient set of loci for gene mapping (Dib *et al.* 1996; see also Stephens *et al.* this volume), for forensic identification of individuals (NRC 1996), and for studying relatedness between individuals (Chakraborty and Jin 1993). Their abundance in the human genome, and higher rate of mutation, compared with other types of polymorphism, also make them useful in microevolutionary studies (Bowcock *et al.* 1994; Deka *et al.* 1995a; Mountain and Cavalli-Sforza 1997). Within a short evolutionary time period, a sufficient number of accumulated mutation events may be observed in divergent populations, and the use of a larger number of loci helps in reducing the stochastic errors of evolutionary inference. To exploit these advantages of microsatellite loci, their unique features should be explored while suggesting appropriate summary statistics of microsatellite variation. Recent work has addressed such properties of microsatellites, including the generalized forward–backward changes of allele sizes produced by mutations (Goldstein *et al.* 1995b; Slatkin 1995; Kimmel and Chakraborty 1996; Kimmel *et al.* 1996). However, most applications of microsatellites are based on the assumption that genetic variation at these loci is maintained by a mutation–drift equilibrium in populations of constant effective size.

The purpose of this chapter is to discuss the consequences of deviations from this assumption, and to suggest means of using selected summary statistics of microsatellite polymorphisms to detect signatures of non-equilibrium conditions. These latter may be introduced by population bottlenecks and/or expansion, by differences in mutation rates between loci, and by constraints of allele size. The theories and methods of inference are suggested by patterns of non-equilibrium detected in the data on microsatellite variation in major human populations. For brevity we exclude detailed derivations in this presentation. These can be found in the publications cited.

## 11.2 Model of microsatellite variability

### 11.2.1 Estimators of allele-size variance and homozygosity

Consider a sample of $n$ chromosomes and a locus with alleles indexed by integer numbers. The expectation of the estimator of the within-population component of genetic variance

$$\hat{V}/2 = \sum_{i=1}^{n} (X_i - \bar{X})^2 / (n-1), \tag{11.1}$$

where $X_i$ is the size of the allele at the locus in the $i$th chromosome present and $\bar{X}$ is the mean of the $X_i$, is equal to $V(t)/2$, where

$$V(t) = E(\hat{V}) = E[(X_i - X_j)^2], \tag{11.2}$$

and $X_i = X_i(t)$ and $X_j = X_j(t)$ are the sizes of two alleles from the population (Kimmel et al. 1996). If $p_k$ denotes the relative frequency of allele $k$ in the sample, then an estimator of homozygosity has the form

$$\hat{P}_0 = \left( n \sum_{k=1}^{n} \hat{p}_k^2 - 1 \right) / (n-1). \tag{11.3}$$

### 11.2.2 The time-continuous Fisher–Wright–Moran model with stepwise mutations

The joint distributions of allele sizes in a stepwise mutation model with sampling from the finite allele pool can be studied by considering that: (i) the population is composed of a constant number of $2N$ haploid individuals or chromosomes. Each individual undergoes death/birth events according to a Poisson process with intensity 1. Upon a death/birth event, a genotype for the individual is sampled with replacement from the $2N$ chromosomes present at this moment (Ewens 1979); and (ii) each individual independently is subject to a mutation which replaces an allele of size $X$ with an allele of size $X + U$, where $U$ is an integer-valued random variable with a given probability generating function (pgf) $\varphi(s) = \sum_{u=-\infty}^{\infty} s^u \Pr[U = u] = E(s^U)$. Mutations occur according to a Poisson process with intensity $v$.

Kimmel et al. (1998) showed that the pgf of the difference between these two allele sizes, $X_1(t) - X_2(t)$, satisfies:

$$R(s, t) = E\left[ s^{X_1(t) - X_2(t)} \right] = R(s, 0) \exp\left[ -\int_0^t a(s, \tau)\, d\tau \right]$$

$$+ \int_0^t [1/2N(\tau)] \exp\left[ -\int_\tau^t a(s, u)\, du \right] d\tau, \tag{11.4}$$

where $a(s, t) = 1/[2N(t)] + 2v[1 - \psi(s)]$ and $\psi(s) = [\varphi(s) + \varphi(1/s)]/2$. Equation (11.4) can be obtained using the coalescent-based approach. Consequently, the expected allele-size variance and homozygosity are given by:

$$V(t) = \partial^2 R(1, t)/\partial s^2 = V(0) \exp\left[-\int_0^t [1/2N(\tau)] d\tau\right]$$
$$+ 2v\psi''(1) \int_0^t \exp\left[-\int_\tau^t [1/2N(u)] du\right] d\tau,$$
(11.5)

and

$$P_0(t) = \pi^{-1} \int_0^\pi R(e^{i\omega}, 0) \exp\left[-\int_0^t a(e^{i\omega}, \tau) d\tau\right] d\omega$$
$$+ \pi^{-1} \int_0^\pi \int_0^t [1/2N(\tau)] \exp\left[-\int_\tau^t a(e^{i\omega}, u) du\right] d\tau \, d\omega, \quad (11.6)$$

under the mutation–drift equilibrium

$$V(t) = V(\infty) = 4vN\psi''(1) = \theta\psi''(1). \quad (11.7)$$

If the single-step SMM is assumed, i.e. if $\psi(s) = (s + s^{-1})/2$ and consequently $\psi''(1) = 1$, we obtain $V(\infty) = 4vN = \theta$ and

$$P_0(t) = P_0(\infty) = (1 + 8vN)^{-1/2} = (1 + 2\theta)^{-1/2}. \quad (11.8)$$

Thus, eqns (11.7) and (11.8) provide expectations of the sample statistics $\hat{V}/2$ and $\hat{P}_0$ (eqns (11.1) and (11.3) respectively), which summarize the within-population variation at a microsatellite locus. These expectations hold under the condition that allele sizes are unrestricted. However, eqns (11.7) and (11.8) differ with respect to the patterns of stepwise mutations. While the expected within-population variance of allele size (eqn (11.7)) holds for any arbitrary Generalized Stepwise Mutation Model, even with a directional (i.e. contraction/expansion) bias of mutations, the expected homozygosity $P_0(\infty)$ assumes single-step changes of allele sizes by mutations, but with directional bias allowed.

### 11.2.3 Restricted mutations

The unrestricted stepwise-mutation model is not sufficient in cases when the mutation pattern is more general. We may instead use a Markov model of mutations with transition probabilities $P_{ij}(t)$ and intensities $Q_{ij}$. Conditions on the intensity matrix: (a) $Q_{ij} \geq 0$, $i \neq j$, (b) $\sum_j Q_{ij} = 0$, all $i$. In the finite-dimensional case $P(t) = \exp(Qt)$.

Let $R_{jk}(t) = \Pr[X_1(t) = j, X_2(t) = k]$, where $X_1(t)$ and $X_2(t)$ are randomly selected chromosomes. From the coalescence theory, we obtain (Kimmel and Chakraborty 1997):

$$R_{jk}(t) = \int_0^\infty \sum_i \pi_i(t - \tau) P_{ij}(\tau) P_{ik}(\tau) \left[ e^{-\tau/(2N)}/(2N) \right] d\tau$$

This is equivalent to the following matrix Lyapunov Differential Equation (Gajič and Qureshi 1995),

$$dR(t)/dt = [Q^T R(t) + R(t)Q] - \frac{1}{2N} R(t) + \frac{1}{2N} \Pi(t), \qquad (11.9)$$

where $dR(t)/dt$ is the time-derivative of $R(t)$, $\Pi(t) = \mathrm{diag}[\pi_i(t)]$, $\pi_i(t) = \Pr[X_1(t) = i]$. If $\Pi(t) \to \Pi$, as $t \to \infty$, then $R(t) \to R$, and the limit (mutation–drift equilibrium) value can be evaluated numerically.

One important special case of Markov mutation pattern is a random walk with reflecting boundaries on states $k = 0, 1, \ldots, K$. In this case, intensities of a backward and a forward step are equal to $vd$ and $vb$, $d + b = 1$ (while the overall mutation rate is $v$) except at the boundaries. This model can be named the Restricted Stepwise Mutation. Under this $K + 1$-state model, the allele-size variance and homozygosity can be expressed as $V^{(K)} = \sum_{ij}(i - j)^2 R_{ij}$, and $P_0^{(K)} = \sum_i R_{ii}$.

Thus, even though an analytically closed-form solution of eqn (11.9) is not possible, its iterative numerical solution provides evaluation of the expectations of the two sample statistics ($\hat{V}/2$ and $\hat{P}_0$) measuring the within-population genetic variability at any microsatellite locus in the presence of possible allele-size constraints (see also Feldman *et al.* 1997).

## 11.3 Inference regarding mutation rates, population-size expansion and allele-size constraints

### 11.3.1 Estimation of relative mutation rates at different microsatellite loci

With allele-size data available from a set of $I$ loci, each studied in $J$ diverse populations, eqn (11.7) can be used to estimate the relative mutation rates in loci of different motif types (Chakraborty *et al.* 1997).

This is so, because, with the assumption that the mutation pattern (allele-size change caused by individual mutations and its distribution) is the same across loci (or groups of loci), the logarithmic variance ($\ln \hat{V}_{ij}$) at the $i$th locus (or the $i$th group of loci) in the $j$th population has the expectation:

$$E(\ln \hat{V}_{ij}) = const + \ln v_i + \ln N_j,$$

so that the standard two-way ANOVA of $\ln \hat{V}_{ij}$ on the same set of microsatellite loci studied in multiple populations can be analysed to estimate the relative mutation rates at different loci. The numerical procedures for ANOVA of $\ln \hat{V}_{ij}$ are exactly the same as described in standard statistical texts (Sokal and Rohlf 1981). The estimator of the logarithm of variance has a distribution similar to normal, as noticed among others by Goldstein *et al.* (1997) and by Kimmel and Chakraborty (1996).

Analysis of the components of variance using the two-way ANOVA can help answer the following questions: (i) is the dependence of mutation rate on locus motif type significant? and (ii) is the dependence on population size significant?

When the dependence on locus motif type dominates in the analysis, the estimated motif type-specific levels of the estimated logarithmic variance are equal, up to an additive constant $C$, to logarithms of motif type-specific average mutation rates. Thus, the motif type-specific average mutation rates themselves can be determined up to a multiplicative constant. This is possible even in the absence of mutation–drift equilibrium, as in eqn (11.5), if it is assumed that $V(0) = \theta_0 \psi''(1)$ represents an older equilibrium, or if $V(0) = 0$, i.e. the evolution started from a monomorphic population.

### 11.3.2 Imbalance index

Equations (11.7) and (11.8) provide two intuitive estimators of the composite parameter

$$\hat{\theta}_V = \hat{V}, \tag{11.10}$$

called the (allele-size) variance estimator of $\theta$, and

$$\hat{\theta}_{P_0} = (1/\hat{P}_0^2 - 1)/2, \tag{11.11}$$

the homozygosity (heterozygosity) estimator of $\theta$. At equilibrium, under the unrestricted single-step stepwise mutation model:

$$E(\hat{\theta}_V)/E(\hat{\theta}_{P_0}) \approx V(\infty)/[1/P_0(\infty)^2 - 1]/2 = 1,$$

which leads to a parametric definition of an index $\beta(t)$, given by

$$\beta(t) = V(t)/\{[1/P_0(t)^2 - 1]/2\}, \tag{11.12}$$

representing imbalance at a microsatellite locus, due to transient states, population size change, or departure from the single-step SMM.

Thus, eqn (11.12) provides another use of the same summary statistics $\hat{V}/2$ and $\hat{P}_0$ of within-population genetic variation at microsatellite loci, signifying whether or not fluctuation of population sizes and/or departure from the single-step SMM has significant impact on the genetic variation at microsatellite loci within any population. However, the estimation of the imbalance index $\beta(t)$

needs some caution. While $\hat{V}$ estimates $\theta$ in an unbiased way (eqn (11.10)), the homozygosity-based estimator of $\theta$ (eqn (11.11)) is not unbiased. Further, single-locus estimators $\hat{\theta}_V$ and $\hat{\theta}_{P_0}$ are subject to rather large stochastic fluctuations. Therefore, when the purpose of estimation of the imbalance index $\beta(t)$ is to study the consequences of fluctuations of population sizes for genetic variation at microsatellite loci, Kimmel et al. (1998) suggest that $\theta$ estimates from the variance and homozygosity be averaged over all loci assumed to conform to single-step SMM, before their ratio is used to compute the imbalance index $\beta(t)$. The sampling error of $\beta(t)$, estimated in this manner, can be evaluated by coalescent-based simulations, as in Kimmel et al. (1998), to make the inference as to whether $\beta(t)$ significantly differs from 1 or, equivalently, $\ln \beta(t)$ significantly differs from 0.

## 11.3.3 Population growth

Coalescence theory was employed by Kimmel et al. (1998) to study the transient behaviour of the imbalance index $\beta(t)$, as defined in eqn (11.12), as a function of time (number of generations), for several patterns of population growth: (i) stepwise population growth, $N(t) = N_0$, $t = 0$, and $N(t) = N$, $t > 0$, (ii) exponential population growth, $N(t) = N_0 \exp(\alpha t)$, $t \geq 0$, and (iii) logistic population growth, $N(t) = K/[1 + (K/N_0 - 1) \exp(-\alpha t)]$, $t \geq 0$, each with three types of initial conditions: mutation–drift equilibrium; initial population monomorphic with only a single allele present; and initial population carrying two alleles differing in size by $k$ repeats, with respective frequencies $p$ and $q = 1 - p$.

Finally, one more complex growth pattern was contemplated, with a population initially of large size $N_{00}$, dropping instantly to a smaller size $N_0$, and then regrowing exponentially to a final size $N$, i.e.

$$N(t) = \begin{cases} N_{00}; & t < 0, \\ N_0 e^{\alpha t}; & t \geq 0, \end{cases} \tag{11.13}$$

where $\alpha = \ln(N/N_0)/T$ has been selected so that $N(t) = N$, if $t = T$.

We used numerical values obtained by Rogers and Harpending (1992), who fitted distributions of pairwise differences of numbers of segregating sites in mitochondrial DNA to the data of Cann et al. (1987). The second row of Table 1 of Rogers and Harpending (1992) contains estimates concerning the world's population expansion. Correcting for the fact that Rogers and Harpending (1992) considered only females while we consider both genders, that is, multiplying all effective sizes by 2, we obtain an expansion from $N_0 = 3254$ to $N = 547\,586$, within $120\,000$ years or $T = 4800$ generations, assuming generation times roughly equivalent to 25 years. We combined these values with mutation rates $\nu = 10^{-4}$ and $5 \times 10^{-4}$ typical for microsatellite loci (Weber and Wong 1993).

For the stepwise and exponential population growth, with equilibrium initial conditions, the imbalance index falls with time to values less than 1, the deviation increasing with the mutation rate $v$. The logistic growth leads to an effect intermediate between those caused by the stepwise and exponential growth. In contrast, under the same model of population growth, but with initial conditions corresponding to a monomorphic population, the index initially is close to 0, but then rapidly, during about 100 generations, increases to a value close to 1 and subsequently follows almost the same trajectory as in the case of equilibrium initial conditions (see Kimmel *et al.* 1998 for further details). The power of the resulting test is yet to be investigated.

Figure 11.1 presents the $\beta(t)$ index values for the bottleneck patterns of eqn (11.13), with the pre-bottleneck population size $N_{00} = 40\,000$ and $N_0 = 3254$, $N = 547\,586$, as above. For an initial period, the index increases from 1 to values higher than 1, the increase being greater for greater mutation rates. After that initial period, an imbalance pattern as in simple exponential growth is restored.

In summary, if before expansion the population is at a mutation–drift equilibrium, the imbalance index deviates downwards from 1 (i.e. $\beta(t) < 1$). In contrast, if the population experiences a bottleneck preceding expansion, there will be a long (e.g. several thousand generations) transient time period during

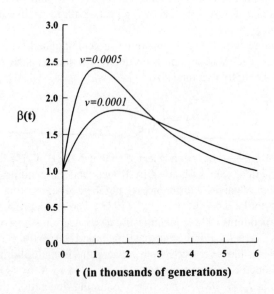

**Fig. 11.1** Values of the $\beta$ index for the bottleneck pattern of eqn (11.13), with the pre-bottleneck population size $N_{00} = 40\,000$, and $N_0 = 3\,254$, $N = 547\,586$, and $T = 4\,800$ generations, with mutation rates $v = 10^{-4}$ and $5 \times 10^{-4}$. Calculations were carried out using eqns (11.5) and (11.6).

**Table 11.1**
Influence of restriction in allele size, mutation rate and
directionality on variance, homozygosity and imbalance index

| $b$ | $K$ | $N$ | $v$ | $V^{(K)}$ | $V^{(\infty)}$ | $P_0^{(K)}$ | $P_O^{(\infty)}$ | $\beta$ |
|-----|-----|-----|-----|-----------|----------------|-------------|------------------|---------|
| 0.5 | 50 | $10^5$ | 0.00001 | 3.79 | 4 | 0.34 | 0.33 | 1.02 |
| | | | 0.00005 | 17.50 | 20 | 0.18 | 0.16 | 1.13 |
| | | | 0.0001 | 32.89 | 40 | 0.13 | 0.13 | 1.15 |
| 0.6 | 50 | $10^5$ | 0.00001 | 2.36 | 4 | 0.39 | 0.33 | 0.83 |
| | | | 0.00005 | 6.14 | 20 | 0.25 | 0.16 | 0.83 |
| | | | 0.0001 | 8.04 | 40 | 0.22 | 0.13 | 0.81 |
| 0.5 | 20 | $10^5$ | 0.00001 | 3.47 | 4 | 0.35 | 0.33 | 0.97 |
| | | | 0.00005 | 13.78 | 20 | 0.19 | 0.16 | 0.96 |
| | | | 0.0001 | 22.62 | 40 | 0.14 | 0.13 | 0.94 |

which $\beta(t) > 1$, before showing the signature of expansion alone ($\beta(t) < 1$). An obvious exception to this general rule occurs when bottleneck is severe enough to make the population monomorphic before expansion, in which case $\beta(t) < 1$ for all times.

### 11.3.4 Allele-size restrictions

We calculated the values of the imbalance index $\beta$, assuming the Restricted Stepwise Mutation model, for a number of combinations of values of mutation rate $v$, allele-size range $K$ and allele-size expansion probability $b$. Examples are presented in Table 11.1. These numerical computations indicate that even in mutation–drift equilibrium, restriction on allele sizes causes deviations of the imbalance index $\beta$ from 1. Further, the factors driving $\beta$ towards lower values are: (i) more severe restriction of allele size (lower $K$); (ii) higher mutation rate (higher $v$); and (iii) directionality of mutation ($b \neq 0.5$).

## 11.4 Data analysis

### 11.4.1 Relative mutation rates

As mentioned above, eqn (11.7), or even eqn (11.5), can be used to estimate relative mutation rate in microsatellites grouped by any biological features such as repeat motif or genomic locations. For example, Chakraborty et al. (1997) used three sets of data on microsatellites grouped by their repeat motifs (di-, tri-, and tetranucleotides) to estimate the relative mutation rates of these three groups of loci. Table 11.2 reproduces the summary results of their analysis. The estimates of natural logarithms of within-population allele-size variances are presented, along with the number of loci of each repeat motif type. These estimates translate into the dinucleotides having mutation rate 1.48–2.16 times higher than the

**Table 11.2**
Summary of the estimated natural logarthims of the within-population allele-size variance

| Data Source | Deka *et al.* (1995*b*) | | Hammond *et al.* (1994) | | GDB | |
|---|---|---|---|---|---|---|
| | $n$ | $\ln V$ | $n$ | $\ln V$ | $n$ | $\ln V$ |
| **Anonymous loci** | | | | | | |
| Dinucleotide | 8 | 1.51 | – | – | 115 | 1.38 |
| Trinucleotide | 1 | 1.00 | 4 | 1.25 | 3 | 1.19 |
| Tetranucleotide | 5 | 0.74 | 8 | 0.57 | 12 | 0.99 |
| **Disease-associated loci** | | | | | | |
| Trinucleotide | 4 | 2.09 | | | 1 | 2.92 |

tetranucleotides, with non-disease trinucleotides in the middle, having mutation rate 1.22–1.97 times higher than the tetranucleotides. In contrast, the disease-associated trinucleotides have a mutation rate 3.86–6.89 times higher than the tetranucleotides. Chakraborty *et al.* (1997) also showed that these differences in mutation rates are statistically significant, using the analysis of variance and the nonparametric Mann-Whitney test (Sokal and Rohlf 1981).

## 11.4.2 Signature of population-size change

Kimmel *et al.* (1998) analysed the allele-size distributions at 60 tetranucleotide loci, in three major human population groups, Africans, Europeans, and Asians (original data reported in Jorde *et al.* 1995, 1997). They sought the signatures of past population-size changes through the estimation of the imbalance index $\beta$ (eqn (11.12)). Adjusting for bias in the ratio estimator of $\beta$, the pooled estimates of $\beta$ were 1.12, 1.33 and 1.82, for Africans, Europeans, and Asians respectively. These estimates were compared to the coalescent simulation-based quantiles of the distribution of $\ln \beta$, under the null hypothesis of constant population size and mutation–drift equilibrium. It was found that the value for Asians exceeded the 0.99 quantile from the null value of $\beta = 1$, the value for Europeans was between the 0.95 and 0.99 quantile, while the value for Africans, residing around the 0.70 quantile, was not significantly different from $\beta = 1$. This behaviour of the imbalance index obtained from the data is consistent with the growth scenarios depicted in Fig. 11.1 that assume a reduced diversity of the population at the time when population expansion begins, representing the consequences of a pre-expansion bottleneck. Further, the trend of sample values of the imbalance index $\beta$ is consistent with the bottleneck being most ancient in Africans, most recent in Asians, and of intermediate age in Europeans. It is not yet clear if this effect is not due to confounding factors such as distributed mutation rates or population substructuring. However, the systematic nature of the imbalance makes such explanations less likely.

**Table 11.3**
Summary statistics for 3 groups of trinucleotide loci averaged over 4 human
populations

| Locus type | Gene-associated | Anonymous | Disease-associated |
|---|---|---|---|
| Number of loci | 6 | 7 | 4 |
| Average heterozygosity | 0.407 | 0.630 | 0.744 |
| Log allele-size variance | −0.459 | 0.437 | 2.034 |
| Coefficient of gene differentiation | 0.314 | 0.135 | 0.131 |
| Relative mutation rate | 1 | 1.35–1.44 | 6.05–24.16 |
| Imbalance index $\beta$ | 2.740 | 1.257 | 2.410 |

## 11.4.3 Allele-size constraints

An application of the Lyapunov Differential Equation (eqn (11.9)) can be illustrated with data reported in Deka *et al.* (1997). The data includes 17 trinucleotide
loci grouped as anonymous (i.e. non-gene-associated and presumably neutral;
seven loci), gene-associated (five loci) and disease-associated (four loci), studied
in four populations, German, Benin, Chinese, and Papua-New Guinea High-
landers. This data set shows that the within-population allele-size variance, as
well as within-population heterozygosity, is lowest for the gene-associated loci,
and highest for the disease-associated loci, with the anonymous ones being
intermediate. In contrast, the coefficient of gene differentiation (a measure of
between-population variability, Nei 1973) shows a reverse trend, namely highest
for the gene-associated, lowest for the disease-associated, and intermediate for
the anonymous loci. These observations are best explained as being mutation-
driven, with the disease-associated loci having the highest mutation rate, and with
the gene-associated ones having the lowest mutation rate. Simulations with the
Lyapunov Differential Equation invoking differential allele-size constraints indicate that the above observations cannot be explained by more severe allele-size
constraints for the gene-associated loci. Indeed, the imbalance indices ($\beta$) for
each population, for the gene-associated and disease-associated loci (summary
in Table 11.3), are more than twice as high than 1, which is inconsistent with the
theoretical predictions. Table 11.1 documents that if constraints are assumed, then
$\beta < 1.15$ in the range of parameters searched, with $\beta < 1$ for most combinations
of data.

# 11.5 Conclusions

In the above descriptions we used two summary statistics of genetic variation
at a microsatellite locus within any population, allele-size variance ($\hat{V}/2$) and
homozygosity ($\hat{P}_0$), and showed that these two statistics offer guidance with
regard to both locus-specific and population-growth related characteristics of
genetic variation. As reflected in eqns (11.1) and (11.3), these two statistics are

readily computed from data on allele sizes, and their frequencies from a sample drawn from the population. Also, it is important to note that eqns (11.1) and (11.3) define symmetric functions of differences of allele sizes in pairs of sample alleles rather than of their absolute sizes. Thus, these two measures of genetic variation within populations are insensitive to directional (contraction/expansion) bias of mutations. The subsequent analyses with the locus–population combinations of estimates of $\hat{V}$ and $\hat{P}_0$ are either carried out according to standard statistical procedures, e.g. ANOVA of ln $\hat{V}$ for estimating the relative mutation rate, or they are based on direct numerical calculations as indicated in the text.

The methods described above are applicable to data on allele-size distributions at any number of microsatellite loci studied in an array of populations. The loci may be grouped by their biological characteristics, for example, by motif type, by their genomic location etc., and the populations by their anthropological affinity and or demographic characteristics.

The theory described above and the analyses of data provided indicate that the unique feature of microsatellites—that the alleles at such loci can be ordered by their repeat sizes—offers an additional advantage for genetic analyses. This provides the scope for developing the theoretical framework for answering questions that cannot be addressed with the classical loci. For example, the population genetic characteristics of microsatellites can be studied by using a general stepwise mutation model that does not necessarily involve only single-step or symmetric changes. This in turn allows estimation of relative mutation rates, examination of signatures of past population size changes, and evaluation of the impact of possible allele-size constraints. Data analyses reviewed above also indicate that although the exact parameters of the mutation process at microsatellites are still not fully known, the models described appear reasonable, and they provide biologically realistic qualitative inference. The observations that the mutation rates at microsatellites are inversely related to the motif size, and the African populations exhibit the indications of the most ancient human bottleneck, are also consistent with other genetic evidence. Since single-locus statistics, used in this theory, involve either higher-order moments of allele sizes (e.g., variance), or ratio-estimators, one limitation of such analyses is that inferences are to be based on statistics averaged over a group of loci. Thus, more experimental data may be required to justify how the loci are to be grouped to avoid the confounding bias of differences of parameter values related to the loci.

# Acknowledgements

This work was supported by grants GM 41399, GM 45861 (R.C.) and GM 58545 (R.C., M.K.) from the National Institutes of Health, and DMS 9409909 (M.K.) from the National Science Foundation and by the Keck's Center for Computational Biology at the Rice University (M.K.).

# 12 Using microsatellites to measure the fitness consequences of inbreeding and outbreeding

J.M. Pemberton, D.W. Coltman, T.N. Coulson and J. Slate

## Chapter contents

## Abstract

Due to the practical difficulties of measuring inbreeding in the field, the fitness consequences of inbreeding and outbreeding have been poorly documented in natural populations. In this chapter we explore the use of microsatellite variation as a tool for measuring how inbred or outbred an individual is. We first point out that average heterozygosity of an individual measured from microsatellite data should more closely reflect inbreeding than previous estimates using allozymes. Second, assuming the stepwise mutation model for microsatellite mutation, we propose that allele lengths can be used to estimate an individual internal distance measure, mean $d^2$, which represents the average time to coalescence for the microsatellites carried by the two gametes giving rise to an individual. Using two vertebrate data sets, we show that mean $d^2$ explains variation in neonatal fitness-related traits which is not explained by individual heterozygosity. We speculate that the most likely source of allele-length variation in natural populations is migration between diverged populations, and that individuals with high mean $d^2$ are fitter because they are heterozygous at more loci throughout the genome. We suggest that mean $d^2$ is a convenient tool for measuring the relative position of individuals within a population on the inbred–outbred continuum.

## 12.1 Introduction

In this chapter we explore the potential advantages microsatellites offer in the study of the fitness consequences of inbreeding and outbreeding in natural populations. Before considering these applications in detail we briefly review the problem.

In diploids the genetic relatedness between the parents of an individual can be viewed as a continuum along which there may be changing fitness consequences for the offspring (Mitton 1993; Shields 1993; Waser 1993). Typically, inbred offspring are less fit ('inbreeding depression') and outbred offspring more fit ('heterosis'), though very outbred matings, for example those found in hybrid zones, may produce offspring of low fitness ('outbreeding depression'). The underlying processes responsible for inbreeding depression and heterosis probably include increased homozygosity for deleterious recessive mutations and overdominance, but the relative importance of each of these processes is not clear (Charlesworth and Charlesworth 1987). Outbreeding depression may be due to disruption of local adaptation or of co-adapted groups of genes or both (Templeton 1986).

The fitness consequences of inbreeding and outbreeding have important ramifications in many areas of biology. Notably, they are a major consideration in commercial plant and animal breeding, they may be responsible for the evolution of many aspects of plant and animal breeding systems, and they are a major concern in conservation biology of species now dependent on captive breeding or intensive management.

Although there is a large literature on inbreeding, outbreeding, and their fitness consequences in domesticated plants and animals, and on non-domesticated organisms brought into the laboratory, it is generally acknowledged that there is a dearth of good studies documenting these phenomena in natural vertebrate populations (Ralls *et al.* 1986; Charlesworth and Charlesworth 1987; Shields 1993). This is unfortunate for various reasons. First, although a species may show inbreeding depression or heterosis in the laboratory, we often do not know how common inbreeding events are in nature, and therefore whether laboratory observations are of any consequence in the wild. Indeed, there is considerable debate in the literature as to what level of relatedness constitutes inbreeding in natural populations (Ralls *et al.* 1986; Shields 1993). Second, there is substantial evidence that inbreeding depression may interact with environmental conditions (Pray *et al.* 1994; Heshel and Paige 1995), so that laboratory measurements may not reflect effects found in the wild. Finally, there is a suite of questions about how inbreeding depression varies with the kind of trait studied (e.g. early-expressed versus late-expressed, sexually selected versus not), which are largely untouched in natural populations.

In general, previous studies of inbreeding in natural populations have adopted one of two routes. In one approach, pedigrees are drawn up for individuals within long-term studies, usually of wild vertebrates (established by observation, or inferred using molecular markers) from which individual inbreeding coefficients ($f$) are estimated and correlated with individual measures of various fitness component or fitness-related traits. Well known examples of this approach include great tits (*Parus major*) in Wytham Wood (Greenwood *et al.* 1978), prairie dogs (*Cynomys ludovicianus*) in South Dakota (Hoogland 1992), and song sparrows (*Melospiza melodia*) on Mandarte Island (Keller *et al.* 1994). This approach is only feasible for some species and in some circumstances. Moreover, the advent of molecular methods for determining parentage has generally reduced the confidence with which parentage can be inferred from behavioural observations, even for apparently monogamous species (Brooker *et al.* 1990). In future it may be possible to use molecularly-determined pedigrees to determine inbreeding coefficients accurately in such studies, though inferring parentage from molecular markers on a sufficiently large scale with high confidence may prove to be a problem for many studies (Marshall *et al.* 1998).

An alternative approach is to analyse individual mean heterozygosity at a sample of codominant molecular markers (hereafter referred to as individual heterozygosity), which should be inversely correlated with inbreeding coefficient. This approach has been investigated extensively with allozymes. Overall, reviews tend to favour a positive association between individual heterozygosity and measures of fitness (Allendorf and Leary 1986; Ledig 1986; Mitton 1993; Britten 1996). However, non-significant results are probably under-reported. A general problem with these studies is the relatively small number of polymorphic loci studied and the rather small number of alleles segregating, so that most loci are homozygous even in the absence of inbreeding.

The microsatellites characterized and screened in natural populations in recent years are commonly far more polymorphic than allozymes, having more alleles and much higher heterozygosity. A first and obvious advantage of microsatellites is that individual heterozygosity measured by microsatellites should be more closely related to the degree of inbreeding than that measured by allozymes. It is perhaps curious that variation in microsatellite-derived individual heterozygosity has not yet been explored in relation to fitness in any natural populations.

The stepwise mutation process of microsatellites (Valdes *et al.* 1993; Weber and Wong 1993) offers a second, more interesting approach to measuring inbreeding and outbreeding. If allele length carries historical information, then an individual internal distance measure can be calculated which reflects the time to coalescence for the two alleles at a locus or, averaged across loci, the mean time to coalescence for the microsatellites studied. More formally, the difference in repeat units between two alleles at a locus is related to their time since coalescence (Goldstein *et al.* 1995*b*; Slatkin 1995). Working at the population level, Goldstein *et al.* (1995*b*) have shown that this distance squared, averaged over many loci, is linearly correlated with the time since two populations diverged. Using the same logic, Coulson *et al.* (1998) have proposed that a similar measure can be estimated for each individual, denoted mean $d^2$. Mean $d^2$ is estimated from the two alleles each individual has at a locus, averaged over loci, and is a measure of the genetic distance between the gametes which formed the individual. Another way of thinking about mean $d^2$ is as a measure of the variance in average allele lengths within an individual. It is assumed that the probability that two gametes differ at loci throughout the genome, including those affecting fitness, is related to mean $d^2$.

Notice that, estimated from the same sample of strongly polymorphic microsatellites, individual heterozygosity and mean $d^2$ put emphasis on different ends of the inbreeding–outbreeding continuum. Individual heterozygosity should reflect recent matings between relatives. Mean $d^2$, though it includes homozygosity, will be strongly influenced by variation in allele length, that is, events deeper in the pedigree. A potent source of intra-population variation in mean $d^2$ will be migration between populations that have diverged at microsatellite allele lengths.

In this chapter, we summarize exploration of individual heterozygosity and mean $d^2$, measured by microsatellites, in relation to two early-expressed fitness measures in two studies of wild vertebrates (Coulson *et al.* 1998; Coltman *et al.* 1998*a*).

## 12.2 Materials and methods

### 12.2.1 Study populations

The two study populations are the red deer (*Cervus elaphus*) population resident on the Isle of Rum, Scotland, UK and the harbour seal (*Phoca vitulina*) population that breeds on Sable Island, Nova Scotia, Canada. Both species are strongly seasonal breeders in which adult females give birth to a single offspring each year (seals) or most years (deer). Both studies concentrate on following individual life histories, and have published extensive previous analyses of non-genetic sources of variation in individual fitness (Clutton-Brock and Albon 1989; Clutton-Brock *et al.* 1982; Bowen *et al.* 1994; Boness and Bowen 1996). Our analyses of individual heterozygosity and mean $d^2$ focus on 670 red deer calves born into the Rum deer population between 1982 and 1996, and on 275 harbour seal pups born into the Sable Island harbour seal population in 1994 and 1995.

Red deer became extinct on Rum in the eighteenth century, and the current population is the result of a series of introductions from at least four source populations since 1845. In contrast, the harbour seals that breed on Sable Island are an unmanipulated natural population. While tagging studies suggest harbour seals are capable of moving large distances at sea (Thompson *et al.* 1994; W.D. Bowen, unpublished data), the level of effective interchange between Sable Island and other harbour seal populations is not known. However, Sable Island harbour seals show some distinct morphological characteristics, such as a reduced number of post-canine teeth and different pelage pattern from mainland populations (Boulva and McLaren 1979), so that it is likely that there are low levels of gene flow among populations in eastern Canada.

## 12.2.2 Study traits

In each population two traits expressed early in life were studied: birth weight and neonatal survival. Newborns were weighed within a few days of birth, and birth weight was estimated after correcting for age at capture. In each population birth weight is related to fitness. In deer, heavy born calves are more likely to survive the neonatal period (see below), more likely to survive their first winter, and if female, likely to have higher fecundity than light-born calves (Clutton-Brock *et al.* 1982; Albon *et al.* 1987; Clutton-Brock and Albon 1989). In harbour seals, birth weight is expected to influence fitness for energetic reasons (Bowen *et al.* 1994)

Neonatal survival is a true fitness component. In red deer, neonatal mortality is defined as survival through the period from birth (median date, 8 June) and 1 October. Most mortality falls close to birth and, where cause can be attributed, is due to birth complications, accidents, predation by golden eagles (*Aquila chrysaetos*) and maternal desertion. On average, neonatal mortality accounts for about 18 per cent of deer calves born and it does not vary significantly from year to year. In harbour seals, neonatal mortality is defined as mortality between birth (median date, 29 May) and day 20, at which time weaning commonly begins (Muelbert and Bowen 1993). Neonatal harbour seal pups swim regularly, and from washed-up remains, a common source of mortality at this time appears to be predation by sharks.

## 12.2.3 Microsatellites, individual heterozygosity and mean $d^2$

In both populations, polymorphic microsatellites were characterized and screened in order to infer paternity for individual juveniles. In the breeding season (the 'rut'), red deer males defend groups of females and mate with those that come into oestrus. However, a previous study using DNA fingerprinting showed that behavioural data are a poor guide to the paternity of most calves (Pemberton *et al.* 1992), so a large-scale paternity analysis using microsatellites was conducted (Marshall *et al.* 1998). Harbour seals belong to a group of seal species which mate aquatically, and microsatellites were employed to infer paternity of Sable Island pups from among the locally resident males (Coltman *et al.* 1998*b*). The loci screened in each population are listed in Table 12.1; they were screened by the conventional methods for the laboratories involved, as described in Bancroft *et al.* (1995) and Coltman *et al.* (1996). Note that in general the harbour seal loci are less polymorphic than the red deer loci, except for SGPV3, which had 22 alleles. All loci are dinucleotide repeats except the seal locus $\beta$-globin, which

**Table 12.1**

Characteristics of microsatellite loci typed in red deer and harbour seals, and influence of $d^2$ measured individually for each locus on birth weight. In deer, the locus-specific tests shown were conducted in a simple model of birth weight with no other variables and no interactions fitted, in which mean $d^2$ including all loci gives $F_{1,669} = 4.34$, $p = 0.038$. In seals, the locus specific tests are conducted in a full birth weight model in which pups sex and maternal age are included, in which mean $d^2$ including all loci gives $F_{3,33} = 7.99$, $p < 0.001$. Significant results ($p < 0.05$, without correction for multiple tests) are shown in bold type

| Species | Locus | Number of alleles | Allele size range (repeat units) | Heterozygosity | Average allele size difference (repeat units) | Effect of removing locus from birth weight model (P) | Effect of locus fitted individually to birth weight model (P) |
|---|---|---|---|---|---|---|---|
| Red deer | OarFCB193[1] | 11 | 13 | 0.76 | 4.6 | 0.062 | 0.173 |
| Red deer | OarFCB304[1] | 9 | 9 | 0.79 | 4.3 | **0.028** | 0.221 |
| Red deer | CelJP15[2] | 10 | 14 | 0.84 | 6.2 | 0.055 | 0.498 |
| Red deer | CelJP27[2] | 6 | 7 | 0.69 | 3.2 | **0.039** | 0.520 |
| Red deer | CelJP38[2] | 8 | 8 | 0.79 | 2.8 | **0.039** | 0.764 |
| Red deer | MAF35[3] | 7 | 6 | 0.67 | 2.7 | **0.032** | 0.481 |
| Red deer | MAF109[4] | 6 | 14 | 0.75 | 7.8 | 0.063 | 0.493 |
| Red deer | OarCP26[5] | 13 | 21 | 0.72 | 11.2 | 0.607 | **0.025** |
| Red deer | TGLA94[6] | 9 | 15 | 0.81 | 8.0 | **0.008** | 0.33 |
| Harbour seal | β-globin[7] | 5 | 4 | 0.73 | 1.9 | **0.005** | 0.667 |
| Harbour seal | Hg8.1[8] | 5 | 4 | 0.53 | 1.0 | **0.002** | 0.951 |
| Harbour seal | Pvc19[9] | 2 | 2 | 0.44 | 2.6 | **0.003** | 0.240 |
| Harbour seal | Pvc43[10] | 4 | 4 | 0.56 | 1.8 | **0.001** | 0.983 |
| Harbour seal | SGPV3[11] | 22 | 26 | 0.89 | 4.6 | **0.015** | **0.024** |
| Harbour seal | SGPV11[11] | 5 | 5 | 0.45 | 2.1 | 0.019 | **0.002** |

[1] Buchanan and Crawford 1993; [2] J. Pemberton, unpublished data; [3] Swarbrick et al. 1991; [4] Swarbrick and Crawford 1992; [5] Ede et al. 1995; [6] Georges and Massey 1992; [7] Gemmel et al. 1997; [8] Allen et al. 1996; [9] Coltman et al. 1996; [10] D.W. Coltman, unpublished data; [11] Goodman 1997.

has a pentanucleotide repeat, and there is no evidence for null alleles segregating at high frequency at any locus.

For each deer calf or seal pup, individual heterozygosity was calculated as the number of loci at which the individual was heterozygous, divided by the total number of loci at which an individual was scored.

For each calf or pup, mean $d^2$ was calculated as:

$$\text{mean } d^2 = \frac{1}{n} \Sigma^n (i_a - i_b)^2,$$

where $i_a$ and $i_b$ are the lengths in repeat units of alleles a and b at locus $i$, and $n$ is the total number of loci at which an individual was scored.

## 12.3 Results

### 12.3.1 Individual heterozygosity and mean $d^2$

Figure 12.1 shows the distribution of individual heterozygosity and mean $d^2$ for the study deer calves and the study seal pups, and Fig. 12.2 shows the correlation between these measures for each population, illustrating the extensive variation in mean $d^2$ found within each individual heterozygosity class.

The component graphs of Fig. 12.1 show various degrees of skew in heterozygosity and mean $d^2$. In particular, mean $d^2$ for harbour seals is heavily skewed, which is partly a consequence of the single hypervariable locus SGPV3 within a sample of less polymorphic loci. In their analysis of birth weight and neonatal survival in deer calves, Coulson et al. (1998) chose not to transform individual heterozygosity and mean $d^2$, because they used generalized linear models which are relatively robust to non-normal data (Genstat 5 Committee 1995), and because residuals in the regressions performed (see below) were normally distributed. However, in their analysis of birth weight and neonatal survival in harbour seal pups, Coltman et al. (1998a) used a logarithmic transformation of mean $d^2$ to normalize their data.

### 12.3.2 Birth weight

In each study population, a model of birth weight was constructed which took into account variables that explain significant variation in birth weight before testing individual heterozygosity or mean $d^2$. In deer calves, these variables include sex of calf (males heavier than females), mother's status (high reproductive effort in previous year depresses birth weight), mother's age as quadratic function (young and old mothers produce lighter calves), and mean April temperature (warm Aprils result in greater foetal growth). In seal pups, explanatory variables for birth weight include sex (males heavier than females), and mother's age as a linear function (young females produce lighter pups).

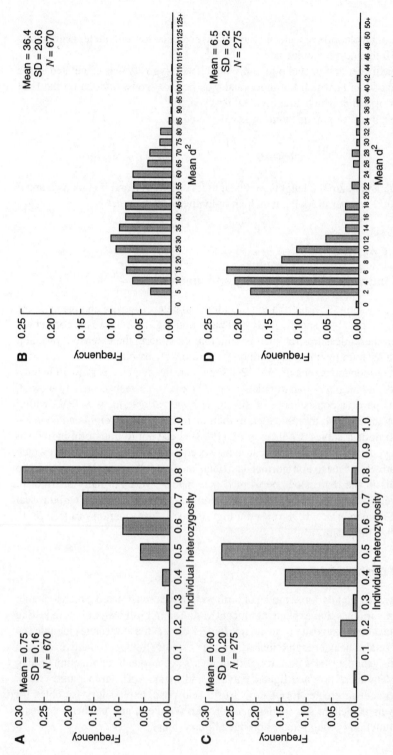

**Fig. 12.1** Frequency distributions of microsatellite-based estimates of inbreeding in red deer and harbour seals. (a) Individual heterozygosity and (b) mean $d^2$ (in repeat units) for all 670 calves born in the Rum study area between 1982 and 1996. Individuals only entered the data set if genotyped at a minimum of seven of the nine loci shown in Table 12.1. (c) Individual heterozygosity and (d) mean $d^2$ for 275 harbour seal pups born on Sable Island in 1994 and 1995. Individuals only entered the data set if genotyped at a minimum of four of the six loci shown in Table 12.1.

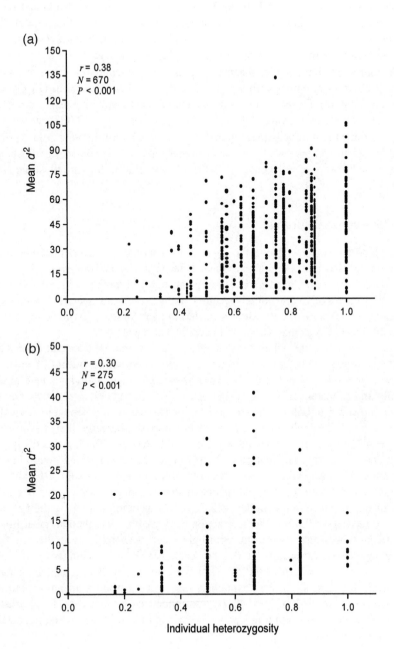

**Fig. 12.2** Relationship between individual heterozygosity and mean $d^2$ in (a) red deer calves and (b) harbour seal pups.

When tested in the models of birth weight described above, individual heterozygosity did not explain variation in deer calf birth weight ($F_{1,635} = 0.72$, $p = 0.40$) (Coulson et al. 1998). In seal pups, however, heterozygous pups weighed more at birth ($F_{1,35} = 8.82$, $p < 0.01$) (Coltman et al. 1998a).

In contrast, in both studies mean $d^2$ explained significantly more variation in birth weight. Deer calves with higher mean $d^2$ were heavier at birth ($F_{1,635} = 4.13$, $p = 0.042$). Furthermore, there was a significant interaction between mean $d^2$, April temperature, and birth weight ($F_{1,634} = 7.2$, $p = 0.007$), in which the prominent feature was that low mean $d^2$ calves born following cold Aprils had low birth weights. Seal pups with higher mean $d^2$ were also significantly heavier at birth $F_{1,35} = 10.64$, $p < 0.001$). These relationships are illustrated in Fig. 12.3.

### 12.3.3 Neonatal survival

In each population, variables affecting neonatal survival were investigated before examining associations with individual heterozygosity and mean $d^2$. In deer calves, a logistic model of neonatal survival included mother's age as a linear function (calves of young mothers are more likely to die), and mean June temperature (calves are more likely to survive in warm Junes). In seal pups no preliminary variables tested explained variation in neonatal survival.

As for birth weight, when fitted to the logistic model of deer calf neonatal survival, individual heterozygosity was not associated with survival ($\chi^2 = 0.1$, d.f.$= 1$, $p > 0.05$). Similarly, there was no association with individual heterozygosity and neonatal survival in seal pups ($\chi^2 = 2.55$, d.f.$=1$, $p > 0.05$). Once again, however, in both studies mean $d^2$ was associated with neonatal survival. High mean $d^2$ deer calves and seal pups were more significantly likely to survive ($\chi^2 = 4.7$, d.f.$=1$, $p < 0.05$ and $\chi^2 = 4.18$, d.f. $= 1$, $p < 0.05$, respectively).

In deer calves, but not seal pups, birth weight was an important determinant of neonatal survival, with low birth-weight calves more likely to die. When birth weight was fitted to the deer calf neonatal survival model (above) it explained significant deviance in the model, as light calves were more likely to die ($\chi^2 = 12.9$, d.f.$=1$, $p < 0.001$). The interaction between birth weight and mean June temperature was also significant, with lighter calves more likely to die in cold Junes ($\chi^2 = 8.5$, d.f. $= 1$, $p < 0.01$); however, mean $d^2$ was no longer significant ($\chi^2 = 0.5$, d.f. $= 1$, $p > 0.05$). Thus the best model of deer calf neonatal survival, in terms of variance explained, does not incorporate mean $d^2$, but it does incorporate birth weight, which was influenced by mean $d^2$. In contrast, mean $d^2$ appears to explain independent variation in birth weight and neonatal survival in seal pups.

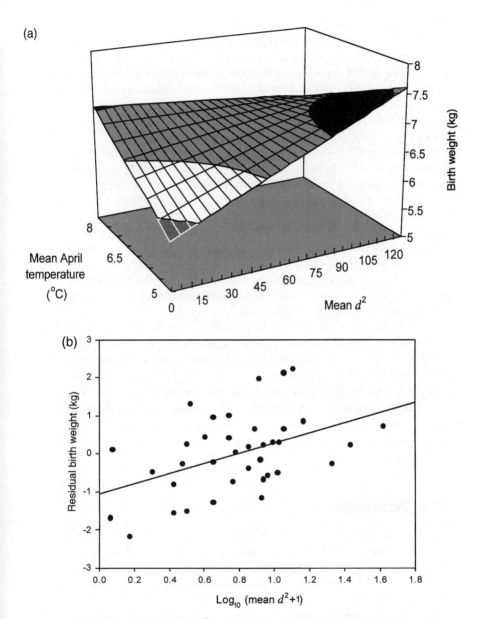

**Fig. 12.3** The relationship between mean $d^2$ and birth weight in (a) red deer calves and (b) harbour seal pups. In red deer calves (a) there is an interaction between mean $d^2$, mean April temperature and birth weight in which the prominent feature is reduced birth weights among low mean $d^2$ calves born following a cold April. In harbour seal pups (b) there is a linear relationship between log mean $d^2$ and birth weight.

### 12.3.4 Influence of individual loci

The importance of each single locus to the overall mean $d^2$ effect was evaluated by testing birth weight models after dropping each locus from the mean $d^2$ calculation in turn, and by fitting the mean $d^2$ of each locus individually as the genetic term. In deer, these tests were conducted in a simple model without other explanatory variables and without interaction terms, while in seals these tests were conducted in a full model including maternal age and pup sex. Results are shown in Table 12.1. In deer the mean $d^2$ effect remains statistically significant or close ($p < 0.10$; Table 12.1) after dropping any one locus, with the notable exception of OarCP26 ($p = 0.607$ following removal), which had the greatest contribution to mean $d^2$. In seals, the mean $d^2$ effect is strongly statistically significant following the removal of any single locus, although dropping SGPV3 (which has the greatest influence on mean $d^2$) or SGPV11 substantially weakens the effect (Table 12.1). Fitted individually, the OarCP26 locus in the red deer analysis, and the loci SGPV3 and SGPV11 in the harbour seal analysis, are significantly associated with increased birth weight. As OarCP26 and SGPV3 are the most polymorphic loci in each study in terms of heterozygosity, number of alleles, and allele size range, these results suggest that more polymorphic loci, and those with greater variance in allele sizes, contribute a disproportionate amount of information to the mean $d^2$ signal.

The analysis presented in Table 12.1 might suggest that the mean $d^2$ effect in deer is entirely due to OarCP26, but this is not correct. When fitted to the full birth weight model, including other explanatory variables and interaction terms, OarCP26 $d^2$ is significant ($F_{1,615} = 4.53$, $p = 0.034$) and so is the interaction term between OarCP26 $d^2$ and mean April temperature ($F_{1,614} = 4.27$, $p = 0.039$), but the effect is substantially weaker than for the multilocus mean $d^2$-mean April temperature interaction ($F_{1,634} = 7.2$, $p = 0.006$ (above)), suggesting that other loci are contributing to the mean $d^2$ effect in the full birth weight model for deer.

## 12.4 Discussion

The analyses of red deer and harbour seal juvenile traits summarized above give strikingly similar results for two individual-specific measures of genetic diversity measured from microsatellite data in two very different field studies. In each case we have shown that mean $d^2$ explains variation in birth weight and neonatal survival not detected by individual heterozygosity.

One inference from the contrast between individual heterozygosity and mean $d^2$ results is that variation in microsatellite allele length among heterozygotes rather than homozygosity generates much of the detected signal. 'Wide' heterozygotes signal greater fitness than 'narrow' heterozygotes. As indicated in the Introduction, a common source of extreme alleles will be successful immigration from populations which have diverged with respect to allele lengths, that is, by individuals which are relatively unrelated to members of the population into which they immigrate. In this case, fitness benefits presumably accrue because progeny are more heterozygous throughout the genome, including loci affecting fitness.

The analysis of the effects of individual locus $d^2$ contributions on birth weight reveals an interesting pattern. In each species the locus with the greatest number of alleles and greatest allele size range (OarCP26 in deer and SGPV3 in seals) is responsible for a substantial proportion of the multilocus mean $d^2$ result, and is significant when fitted on its own. In addition, in harbour seals SGPV11 has a significant contribution when fitted on its own, and in deer there is evidence, from the full birth weight model, that other loci contribute to the effect. The fact that $d^2$ values for particular loci influence birth weight might be taken to suggest that the multilocus mean $d^2$ results presented are consistent with the idea that we have revealed birth weight effects at loci linked to particular microsatellites. However, the fact that the most polymorphic loci are involved in each species instead suggests to us that there is a correlation between microsatellite polymorphism and efficacy as a marker of genome-wide heterozygosity. Future studies, involving larger numbers of markers, will investigate this point and methods for standardising $d^2$ across loci, such as weighting the mean $d^2$ calculation by locus-specific variance.

The hypothesis that associations between mean $d^2$ and fitness-related traits arise predominately from population mixing, or outbreeding, requires further investigation. If accurate, then associations might be expected to occur in the red deer population on Rum, as it was formed by recent introductions from deer populations in England and Scotland, and individuals are likely to vary greatly in their recent coancestry. The results from the harbour seal population at Sable Island are somewhat more surprising. As this population is relatively isolated, and individuals show relatively low levels of genetic variability both at microsatellite loci (Coltman et al. 1996) and in DNA fingerprints (Harris et al. 1991; Kappe et al. 1997; C. Schaeff unpublished data), it seems unlikely that there are high levels of immigration from other populations in Eastern Canada, and that the population is more genetically homogenous. The fact that mean $d^2$ was positively associated with birth weight and neonatal survival in this harbour seal population may suggest that enough population mixing has taken place to generate differences in heterozygosity at fitness-affecting loci. On the other hand, the positive associations with heterozygosity were also statistically significant, which, coupled with low genetic variability overall, suggest that individuals in this population may also suffer from increased homozygosity due to genetic drift, given the small population size. Mean $d^2$ requires further study at both the theoretical and empirical level to determine its usefulness under different levels of population isolation and panmixia.

Thus far, we have carefully avoided describing the effects noted in deer and seals as evidence for either inbreeding depression or heterosis or both. This is because we cannot really determine whether we are detecting reduced fitness in particularly homozygous individuals, or enhanced fitness in particularly heterozygous individuals, although the difference in signal between individual heterozygosity and mean $d^2$ perhaps suggests that heterosis is a more likely explanation. The problem is that we do not know how heterozygous the loci affecting fitness are—we only know that their heterozygosity is likely to be correlated with mean $d^2$. By the same token, we cannot distinguish whether the positive associations result from increased homozygosity at deleterious recessive loci, or from increased homozygosity at loci with overdominance.

In conclusion, we suggest that mean $d^2$ offers a practical tool for investigating the fitness consequences of inbreeding and outbreeding in natural populations. In particular, it offers great advantages over methods that depend on pedigrees, since each individual's status can be assessed from a single sampling event, which may also, as in the case of the deer and seals described here, be an opportunity for collecting data on important fitness-related traits such as birth weight.

## Acknowledgements

We thank all those involved with the red deer project over many years, especially Tim Clutton-Brock, Steve Albon, Fiona Guinness, and all those involved with the harbour seal project, especially Don Bowen and Jonathan Wright. The projects are funded by NERC and BBSRC (UK) and NSERC and the Department of Fisheries and Oceans (Canada). Tristan Marshall made helpful comments on an early draft of this chapter.

# 13 Microsatellites in conservation genetics

Mark A. Beaumont and Michael W. Bruford

## Chapter contents

## Abstract

In this chapter we will examine the role that microsatellite markers have played, and could potentially fulfil in the future, in the emerging field of conservation genetics. We first briefly discuss the applications, advantages, and disadvantages that microsatellite analysis brings to conservation genetics. We then highlight some current applications through the use of case study examples, ranging from the detection of inter-specific hybridization, through inferring population history, to distinguishing among demographic factors affecting present day allele frequencies, and include studies of inbreeding and reproductive success. In the final section we concentrate on the actual and potential additional information that microsatellites can give when compared with other quasi-neutral genetic markers, and explore the possible theoretical advantages and limitations that microsatellites have in conservation studies.

## 13.1 Introduction

Knowledge of the demographic history of populations is important when making informed decisions about population management. For example, there may be concerns about the level of inbreeding within a population, the ratio of effective population size to census size, and the effective population size itself (Lande and Barrowclough 1987). To ameliorate possible deleterious consequences of inbreeding depression, and the risks posed by maintaining monocultures, it might be decided to move animals from different populations. However, this raises questions of which populations should be used to exchange individuals, and which should be maintained as genetically distinct. Microsatellites can be used to answer a number of question posed by the above considerations, and we will consider these below.

Although the application of microsatellites to conservation genetic studies is relatively recent, it is now becoming routine. Since the first publications, which appeared in a special conservation genetics edition of the journal *Molecular Ecology* three years ago (e.g. Gottelli *et al.* 1994; Taylor *et al.* 1994) and shortly thereafter (Morin *et al.* 1994*a,b*; Roy *et al.* 1994), studies have proliferated over the last two years (e.g. Paetkau *et al.* 1995; Spencer *et al.* 1995; Coote and Bruford 1996; Garcia-Moreno *et al.* 1996; Mundy *et al.* 1997; Smith *et al.* 1997). Our aim is not to review large numbers of these studies; rather we attempt to highlight specific conservation-relevant applications through a few case studies, and speculate on how these markers may be used in the future.

The range of studies which can usefully be tackled at the molecular level within a conservation genetics framework has been well documented (e.g. Hedrick and Miller 1992; Bruford and Wayne 1993*a,b*; Frankham 1995; Smith and Wayne 1996). Microsatellites are, however, slightly restricted in their potential application, largely due to their inappropriateness for phylogenetic studies. The main areas where they have been used are in studies of hybridization, population history, and phylogeography, in detecting population bottlenecks and inbreeding, and assessing the impact of reproductive behaviour, social structure, and dispersal on the genetic structure of endangered populations (see case studies).

## 13.2 Advantages and drawbacks

Perhaps the major reason why microsatellites are being so widely applied in the study of endangered species is that they offer advantages which are particularly appropriate in conservation. First, in many species they are relatively easy to obtain, either directly through the isolation of species-specific markers, which involves constructing a genomic DNA library (which can be enriched for the sequences sought, e.g. Armour *et al.* 1994; Kandpal *et al.* 1994; Hammond *et al.* 1998), or by the application of markers originally isolated from related species (e.g. Moore *et al.* 1991; Schlötterer *et al.* 1991; Coote and Bruford 1996; Rico *et al.* 1996). Second, different loci can be used according to their level of variation, which ranges from heterozygosity values a little greater than that of normal allozyme systems to those as variable as many minisatellite loci. Third, as genetic systems, they are comparatively easy to automate, with multiplex amplification of up to five loci possible in a single PCR reaction, and multiple loading of up to fifteen loci per individual being carried out in some highly optimized systems. Most of the automated DNA sequencing systems available on the market have specific fragment analysis software available, making microsatellite analysis comparatively easy, with high throughput possible. Finally, unlike the minisatellite sequences which are detected in traditional DNA fingerprinting methods, microsatellites are amplified by PCR and thus can be used on non-invasively sampled material. This raises the possibility of tracking the demography, movements, and social structure of populations without needing to come into direct contact with the animals themselves (e.g. Morin and Woodruff 1996). Although the technical challenges facing those attempting microsatellite analysis from saliva, hair, or faecal material are considerable (e.g. Gerloff *et al.* 1995; Taberlet *et al.* 1996, 1997; Gagneux *et al.* 1997), they may in many cases be surmountable, and laboratories are beginning to employ techniques which until now have only been used in forensic science in an attempt to understand endangered populations in previously unattainable detail.

However, it is neither justifiable nor realistic to give the impression that microsatellite analysis is without its drawbacks. First, there is currently undocumented anecdotal evidence emerging from a number of laboratories that certain groups of organisms can be extremely difficult to obtain microsatellites from. Many plant species, several invertebrate groups, such as lepidopterans (where micro- and minisatellites seem generally scarce, e.g. Saccheri and Bruford 1993), some dipterans and gastropods, and a number of avian groups have proved difficult to work with, and it is unlikely that this can be entirely explained by general laboratory-to-laboratory variation in competence. Certainly there seems to be great variability in the amount of microsatellite sequence present in the genomes of different species, and these differences can accumulate over relatively short periods of evolutionary time (e.g. Rothuizen *et al.* 1994). As stated previously, evidence is currently at best anecdotal, but some consistent patterns are emerging in these studies, and before embarking on a potentially protracted and expensive procedure, the advice of others working on similar taxa should ideally be sought.

Second, a variety of problems associated with the PCR process are becoming apparent. Non-amplification of certain alleles due to substitutions, insertions, or deletions within the priming sites can lead to apparent null alleles appearing in population studies (e.g. Callen *et al.* 1993; Paetkau and Strobeck 1995; Pemberton *et al.* 1995). *Taq* polymerase-generated slippage products are routinely seen, especially in mono- and dinucleotide microsatellite loci, which can sometimes make allele scoring problematic (e.g. Schlötterer and Tautz 1992; Gill *et al.* 1997), and finally the tendency of *Taq* polymerase to add an additional dATP to PCR products can cause single-base shifts and additional sizing problems (e.g. Ginot *et al.* 1996). Third, although automation has proven to be a powerful advance in terms of genotype throughput, the once generally held view that this would lead to the elimination of problems when comparing data generated among different laboratories seems to have been over-optimistic. Although these problems are not generally documented in the literature, they seem to be general, and a constraining force when analysing comparative data sets, especially when comparing data generated using different chemistries and hardware.

The fourth problem associated with microsatellite genotyping concerns typing from non-invasively sampled material. In particular, hair and faecal material have proved to be difficult sources for analysis. Because each locus is usually in just one copy per cell (as opposed to the multiple copies in mitochondrial DNA typing), stochastic amplification problems can arise, especially where alterations of stringency conditions have been made. Several artefacts have been observed in these situations, including allelic 'dropout' (the stochastic non-amplification of one or two alleles), non-specific amplification fragments obscuring analysis of the locus being studied, and incorrect genotypes can be generated (due to slippage very early in the amplification process leading to the predominance of artefactual fragments in the final product, or simply by contamination; Gerloff *et al.* 1995; Taberlet *et al.* 1996, 1997; Gagneux *et al.* 1997). Very rigorous approaches are therefore required to eliminate these problems and these are both time-consuming and expensive (Taberlet *et al.* 1996, 1997).

Finally, a major problem associated with microsatellite analysis, and one explicitly dealt with in the final section of this chapter, concerns the fact that we do not yet have a well-substantiated evolutionary model which can be applied universally, or even to a subset of markers used in most studies. Without such a model, inference is difficult using allele frequency distributions, accurately quantifying genetic differentiation is problematic, and it is hard to explain the differences we see across species in terms of allele length and variability. In conservation genetics this problem is often not as relevant as elsewhere, since in many cases genetic divergence is associated with the effects of recent drift, bottlenecks and inbreeding, as opposed to mutation. Many creditable attempts have been and are currently being made to study mutation processes, develop algorithms to help analyse size-sensitive data, and to apply likelihood-based coalescent modelling to make fuller use of the data contained within microsatellites, which can enable the testing of alternative hypotheses to explain observed data. These will be reviewed and explored further.

# 13.3 Case studies

## 13.3.1 Hybridization

Several studies have used allele size or frequency differences to examine the effects of intra-specific or subspecific hybridization in endangered species. Perhaps two of the best examples published to date centre around work on wolf-like canids, a group with a great propensity to hybridize, and where offspring are very often perfectly fertile. This poses particular problems for endangered populations, and in the case of both the Ethiopian and red wolf, hybridization has had a significant impact.

Currently, the world's most endangered canid is the Ethiopian wolf *Canis simensis*. The species has a very narrow range, being found in only six isolated populations in the Ethiopian highlands, with a total population of approximately 500 individuals (Sillero-Zubiri and Macdonald 1997). Ethiopian wolves have been in decline due to a combination of habitat loss and extermination by humans. Additionally the wolves are in contact with feral dogs, resulting in extensive hybridization with the dog population also acting as a disease reservoir for both distemper and rabies, outbreaks of which have led to further declines. Using microsatellites, Gottelli *et al.* (1994) addressed a number of questions concerning Ethiopian wolves that have conservation implications. Nine canine microsatellite loci were analysed in hybridizing populations in the Upper Web Valley and Sanetti Plateau, and the results showed that hybridization was occurring at a significant level, primarily between female Ethiopian wolves and male domestic dogs. Interestingly, a high level of precision could be ascribed in the hybridization studies, because many of the Ethiopian wolf alleles were not found in the local dogs. Such an observation could be interpreted as reflecting their phylogenetic distinctiveness, through species-specific allelic differences. Interestingly, Ethiopian wolf alleles show almost completely overlapping size ranges with dog alleles. This is perhaps surprising, given that non-overlapping allele size ranges have accounted for a number of the observations of species-specific alleles in previous hybridization studies (e.g. in taurine and zebu cattle, MacHugh *et al.* 1997). This observation may suggest that the power that microsatellites apparently have to distinguish between Ethiopian wolves and dogs could simply be a function of the disparity of genetic diversity in the two populations. Ethiopian wolves only show about 30–40 per cent of the heterozygosity and allelic diversity of outbred canids, and on average only possess two to three alleles per locus, in contrast to the much higher numbers of alleles found in dogs. An alternative explanation is simply that the loci chosen are under some functional constraint that keeps their length within certain size ranges.

In a controversial and now well documented study in 1991, an extensive mitochondrial (mt)DNA analysis cast doubt on the genetic integrity and identity of the endangered red wolf of the south-eastern region of the United States (Wayne and Jenks 1991). In order to explore this problem further, and to avoid arriving at erroneous conclusions based only on female-mediated introgression (mtDNA is maternally inherited), Roy *et al.* (1994, 1996) analysed red wolves using ten microsatellite loci in both extant and historical specimens. This analysis was carried out within the context of a wider ranging study of grey wolf and coyote hybridization across their respective ranges in the USA. In the case of the red wolf, the prevailing evolutionary hypothesis prior to Wayne and Jenks' study was that the species had an early Pleistocene origin, and could potentially have been the predecessor of modern coyotes and gray wolves. Unfortunately, during the 1940s, the red wolf almost vanished from the wild, and the few remaining individuals were seen to be hybridizing with coyotes. By analysing pre-1940 red wolf pelts, the hybridization question could be fully explored, and here the microsatellite data substantiated the mtDNA findings, that the red wolf originated through hybridization between gray wolves and coyotes. Whilst hybridization may have occurred repeatedly over time prior to European settlement in the USA, most grey wolf × coyote hybridization has been recently induced by human-mediated habitat alteration. In contrast to the Ethiopian wolf study, the major technical challenge in this analysis was to assess the significance of the lack of unique (or private) alleles in red wolves. Specifically, because modern populations of red wolves lack unique alleles, this does not mean that such alleles did not originally exist, because they could quite readily have been lost due to genetic drift and/or hybridization in the last two centuries. Such problems can be modelled, and current data clearly show that the red wolf has a hybrid origin, even when testing for such effects.

### 13.3.2 Divergence among populations

Still largely absent from the literature are studies where microsatellite distributions have been described in populations known to have been isolated for long periods of evolutionary time. This is unfortunate (though likely to change in the near future), because it is only under these circumstances that empirical data can be used to test the hypothesized superior performance of the so-called 'step-wise' distance measures (Feldman *et al.* this volume) recently derived specifically for use in populations where microsatellite mutational processes might account for a proportion of the differentiation seen (e.g. Goldstein *et al.* 1995*b*; Slatkin 1995). The two examples highlighted below both used microsatellites in the context of measuring gene-flow, in the first to test hypotheses of selection-induced quantitative genetic differentiation in unstable habitats, and in the second study to identify population units in a highly seasonally mobile carnivore.

In a study designed to examine the importance of rainforest–savannah ecotone forest patches as sources of evolutionary novelty in the tropics, Smith *et al.* (1997) measured gene flow and morphological divergence in 12 populations of a common rainforest passerine bird, the little greenbul (*Andropadus virens*). Eight microsatellite loci were used to derive gene-flow estimates indirectly through *F* statistics, and contrary to expectations, populations in the forest and the ecotone (the transition zone between the African rainforest and savanna) were found to be morphologically divergent, despite the fact that the microsatellites revealed quite high levels of gene flow. Although this particular species was chosen because it is both common and relatively sedentary, different forest patches did not show the high levels of stochastic differentiation expected under a drift model of divergence, and there were no private alleles found at high frequency in any population, results concordant with those generated using mtDNA control region. This disparity between morphologic and quasi-neutral genetic divergence has also been observed in a similar study of greenfinches in Europe by Merila (1997), and the combination of neutral molecular markers with quantitative genetic analysis in a biogeographical context is a potentially important development in understanding population differentiation and evolutionary processes.

The genetics of some large carnivore populations have been characterized in the past by relatively low levels of variation, especially when using allozyme markers. Paetkau *et al.* (1995) investigated this phenomenon in polar bears, which demonstrate particularly low variation and consequently intractable population genetics. Eight loci were used to study the relationships among four bear populations in Canada. Using microsatellites they uncovered considerable genetic variation, with an average heterozygosity of approximately 60 per cent. All populations were significantly differentiated, including two populations in the Beaufort Sea, and while the levels of differentiation generally reflected geographic location, they also suggested patterns of gene flow not obvious from geography, which may reflect unusual dispersal routes. The authors also introduced an assignment test whereby the origin for a given sample could be predicted, based on its expected genotype frequency, a method which is now being used more widely.

### 13.3.3 Inferring past demographic processes

It is a widely held, if inaccurate, view that conservation genetics is primarily concerned with measuring levels of genetic diversity within populations, and attaching some notional risk of extinction to those populations exhibiting low variation because of a presumed cost of inbreeding (e.g. Brookes 1997). In reality, studies of genetic variation which are carried out in the absence of a known historical context, and which use genetic diversity as a surrogate for fitness and evolutionary potential, are nowadays rare, as the following examples clearly show. These studies have placed considerable emphasis on interpreting the genetic diversity they measure in a sound evolutionary context. Later in this chapter we will focus on a similar study recently carried out by ourselves where we use historically accurate demographic data to understand the evolutionary origins of fragmented populations of South African buffalo (O'Ryan *et al.* 1998).

In perhaps the first study of genetic bottlenecks using microsatellites in a natural population, Taylor *et al.* (1994) investigated the utility of these markers to measure and interpret the genetic variation present in the northern hairy-nosed wombat (*Lasiorhinus kreftii*), one of Australia's most endangered marsupials. This unusual and biologically interesting species has undergone a major decline, especially since agriculture started to impinge in its habitat in the last 120 years. The species now exists in one location (Epping Forest National Park, Queensland) as a single colony. Comparisons with a closely related species which has a similar life history, the southern hairy-nosed wombat (*Lasiorhinus latifrons*), showed that the northern species demonstrates approximately 40 per cent of the heterozygosity found in the southern species. Further, using museum specimens collected in 1884, the authors were also able to assess microsatellite variation in an extinct population of the northern hairy-nosed wombat, from Deniliquin, New South Wales, 2000 km to the south of the extant population, and found that the apparent loss of variation in the Epping Forest colony was consistent with an extremely small effective population size in the last 120 years.

In a study of an avian island endemic, Mundy *et al.* (1997) used microsatellites to characterize variation in the present-day and historic population of the San Clemente loggerhead shrike, a Californian Channel Island subspecies currently endangered with a population of about 130 individuals. The authors characterized this variation in the context of studies of two populations of an abundant, related subspecies found 120 km away on the mainland. They found that the current population on San Clemente has 60 per cent of the variation of the mainland populations and is strongly differentiated. Comparison of the present-day population with birds from a 1915 sample showed most of the variation to have been lost before recent population crashes, and only a 20 per cent decrease in variation was measurable in the extant samples. Measurements of gene flow with the mainland were very low, despite limited vagrance from populations of the mainland subspecies onto San Clemente. The authors concluded that the island subspecies is sufficiently differentiated to justify conservation efforts aimed at avoiding further demographically mediated decline.

### 13.3.4 Reproductive success, social structure and inbreeding

Microsatellites have become the tool of choice in recent years for analysing reproductive success, social structure, and relatedness. Particularly within the framework of studying social evolution and behaviour in the Hymenoptera (where microsatellites appear to be both abundant and variable), microsatellite analysis has made studies of relatedness and reproductive skew possible where previously it was impossible (e.g. Hughes and Queller 1993, Queller *et al.* 1993; Bourke *et al.* 1997). Of particular importance in conservation biology is how, in socially structured populations, fragmentation, demographic contraction, and inbreeding might affect and potentially effect changes in the distribution and abundance of genetic variation and relatedness. Within this context several studies have examined the link between social and genetic structure in endangered natural populations.

In primates, a variety of different social organizations are found among species, many of which are threatened in the wild. In 1994, in two separate analyses, Morin *et al.* examined hypotheses about chimpanzee social behaviour by non-invasive genotyping of free-ranging individuals from 20 African sites. Relatedness among individuals in one community was extrapolated from allele-sharing data from eight microsatellites. Males (in chimpanzees, the non-dispersing sex) were found, within groups, to be related on the order of half-siblings, supporting the kin-selection hypotheses for the evolution of cooperation among males. Morin *et al.* also carried out paternity exclusions for 25 chimpanzees including ten for whom the mother was also genotyped. Twelve to twenty males were potential fathers, based on age and/or sexual behaviour. Of these 25 offspring, the data permitted the identification in four cases of a previously undetermined father. In an additional four cases all but two to five potential fathers could be excluded, and in the remaining cases all living males could be excluded. This lack of conclusiveness in paternity determination in chimpanzees is likely to be due to a combination of insufficient loci having been utilized in a species with a male-biased relatedness structure within social groups, and the effects of reproductive male turnover coupled with long generation times. These results are in contrast to the study of Altmann *et al.* (1996) who carried out paternity analysis in savannah baboons, a species whose social structure is very different. Baboon social groups include multiple adults of both sexes but, as in many mammals, males are the dispersing sex. Females stay in their natal groups throughout their life and form dominance hierarchies where status is inherited. Using ten microsatellite and two allozyme loci, Altmann *et al.* were able to determine paternity for 27 offspring over a four-year period, and in so doing confirm the reproductive priority of dominant males. However, the high short-term variance in reproductive success did not translate into equally high long-term variance, because male dominance status was unstable over time. Relatedness analysis also confirmed that group females had considerably closer and more variable genetic affiliations than group males, as expected from behaviour and demography in this species.

## 13.4 Alternative methods of data analysis in conservation biology

### 13.4.1 Introduction

One feature of microsatellites is that they have the potential to exhibit a high degree of polymorphism. We can view the frequency distribution of length variants from a genealogical perspective (Hudson 1990; Donnelly and Tavaré 1995; Donnelly this volume). Changes in demographic history alter the expected shape of gene genealogies. The genealogy of a sample at a particular locus is a random deviate from the distribution of possible genealogies with this expected shape. Thus, even if we can see the genealogy directly, there is statistical noise that dilutes our powers of inference. However, the actual gene tree is invisible to us. Mutations act as spotlights illuminating certain parts of the tree, and from this we try to infer its shape. Clearly, there are many veils that separate the genetic data from the underlying history of the populations. To some extent, this can be ameliorated by exploiting another feature of microsatellites: we are often in a position to use many loci. A working assumption is that the genealogies of unlinked loci may be regarded as uncorrelated. Thus they can be viewed as independent realizations of the same process, and we can usefully combine data across many loci to infer the expected shape of the genealogy, and thence the demographic history of the population.

There are still additional caveats to add to even this heavily-hedged view-point. In particular, selection will also be a powerful force affecting the shape of genealogies. However, once again, by studying many loci we may hope to disentangle selection from demographic effects. In particular, the demographic history experienced by each locus is the same, and therefore neutral loci should tell a consistent story. By contrast, selection may act in arbitrary ways, and, provided it is not widespread, may be detected by comparing loci.

## 13.4.2 Parameter estimation: expectation-matching and likelihood

There are two primary ways of using gene frequency data to infer mutational and demographic parameters: those based on method-of-moments, or expectation-matching; and those based on likelihood.

By 'expectation-matching' we mean a procedure such as the following (although the logical development may not be in the order given!). A statistic is devised that can be measured from the data—for example, heterozygosity, number of alleles, variance of allele length, $F_{st}$, $R_{st}$, etc. A theory is developed that relates the expected value of this statistic to a parameter that cannot be directly measured—mutation rate, population size, growth rate, migration rate, etc. By expected value is usually meant the average value of the statistic estimated from a large number of individuals at a large number of loci. Our estimate of the statistic (from relatively few individuals at relatively few loci) is then matched with the expected value, the relationship is inverted, and the parameter is inferred. Ideally, a sampling theory is developed which enables us to obtain confidence limits on our inferred parameters. This is often very difficult, and, with the cheap availability of fast computers, a more recent development is to obtain confidence limits directly from simulation.

Expectation-matching is by far the most highly developed method of inference in population genetics. The wide-ranging results and applications, particularly with respect to microsatellites, are described in the chapters by Feldman et al. and by Chakraborty and Kimmel in this volume, and will not be discussed further here.

Methods based on likelihood, although they have a long history of development for k-allele and infinite allele markers (Cavalli-Sforza and Edwards 1967; Ewens 1972; Felsenstein 1981a; Barton et al. 1983) have only very recently become more generally applied to molecular markers (Griffiths and Tavaré 1994a,b,c; Kuhner et al. 1995). By 'likelihood' is meant the following (see Weir 1996). A statistical model is devised which gives the relative probability of obtaining a gene-frequency configuration, given some demographic and mutational parameters. By 'relative' we mean that this probability is known up to a multiplying constant, which may, or may not, be 1. Thus, for a microsatellite locus studied in one population, we might code the gene frequency information in a matrix, **a**, which associates each microsatellite length with the number of times it occurred in the sample. We may wish to make inferences about the parameter $\theta = 4N_e\mu$, assuming the population size remains constant. We could write a function $d(\mathbf{a}, \theta)$ (practically, as a computer program: Nielsen 1997), which gives us the probability of obtaining the configuration given the parameter $\theta$. Intuitively it is reasonable that values of $\theta$ that give higher probabilities for configuration **a** to have occurred are in some sense 'better' than others that give lower probabilities, and the value of $\theta$ that gives the highest value, the maximum likelihood estimate (mle), is the 'best'. For many probability models it is possible to show that parameter estimates based on the mle tend to the 'true' value as the sample size becomes large, a property known as consistency. It is interesting to note, however, that genealogical models do not necessarily have this property (Joyce 1994).

There are a number of potential advantages of using likelihood. It enables us to use the information contained in the full haplotype-frequency distribution rather than statistics that are simple functions of it. For example, the statistic $F_{st}$ can be related to migration rate in an equilibrium model of a large number of islands, and it can also be related to time since divergence in a non-equilibrium case when populations start out with identical gene frequencies but evolve independently through drift without exchange of migrants. The statistic itself cannot be used to distinguish between these two very different models of demographic history. Yet, for example, the probability function of gene frequencies is known to be different in each case (Balding and Nichols 1995; Rannala and Hartigan 1996; O'Ryan *et al.* 1998), and therefore the two scenarios can in principle be distinguished, as discussed later, but this difference is not captured by the statistic $F_{st}$.

Once a likelihood function can be evaluated, it is then possible to apply Bayesian approaches to inference. The two modern traditions in statistical inference are the frequentist and Bayesian approaches. In the frequentist approach, the unknown parameter is assumed fixed and the data are regarded as some instantiation of a probability model with that parameter. By contrast, in the Bayesian approach, statements of probability are subjective (Gelman *et al.* 1995): the data are assumed fixed (they are observables), and the parameters have a probability distribution that depends on our initial beliefs and how these are changed by the data. From Bayes' theorem the probability of a parameter, given the data (the posterior), is proportional to the probability of the data given the parameter (the likelihood) multiplied by the probability of the parameter (the prior). The likelihood multiplied by the prior needs only to be known up to a multiplying constant: a normalizing constant ensures that the posterior distribution of the parameter sums (integrates) to one. Thus the function $d(\mathbf{a}, \theta)$, introduced earlier, which gives the probability of the microsatellite configuration **a** for a value of $\theta$ is normalized so that the sum of the probabilities of the possible values of **a** is 1. However, we could normalize it for a particular value of **a** so that the sum (integral) of possible values of $\theta$ sums to 1. This would be the posterior distribution of $\theta$ if we had believed that all possible values were equally probable. In reality we would not have such a belief: for example, we would think it very improbable that the mutation rate was higher than 10 per cent for any microsatellite locus, or lower than $10^{-8}$; and we would have corresponding beliefs about $N_e$. While this subjective aspect of Bayesian inference is discomforting to some, it cannot be denied that in population genetics there is often a wide choice of models that could be used, and the particular choice that is made reflects the prior beliefs of the modeller. Bayesian reasoning puts this, often tacit, process on a firm logical footing.

A problem that is often raised with likelihood-based inference is that the increased power over expectation-matching methods is associated with increased sensitivity to model assumptions. For example, Ewens (1972) showed that his ml estimator for $\theta$ based on sample size and number of alleles was considerably more powerful than the corresponding expectation-matching estimator based on observed heterozygosity. Kimura (1983) criticized Ewens estimator on the (rather ironic) grounds that variation at allozyme loci was seldom strictly neutral, and rare deleterious variants would inflate the number of alleles observed, whereas heterozygosity is little affected. However this rather begs the question of what is the purpose of inference. Model-testing procedures, which should be an important part of likelihood and Bayesian analyses (Gelman *et al.* 1995), can be used to show that a particular model does not fit very well, and therefore it could be argued that sensitivity to the modelling assumptions is an advantage, because it stimulates further studies.

The current drawbacks are mainly operational: (1) they are highly dependent on time-consuming computer simulation; (2) they generally require substantial user-input/insight to use correctly and thus, although programs have been made available, they do not appear to have been widely used; and (3) they are technically demanding to implement, and thus their development is lagging behind that of expectation-matching approaches. Nonetheless, it is important to consider these methods because they provide us with particular insights into the problems that are the concern of this chapter.

### 13.4.3 Example of the use of likelihood in the case of drift

The model we discuss here is that of a number of populations that are diverging through drift. This has been discussed by Cavalli-Sforza and Edwards (1967), and Felsenstein (1981*a*) (see also Nielsen *et al.* 1998). We will discuss the advantages and limitations of this model, and describe its application to a specific conservation/management problem (O'Ryan *et al.* 1998).

## Theory

We can calculate the probability of a sample configuration as a function of gene frequencies at a certain time, $T$, in the past, and the harmonic mean effective population size, $N_e$. Using some suitable generation time, $G$, the probability can be given in terms of a scaled time, $T_s = T/(2N_eG)$, and the ancestral gene frequencies $\mathbf{x}$ (see Appendix, p. 275). In this model we condition on the number of alleles observed, $k$. An approximation for this probability has been given by Felsenstein (1981a). However, it is possible to calculate probabilities without this approximation using coalescent theory (Donnelly, this volume). In general, for most mutational and demographic models, the probability of a sample configuration can be written as a recursion (described in Griffiths and Tavaré 1994a,b,c). This has been used by Nielsen (1997) to calculate likelihoods in some models of microsatellite evolution. The relevant recursion for the case of drift considered here is given in O'Ryan et al. (1998), and also in the Appendix. In general, when there are mutations, the recursion cannot be solved explicitly. However, in the case of drift an explicit solution can be obtained, and is given in the Appendix. This solution involves a very large number of terms, and a more convenient approximate solution can be obtained using the numerical method introduced by Griffiths and Tavaré (1994a), which is also described in the Appendix (p. 276).

This model is particularly useful when we have more than one sample, and where we can assume that at some stage in the past the lineages of each sample are drawn randomly from the same gene-frequency distribution. The samples could be from the same population taken at different times, or from two different populations. The samples then act as 'triangulation' points to enable us to make inferences about $T_s$ without the need to make any assumptions about the distribution of ancestral gene frequencies. For example, we do not need to make the assumption that the ancestral population is in mutation–drift equilibrium, or that a particular mutation model pertains.

Clearly, the model is only applicable to situations where there is good historical evidence that the effects of mutations can reasonably be ignored. In this sense, therefore, the model is completely general and is not specific to microsatellites. However, it is likely that the majority of data sets for which the method could be applied would be composed of microsatellites. One possible additional advantage of microsatellites is that, with a stepwise mutation process, the occasional mutation to an already occupied length class will have negligible effects on the expected gene-frequency distribution under drift. Another advantage is that no assumptions need to be made about changes in population size prior to sampling. A consequence of the theory of the coalescent in fluctuating populations is that the distribution of the number of coalescences in a certain period is equal to that of a stable population whose effective size is the harmonic mean of the effective size of the variable population over that period (Griffiths and Tavaré 1994b).

We can calculate the probability of obtaining a sample configuration from one population and one locus given $T_s$ and $\mathbf{x}$. When there are a number of different populations and a number of different loci it is assumed that the ancestral **x** is the same for all populations, but different for each locus, and that $T_s$ is the same for each locus but different for each population. Otherwise they are assumed to have evolved independently. Joint likelihoods for **x** and the separate values of $T_s$ in each population can then be calculated by multiplying the probabilities estimated separately for each population and each locus. Obviously these are strong assumptions: in particular the loci may not be independent due to effects of small population size and/or linkage, and $T_s$ may not be the same for each locus (because of selection on that segment of chromosome, for example). The demographic history modelled in this case is a multiway split. However, as discussed in the example below, we may wish to consider additional bifurcations of populations after the initial split. The method for doing this is described in O'Ryan *et al.* (1998).

For many problems, this approach is clearly very highly dimensional. Usually we are not interested in the ancestral gene frequencies **x** but wish to make statements about the different $T_s$. Because it would be impossible to look at the likelihood surface of all the parameters, it is more convenient to look at the marginal likelihood surfaces for each parameter separately (Tanner 1983; Gelman *et al.* 1995). By 'marginal' is meant the likelihood function of one parameter averaged over the likelihoods of all the others. In the case discussed here, direct integration would not be practicable. As described earlier, in principle the likelihood function of all the parameters could be normalized (by dividing the likelihood function by its volume) to be a probability distribution. Random draws of sets of the parameters could be made. The observed distribution for a particular parameter would then be its marginal likelihood. The most practicable way of doing this is to use a method known as Metropolis–Hastings simulation (Tanner 1993; Gelman *et al.* 1995), discussed further in the Appendix (p. 278).

### Example: demographic history of South African buffalo populations

A total of 105 African buffalo from four managed populations were genotyped at seven microsatellite loci (O'Ryan *et al.* 1998). One sample came from Kruger National Park (KNP), which has the largest number of buffalo. Smaller populations are in the Addo National Park (ANP) and the Umfolozi–Hluhuwe Complex (UHC). There is good historical evidence that up until around a hundred years ago these populations were part of a network of populations through which gene flow could occur (Smithers 1983). An outbreak of foot-and-mouth disease in 1894, coupled with a rinderpest epidemic in 1896, resulted in the loss of approximately 95 per cent of the buffalo in South Africa (Smithers 1983). The fourth population, St Lucia (St L), was founded with individuals from UHC 16 years before sampling. For all these populations we can reasonably assume that there has been little gene flow between them during the last hundred years. Good estimates of census sizes are available for ANP and UHC over a period from 1930 to the time of sampling. There are also good census-size estimates for St Lucia since the time of foundation. In the following, $N_e$ refers to the harmonic mean effective population size in a time interval, and $N_c$ refers to the harmonic mean census size in the same interval.

These data allow us to pursue a number of lines of enquiry. For example, it is possible to obtain accurate estimates of the ratio of $N_e$ to $N_c$ for the St Lucia and UHC populations. We can also estimate $N_e$ separately for KNP, ANP, and UHC, assuming a hundred-year interval of independent divergence. Of great interest, however, is that these data allow us to test the hypothesis that currently observed gene-frequency differences have arisen purely from drift. The alternative hypothesis would be that the populations were originally differentiated. This assumes negligible effects of selection at linked sites. Using the inferred ratio of $N_e$ to $N_c$ from the UHC/St Lucia split, we can make estimates of harmonic mean census size for the three older populations, and compare these with our historical knowledge of the census sizes to see if they are reasonable. We used the methods described above and in the Appendix to calculate likelihoods for the data. It is assumed that the loci are independent and the likelihoods can be multiplied over loci. The demographic history that is modelled can be regarded as a tree with an initial three-way split and one further bifurcation.

The main points to come out of the analysis were as follows. The effective population size as a proportion of the census size was estimated to be in the range 7.3–30 per cent, with the mle at 13 per cent for the St Lucia population. Interestingly, the comparable estimate for UHC appeared to rise to an asymptote, and therefore only a lower limit (where the log-likelihood is 2 units less than the apparent maximum) of 3.6 per cent could be made. The interpretation here is that the degree of drift inferred for UHC is so small that the UHC sample could have been made directly from the ancestral population. The likelihood curve for the ratio of $N_e$ to $N_c$ estimated for St Lucia can be combined with the likelihood curves for $N_e$ to estimate $N_c$ in the case of the three-way split. We estimated $N_c$ for Kruger, Addo, and UHC for the hundred-year period to be $<1600$, 46–256, and 101–462, respectively. Historical census estimates gave $\gg 1600$ for KNP, $<132$ for Addo, and $<492$ for UHC. Genetic estimates only give a lower bound for Kruger, as occurred with UHC in the two-way split. The historical estimate of the Kruger census size is anecdotal: the park is very large, the current population size is 35 000, and is always believed to have been fairly large. The other two estimates are based on detailed census counts since 1930, with the additional assumption that $N_c$ was smaller for the forty-year period beforehand. If the historical census estimates had been considerably larger than the genetic estimates, then we would be forced to conclude that the assumption of a common ancestry was violated, and that the original populations had already been differentiated from each other.

## Caveats

There are a number of caveats that have to made about the analysis described above. Clearly the test is quite weak. There are wide bounds on the genetic census estimates. In the case of the two populations for which we have detailed census information, we are only able to make upper bounds on the historical estimates of population size, when lower bounds are needed to test the hypothesis. In addition, the estimates of the ratio of effective population size to census size depend on generation time. For reasons given in O'Ryan *et al.* (1998), the genetic estimates of historical $N_c$ do not depend on generation time, but do depend on the assumption that generation time is the same in all populations.

In many ways the historical information would have been better used to carry out a fully Bayesian analysis of the data (Gelman *et al.* 1995). For example, rather than assuming the ancestral lineages were multinomial random samples from the same ancestral frequencies, we might have assumed that the ancestral frequencies could have been distributed according to, say, an island model parametrized by an ancestral inbreeding coefficient, $F_{st}$ (Balding and Nichols 1995; Rannala and Hartigan 1996). Tight prior distributions could have been made for $N_c$ using the historical information. Priors could have been made for the ancestral $F_{st}$ (i.e. low probability density for $F_{st} \gg 0.05$, Peter Arctander personal communication). A prior could also have been made on the generation time, $G$. We could then obtain a marginal posterior distribution for ancestral $F_{st}$ directly. This would give us some estimate of the power of the analysis: it may well be that quite high historical values of $F_{st}$ are consistent with the data.

## 13.5 Future trends in genetic analysis for conservation

It is likely that expectation-matching methods will continue to be dominant in the medium term, for the reasons discussed earlier. However, it may be desirable to consider more Bayesian approaches to genetic data analysis, because there is often a great deal of historical information about the populations of interest, and this can be used to construct prior distributions. It will be possible to combine these priors with new general methods of calculating likelihoods of demographic and mutational parameters with genetic data (Griffiths and Tavaré 1994*a*,*b*,*c*; Kuhner *et al.* 1995; Wilson and Balding, 1998; Beaumont, submitted). Using Metropolis–Hastings simulation, joint and marginal posterior densities can be estimated for multiparameter models (Tanner 1993; Gelman *et al.* 1995). In the context of Bayesian or likelihood-based analyses, we will briefly discuss extensions of the drift model described above, and then give a description of methods that incorporate mutation models.

### 13.5.1 Non-mutation models

One simple extension is to use the method to infer $N_e$ by the temporal method. In the temporal method a population is sampled at a number of different time points. For modelling, one would assume that the initial time sample is a random (multinomial) sample from some unknown gene frequency distribution, and that subsequent samples have undergone increasing degrees of drift, parametrized by

$T_s = T/(2N_e G)$. Since $T$ is known with little error, posterior distributions for $N_e$ can be obtained for the different time points.

One approach that may not be useful for microsatellites, but is worth describing nonetheless, is that if one is willing to assume an infinite allele model (IAM) it is possible to detect bottlenecks in single populations. The effect of a bottleneck is to produce a transient departure from the IAM mutation–drift equilibrium. The situation can be modelled by assuming that the founder lineages are drawn from a population at equilibrium, where the probability distribution of the founder alleles can be calculated using the Ewens sampling formula. The subsequent period can be regarded as sufficiently short that it can be modelled by the drift model described above. It is then possible to integrate (average) over all possible values of ancestral $\theta = 4N_e\mu$ (where $N_e$ is ancestral), and obtain marginal distributions for $T_s = T/(2N_e G)$. If $T_s$ has very low likelihood at zero, then this would be good evidence that a bottleneck has occurred.

Another application is in trying to reconstruct breed or population 'trees' in the case where populations are continually splitting. As discussed above, with the buffalo data, it is possible to calculate the likelihood for a particular tree. The branch-swapping Metropolis–Hastings approach used by Rannala and Yang (1996) and Yang and Rannala (1997) could be applied to populations, and one could then obtain the posterior probabilities of particular trees. This approach avoids the problems of bootstrap resampling (Yang and Rannala 1997).

Another drift model that is based on different assumptions is one where the populations are in migration–drift equilibrium. In this case, rather than having a particular founder time, it is assumed that the populations are subject to a constant stream of immigration from some outside gene pool with unchanging gene frequencies. Wright's (1951) equation gives us the pattern of gene frequency variation (in a k-allele model) that we might expect among different islands for a given level of inbreeding. In practice we will have a sample from this distribution (Rannala and Hartigan 1996). A two-allele version was introduced by Barton *et al.* (1983). The multi-allele versions have been studied by Balding and Nichols (1995) and Rannala and Hartigan (1996). Because the migration–drift model and the founder–drift model each predict different probability distributions for allele frequencies (as discussed earlier in the context of Fst), it is possible to use Bayesian methods of switching between two models (Carlin and Chib 1995) to obtain the posterior probability of each model. It appears that the power of the method is weak (Beaumont, unpublished observations; Ian Wilson, personal communication), but if a sufficiently large number of loci are used it may well be possible to distinguish migration–drift equilibrium versus non-equilibrium explanations of population structure.

## 13.5.2 Mutation-based models

Once the assumption of no mutations becomes untenable, then we have to use more sophisticated models of the mutation process at microsatellites. The two currently published methods of calculating likelihoods by simulation are the method of Griffiths and Tavaré (1994$a$,$b$,$c$) (described in this chapter) and the method of Kuhner $et$ $al.$ (1995).

The method of Griffiths and Tavaré has been successfully applied by Nielsen (1997) to microsatellites evolving according to a stepwise model to obtain likelihood estimates of $\theta$, and also to detect departures from the single-step mutation process. The method explores a large number of possible genealogies consistent with the data, largely independent of their probability of occurring. Because of this it can be inefficient to use on genetic data with a large number of mutational states, such as microsatellites under a stepwise model, or sequences. For example it can be difficult to calculate likelihood surfaces for large values of $\theta$ in microsatellites, even if a truncation method is used to restrict the method to genealogies with a manageable number of mutations (Nielsen 1997).

Kuhner $et$ $al.$ (1995) have devised a different approach for obtaining likelihood-based estimates of $\theta$ from sequence data using Metropolis–Hastings sampling. For a given value of $\theta$, the probability of obtaining a particular tree shape can be calculated. Given this tree shape, the probability of obtaining the data can be calculated (Felsenstein 1981$b$). The problem is that there is a very large number of trees compatible with the data, and the probability of obtaining the data must be summed over all these possibilities. Kuhner $et$ $al.$ use Metropolis–Hastings sampling to produce a sequence of trees in proportion to their probability of occurring given theta, multiplied by the probability of obtaining the data given the tree. By itself, this gives the distribution of trees conditional on the data and the value of $\theta$, but not the likelihood. However the likelihood ratio for two values of $\theta$, say $\theta_1/\theta_0$, can be estimated using a method called Importance Sampling. For example, if the sequence of trees was generated for some value of $\theta$, say $\theta_0$, we can compute the probability of obtaining a particular tree with $\theta_1$, and also the probability with $\theta_0$. The average value of the ratio of these two probabilities over all the sampled trees is an estimate of the likelihood ratio. The rationale behind the method is given in Kuhner $et$ $al.$ (1995), and an explanation of importance sampling is given in Tanner (1993) and Gelman $et$ $al.$ (1995).

Two other approaches currently under study were both introduced to the workshop. In one case, Wilson and Balding (1998) calculate the probability of a tree conditional on the data and the states of the interior coalescent nodes. The shape of the tree, the states of the interior nodes, and the demographic/mutational parameters (for example, $\theta$) can be updated using Metropolis–Hastings sampling, and thereby marginal posterior distributions for demographic parameters can be obtained. The approach presented by Beaumont (submitted) is very similar, but explicitly places mutations along the lineages. These methods are more flexible than the previously described approaches, because they allow the full marginal and joint posterior distributions for parameter values to be estimated. Thus they permit a more Bayesian approach to inference.

# 14 Microsatellites and the reconstruction of the history of human populations

Andrés Ruiz Linares

## Chapter contents

## Abstract

Due to the very large number of microsatellite markers currently available as a result of the human genome project, microsatellites are presently the most powerful way of probing the diversity of the whole human genome. Here I summarise, within the context of other genetic studies, some of the principal inferences drawn from currently available microsatellite data sets relative to the analysis of human evolution. Consistent with various previous studies, microsatellites indicate that the greatest genetic distance is seen between African and non-African populations. In agreement with the recent African origin for modern humans, microsatellites also show that African populations are genetically the most diverse. Furthermore, estimates of time since population splits based on the step-wise mutational model place the African–non-African separation at about 156 000 years ago, a figure inconsistent with the multi-regional model of human evolution. One of the principal problems for the use of microsatellites in human evolutionary studies is range constraint, which seems to affect more seriously tri- and tetranucleotide repeats, having a higher mutation rate, than dinucleotide repeats. The more rapidly evolving microsatellites are likely to be mainly informative for regional population studies, particularly those examining compound Y chromosome haplotypes.

## 14.1 Introduction

The potential utility of microsatellites as polymorphic genetic markers was evidenced almost simultaneously in *Drosophila* and humans about ten years ago (Tautz 1989; Weber and May 1989). Since then, microsatellites have quickly taken over as the preferred marker for genetic mapping in humans, and will most likely remain as such for some time. As will be reviewed here, applications to human evolution studies have provided some of the initial evidence showing that microsatellites are extremely informative in population genetics. These studies have also helped define more precisely some of the key factors determining the range of situations in which (with current technology) microsatellites provide the most cost-effective and at times most informative genetic marker.

Thus far microsatellite population surveys have mostly approached the problem of modern human origins and the differentiation of major human groups. More recently, attention is being paid to their utility for the detailed study of particular human populations. These applications of microsatellites to human evolutionary studies will be presented in turn.

## 14.2 Genetics and modern human origins

Controversies regarding modern human evolution revolve to a considerable extent around the issue of the place of origin and age of our species, and the possibility and extent of a genetic contribution of archaic to modern humans (Cavalli-Sforza *et al*. 1994; Relethford 1995; Ruvolo 1996; Hammer and Zegura 1996). Palaeoanthropological evidence has established Africa as the region where hominization gradually took place, with *Homo erectus* representing the first hominid migrating out of the African continent around 1–2 million years ago. Some palaeontologists have proposed that anatomically modern humans also appeared initially in Africa about 100 000 years ago, and from there populated the rest of the world without intermingling with *H. erectus* or other archaic *Homo sapiens* native to the regions being colonized by the now fully modern humans (Stringer and Andrews 1988). At the other extreme it has been proposed that there was a gradual transformation of descendants of *H. erectus* world-wide into modern humans (Wolpoff *et al*. 1984). Various intermediate scenarios have been proposed that differ mostly in the extent of the genetic contribution of local archaic forms to a modern *Homo sapiens* migrating out of Africa, and the level of gene flow between ancient human populations (Relethford 1995).

## 14.2.1 Classical marker studies

Genetic data have been used to approach questions of human evolution for several decades (Cavalli-Sforza *et al*. 1994). The most comprehensive database available so far consists of the so-called classical markers (blood groups and proteins) which have been typed for large population samples world-wide. Data for these markers have been summarized recently, and provide a broad picture of human evolution from a genetical perspective, which complements descriptions based in other sources such as archaeology and linguistics (Cavalli-Sforza *et al*. 1988, 1994; Nei and Roychoudhury 1993; Cavalli-Sforza 1997).

One important conclusion from the study of classical markers is that there is a relatively small degree of genetic divergence between existing human populations. Most estimates of population structure between continental populations based on these data give values of $F_{st}$ of about 10 per cent, indicating that the brunt of human genetic variation is present within continents (Lewontin 1972; Nei and Roychoudhury 1974; Cavalli-Sforza *et al*. 1994). Another interesting result from these studies is that, on average, the greatest genetic divergence is seen between African and non-African populations (Fig. 14.1), suggesting that the initial population split occurred between them.

Both the low level of inter-population divergence, and an initial split between African and non-African populations, are consistent with the model of a recent African origin for modern humans. However, confidence in these observations is undermined by the fact that classical markers are likely not to evolve neutrally, and that they do not reflect the full extent of underlying genetic variation at the DNA level.

## 14.2.2 DNA markers: single locus studies

The first large database available on polymorphisms at the DNA level consisted of mitochondrial DNA (mtDNA) haplotypes, which started being collected in the early 1980s. Currently, mtDNA also represents the largest sequence database available for human evolutionary studies. The interpretation of the mtDNA data with regard to human origins has been the subject of considerable debate (Cann *et al*. 1987; Stoneking 1993; Templeton 1993; Ayala 1995). These data show a greater genetic diversity of African populations, and are consistent with a fairly small evolutionary effective population size for humans (of the order of 10 000 individuals), with most coalescent time estimates for mtDNA producing values around 200 000 years ago. These observations are compatible with the model of a recent African origin for modern humans, although alternative explanations have been put forward (Relethford 1995; Hey 1997).

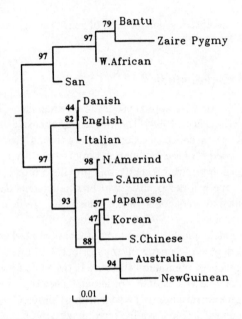

**Fig. 14.1** Neighbour-joining tree relating 14 native populations based on information for 120 classical markers (Cavalli-Sforza *et al.* 1994) and Nei's standard genetic distance (Nei 1987). The root separates African (Bantu, Zaire Pygmy, West African, San) from non-African populations. Numbers at the nodes refer to bootstrap percentages for 100 bootstraps. The tree was rooted using the midpoint method.

One of the best studied nuclear regions at the DNA sequence level is the Human Leukocyte Antigen (HLA) gene cluster, for which sequence variation has been examined in large population samples world-wide (Tsuji *et al.* 1992). However, phylogenetic interpretation of these data is difficult, due to the known selective pressure acting on these genes, and the complex mutational events affecting them. Nevertheless, the presence of HLA transpecies polymorphisms among primate species could indicate that the effective population size of humans might not have been extremely low, thus putting a limit to extreme interpretations of the out-of-Africa model postulating a population bottleneck associated with speciation (Ayala *et al.* 1994; Ayala 1995).

Despite being very informative, both mtDNA and HLA data suffer from the drawback of selection possibly complicating any inferences made from them (Hey 1997). In addition, they both represent a single locus, and as such cannot provide the full picture of evolutionary events as reflected in the human genome as a whole. This is also a problem with other genomic regions, such as the beta-globin gene cluster, for which considerable population data are becoming available (Fullerton 1995; Harding *et al.* 1997).

## 14.2.3 DNA markers: genomic sampling

An initial attempt at sampling variation at the DNA level of the whole human genome was initiated in the late 1980s, employing restriction fragment length

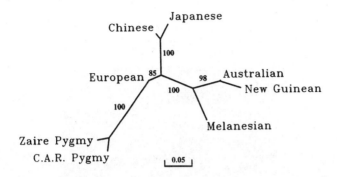

**Fig. 14.2** Neighbour-joining tree based on 87 RFLPs for eight populations (Lin *et al.* 1994, and references therein) and Reynolds *et al.* (1983) genetic distance. Numbers at the nodes refer to bootstrap percentages for 100 bootstraps.

polymorphisms (RFLPs). Analyses of these data produced a pattern of population relationships similar to that obtained with classical markers (Figs 14.1 and 14.2). Interestingly, it was suggested that Caucasians could represent an ancient admixture between African and Asian populations (Bowcock *et al.* 1991). Such an admixture introduces complications in the representation of human evolution as a strictly bifurcating process, as exemplified by the relative sensitivities of phylogenetic reconstruction methods to varying degrees of admixture (Ruiz-Linares *et al.* 1995; Cavalli-Sforza 1997).

For a brief period of time, nuclear RFLP information was collected systematically, but it was rapidly superseded by microsatellites due to their abundance, ease of typing, and high degree of polymorphism. It seems likely that in the future, human population genetic studies will turn again to the genomic screening of scattered nuclear DNA sequence variants, but this time in the form of SNP chips, allowing the rapid collection of large amounts of population data. Technological advances are also likely to permit the comparison of long stretches of genomic sequence across populations in the not too distant future, allowing full exploitation of the methods of analysis developed in single locus studies.

## 14.3 Microsatellites and the problem of modern human origins

### 14.3.1 Genetic diversity

As was seen with other markers, microsatellite data indicate a low level of genetic divergence between continental human populations, with $F_{st}$ values of about 8–11 per cent (Bowcock *et al.* 1994; Deka *et al.* 1995*b*; Perez-Lezaun *et al.* 1997; Barbujani *et al.* 1997; Jorde *et al.* 1997). In agreement with mtDNA data, dinucleotide repeats show the greatest level of genetic diversity in African populations (Table 14.1). This had not been seen previously with the other nuclear markers, possibly because these had been selected preferentially in populations

**Table 14.1**
Observed percentage heterozygosity of 14
native populations (± standard error) based on
the data of Bowcock *et al*. (1994)

| | |
|---|---|
| **Africa** | |
| C.A.R. Pygmy (Biaka) | 83.4 ± 2.5 |
| Lisongo | 81.1 ± 2.3 |
| Zaire-Congo Pygmy (Mbuti) | 77.6 ± 2.6 |
| Mean | 80.7 ± 1.5 |
| **Europe** | |
| N. European | 73.5 ± 2.9 |
| N. Italian | 72.6 ± 1.5 |
| Mean | 73.0 ± 1.6 |
| **Asia** | |
| Japan | 70.7 ± 5.2 |
| China | 68.4 ± 3.2 |
| Cambodia | 66.3 ± 2.6 |
| Mean | 68.5 ± 2.1 |
| **Oceania** | |
| New Guinea | 64.1 ± 3.2 |
| Melanesia | 64.1 ± 2.0 |
| Australia | 62.1 ± 3.8 |
| Mean | 63.6 ± 1.8 |
| **America** | |
| Maya | 65.2 ± 2.6 |
| Surui | 59.3 ± 3.2 |
| Karitiana | 51.8 ± 4.8 |
| Mean | 58.8 ± 2.3 |

of European descent and have a lower level of variability than microsatellites, thus considerably biasing estimates made from them (Mountain and Cavalli-Sforza 1994; Rogers and Jorde 1996).

The finding of a higher genetic diversity of African populations is consistent with the hypothesis that they represent the oldest human population, as postulated by the recent African origin model (Cann *et al*. 1987). This observation has been recently re-evaluated by Jorde *et al*. (1997), who also found a significantly higher heterozygosity in African populations relative to populations from other continents, using different population samples and microsatellite loci.

## 14.3.2 Population and individual relationships

Population trees based on genetic distances calculated from microsatellite allele frequencies have been found to be highly consistent with those obtained with other nuclear markers, and are statistically robust (Figs 14.1, 14.2, 14.3). There is a

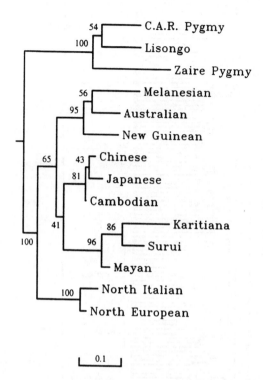

**Fig. 14.3** Neighbour-joining tree relating 14 native human populations based on allele frequencies for 30 microsatellite loci (Bowcock *et al.* 1994) and Nei's standard genetic distance (Nei 1987). The tree was rooted by midpoint. The numbers at the nodes indicate bootstrap percentages for 100 bootstraps.

highly significant clustering of all continental populations and of African and non-African populations. As seen with classical markers and RFLPs, the genetic distance between African and non-African populations is the greatest, as shown by their separation when the tree is rooted using the midpoint method (Fig. 14.3).

The availability of what is essentially a 'DNA fingerprint' for each individual examined also allowed Bowcock *et al.* (1994) to evaluate the genetic relatedness between individuals. When trees were constructed from inter-individual genetic distances it was found that individuals clustered according to their geographic origin (Fig. 14.4). The great majority of individuals examined formed discrete clusters that coincided with the continent of origin of the sample. In addition, there was a tendency to form sub-clusters within continents that corresponded to the population of origin of the sample.

The possibility of evaluating the genetic relationship of individuals sampled from a population based on multi-locus genotypes opens new avenues in population genetic analyses, which are just beginning to be explored. This type of data eliminates the need for prior population definition, which could be helpful in refined studies of recent population movements, as well as facilitate individual admixture estimation. Their potential utility in the detection of recent immigration

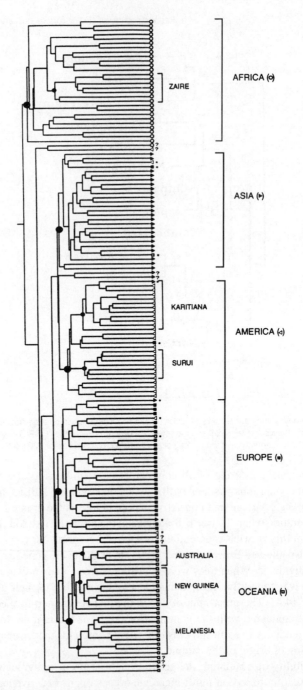

**Fig. 14.4** UPGMA tree relating 148 individuals from 14 native populations (listed in Table 14.1). Inter-individual distances were based on information for 15–30 microsatellite loci (Bowcock et al. 1994) and calculated for each pair-wise comparison as: $-\log(P_s)$ where $P_s$ = total number of shared alleles summed over loci/(2 × number of loci compared). Black dots indicate nodes defining geographic clusters of individuals. Dots refer to individuals present in a cluster not corresponding to their geographic origin. Question marks indicate individuals occupying an undefined position relative to the clusters.

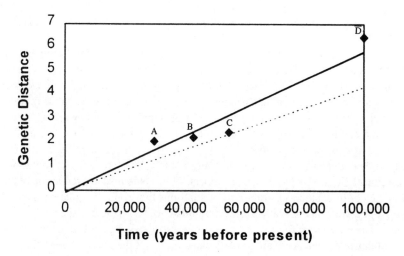

**Fig. 14.5** Regression of the $(\delta\mu)^2$ genetic distance (Goldstein *et al*. 1995*a*) between populations, based on microsatellite data from Bowcock *et al*. (1994), and time since population separation estimated from archaeological evidence. Populations compared are: A, Amerind vs. East Asian; B, European vs. East Asian and Amerind; C, Melanesian vs. Sahulland; D, African vs. Non-African. The dotted line is the expected genetic distance between these populations using an experimentally determined mutation rate of $5.6\times10^{-4}$. The slope of the regression line gives a calibrated mutation rate of $7.96\times10^{-4}$.

and for ethnic affiliation of DNA samples has recently been evaluated (Mountain and Cavalli-Sforza 1997; Rannala and Mountain 1997; Shriver *et al*. 1997).

## 14.3.3 Dating the migration out of Africa

Apart from estimates of genetic diversity and population relatedness, dating the migration out of Africa is a critical piece of information that genetic studies can contribute to theories of human origins. Microsatellites have been particularly helpful in this regard.

Although the population relationships obtained with microsatellites (Fig. 14.3) are consistent with those seen with classical markers and RFLPs, the upper segments of this tree are short when compared with the branch lengths separating populations within continents, which must have separated within a much shorter time span. This distortion in branch lengths suggests that for microsatellites, linearity with time of ordinary genetic distances is lost at greater time depths.

The loss of linearity with time of traditional measures of genetic distance is the result of the mutational mechanism of microsatellites, which seem to evolve mostly through stepwise mutations (Nei 1987; Hancock, this volume). Properties of this mutational mechanism have been recently incorporated into measures of genetic distance seeking to make optimal use of microsatellite data (Goldstein *et al*. 1995*a,b*; Slatkin 1995). These new measures have the added advantage that they are relatively insensitive to variations in population size, and depend mostly on the mutation rate. Furthermore, microsatellite mutation rates are sufficiently high that they can be determined experimentally (Weber and Wong 1993; Heyer *et al.* 1997).

Using this approach, Goldstein *et al.* (1995a) obtained a date for the separation of African and non-African populations of 156 000 years ago (with a 95 per cent confidence interval of 75 000 to 287 000 years ago), which is highly consistent with a scenario of recent human origins in Africa. Interestingly, when the relation of genetic distance to archaeological dating of population splits was examined, an estimate of the mutation rate was obtained that did not differ greatly from that obtained from pedigrees ($7.96 \times 10^{-4}$ vs. $5.6 \times 10^{-4}$, respectively; Fig. 14.5). The difference between these two values could relate to the frequency of mutations changing allele size by more than one repeat. This would lead to a reduction of the estimates of time since separation of populations that could be quite substantial (Minch, Feldman and Cavalli-Sforza unpublished). Nevertheless, the fairly good concordance between the two mutation rate estimates suggests that the recent date obtained for the African–non-African split is not the result of gene-flow, in agreement with the recent replacement hypothesis of modern human origins.

### 14.3.4 Informativeness of dinucleotide repeats vs. microsatellites with larger motifs

Recently, data have been obtained for world populations using microsatellites consisting of tri- or tetranucleotide repeats (sometimes called short tandem repeats or STRs). Qualitatively, the results obtained with dinucleotide repeats and STRs are similar. However, STR data do not seem to reflect population structure to the same extent as dinucleotide repeats. In particular, STRs produce mean $F_{st}$ figures somewhat lower than CA repeats, and result in trees with a less well defined structure, suggesting that STRs are somewhat less informative than CA repeats on the human evolutionary scale (Deka *et al.* 1996; Jorde *et al.* 1995, 1997; de Knijff *et al.* 1997; Perez-Lezaun *et al.* 1997).

The lower informativeness of STRs could relate to an allele size constraint acting on microsatellites (Bowcock *et al.* 1994; Garza *et al.* 1995), particularly considering that tetranucleotide repeats seem to have mutation rates of the order of $10^{-3}$ while dinucleotide repeats mutate perhaps an order of magnitude more slowly (Weber and Wong 1993; Heyer *et al.* 1997).

### 14.3.5 Y chromosome microsatellites in world populations

There has been a long-standing interest in using Y markers for human evolutionary studies (Jobling and Tyler-Smith 1995; Hammer and Zegura 1996). This is due mostly to the particular genetic properties of the greater part of this chromosome, which is inherited as a single unit, and exclusively through the male lineage, thus offering a male counterpart to mtDNA. However, progress on the Y had until recently been slow due to the paucity of polymorphism on this chromosome, particularly of markers that are easy to type and interpret. The recent isolation of microsatellites from the Y chromosome has changed this situation considerably.

An initial analysis of five microsatellites (four dinucleotide repeats and one tetranucleotide repeat) in the same population samples studied by Bowcock *et al.* (1994) revealed a significant level of geographic structuring of the microsatellite Y-haplotypes (Ruiz Linares *et al.* 1996; Fig. 14.6). This reflection of geography in the genetic data is much lower than that observed for the unlinked autosomal microsatellites, but much greater than seen for mitochondrial DNA.

More recent examination of additional microsatellites in various world populations has revealed very little geographic structuring of Y-haplotypes (Deka *et al.* 1996; de Knijff *et al.* 1997). As most of the more recently available Y microsatellites are tri- or tetranucleotide repeats, it is possible that the same problem that affects autosomal STRs is affecting Y-linked ones, namely that the mutation rate is so high that range constraint leads to a rapid loss of the phylogenetic informativeness of these markers (de Knijff *et al.* 1997). This suggests that Y-linked microsatellites are going to be mostly informative for the phylogenetic study of particular geographic populations but not for the human species as a whole. In fact, there has been a recent tendency towards the use of Y microsatellites for regional population studies, as the following section illustrates (Section 14.4; see also Zerjal *et al.* 1997; Karafet *et al.* this volume; Thomas *et al.* in preparation; Ruiz Linares *et al.* in preparation).

## 14.4 A regional case study: the origin of native Americans

Although there is agreement that during the initial peopling of the American continent people migrated through the Bering strait from Asia into America, there is considerable debate among archaeologists as well as among geneticists as to the details of this process. A widely popular 'synthetic' model posits peopling of the New World in three migration waves (although other researchers have postulated from one to six migrations) originating somewhere around southeastern Siberia–Manchuria–Mongolia (Fiedel 1992; Cavalli-Sforza *et al.* 1994). Most archaeological evidence points to an initial peopling around 12 000 years ago, but much older sites have been reported, particularly in South America (Fiedel 1992).

Available Y chromosome data have been interpreted as suggesting that all Native American populations descend from a single migratory wave possibly, originating in a restricted area of East Asia. A specific haplotype defined by a heteroduplex polymorphism and a tetranucleotide repeat was found at very high frequency in Amerind populations from North and South America (Pena *et al.* 1995; Santos *et al.* 1996). In agreement with this observation, Underhill *et al.* (1996) found a C/T transition polymorphism apparently restricted to Amerind Y chromosomes, with the usually predominant T allele having a widespread distribution among Amerind populations. However, it is not clear if the C-bearing

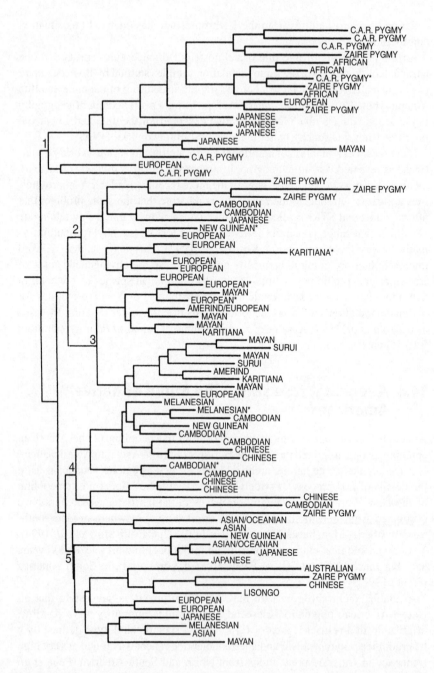

**Fig. 14.6** Neighbour-joining tree relating 78 Y chromosome haplotypes using as genetic distance $1-P_s$ ($P_s$ as defined in Fig. 14.4). Haplotypes seen in more than one copy per population are indicated with asterisks; those shared between populations within a continent have the continent name. The tree was rooted by midpoint. Numbers indicate clusters which tend to include haplotypes of a specific geographic origin: 1 is mostly African; 2, European; 3, Amerind; 4, Asian; 5, Asian/Oceanian. The clustering pattern is statistically significant (Ruiz-Linares *et al.* 1996).

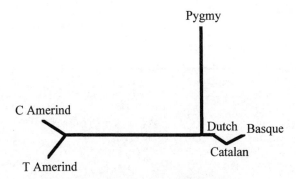

**Fig. 14.7** Neighbour-joining tree relating Amerind C/T Y chromosomes to the Mbuti Pygmy, Basque, and Catalan populations, based on Nei's standard genetic distance and data for six Y microsatellites.

chromosomes detected in Amerinds are the result of admixture as suggested by Underhill *et al.* (1996).

In order to evaluate the origin of Amerind C-bearing chromosomes, Ruiz Linares *et al.* (in preparation), examined seven Y microsatellites in five Colombian Amerind populations. A phylogenetic tree depicting the relationship among Amerind and non-Amerind Y chromosomes (Fig. 14.7) shows a close relationship between C and T Amerind chromosomes, suggesting that most C chromosomes detected in the Colombian populations are not the result of recent admixture. This indicates the existence of two types of Y chromosome in the New World prior to European colonization. As the T allele is widespread and is usually seen in Amerind populations at high frequencies, this suggests that the C/T transition happened before entry into the New World, indicating the presence of at least two types of Y chromosomes among the founders of Amerind populations.

Calculating the date of the C to T transition could be of use in estimating the initial date of entry into the American continent. A simple approximation to estimate this date is provided by calculating the age of the most recent common ancestor (MRCA) of a set of microsatellite haplotypes, using the dynamic for the square of the difference of microsatellite repeat number between haplotypes (Goldstein *et al.* 1995*b*; Slatkin 1995). For the case of the hypothetical ancestor and its descendants this is just $\mu \times t$ ($\mu =$ mutation rate, $t =$ generations). Taking the most frequent allele seen in present T chromosomes as the ancestral allele, one can infer the putative ancestral haplotype on which the C to T transition occurred (Table 14.2). This approach produces a figure for the MRCA of T chromosomes of about 15 000 years; assuming a microsatellite mutation rate of $2.1 \times 10^{-3}$ and a generation time of 27 years. This date estimate has a greater consistency with a relatively recent date of peopling of the New World, although more work is necessary.

**Table 14.2**
Allele frequencies at six microsatellite loci for C/T Colombian Amerind
Y-chromosomes ($n = 129$) defining putative ancestral haplotypes (A1
and A2) on which the C to T transition might have occurred

| | | | | | | | A1 | A2 |
|---|---|---|---|---|---|---|---|---|
| DYS19 | 182 | **186** | 190 | 194 | 198 | | | |
| C-chrom. | 0.017 | **0.525** | 0.271 | 0.169 | 0.017 · | | | |
| T-chrom. | 0.019 | **0.750** | 0.154 | 0.077 | 0.000 | | **186** | **186** |
| DYS388 | **129** | 132 | 135 | 138 | 141 | 144 | | |
| C-chrom. | **0.661** | 0.220 | 0.017 | 0.017 | 0.034 | 0.051 | | |
| T-chrom. | **0.717** | 0.264 | 0.000 | 0.000 | 0.000 | 0.019 | **129** | **129** |
| DYS389 | 247 | **251** | 255 | 259 | | | | |
| C-chrom. | 0.182 | **0.500** | 0.318 | 0.000 | | | | |
| T-chrom. | 0.208 | **0.479** | 0.229 | 0.083 | | | **251** | **251** |
| DYS390 | 203 | 207 | 211 | 215 | **219** | 223 | | |
| C-chrom. | 0.019 | 0.115 | 0.269 | **0.346** | 0.231 | 0.019 | | |
| T-chrom. | 0.029 | 0.000 | 0.118 | 0.235 | **0.618** | 0.000· | **219** | **219** |
| DYS392 | 248 | 251 | 254 | **257** | 260 | | | |
| C-chrom. | 0.186 | 0.093 | 0.070 | **0.465** | 0.186 | | | |
| T-chrom. | 0.200 | 0.000 | 0.000 | **0.400** | **0.400** | | **257** | **260** |
| DYS393 | 120 | **124** | 128 | 132 · | 136 | | | |
| C-chrom. | 0.111 | **0.537** | 0.167 | 0.185 | 0.000 | | | |
| T-chrom. | 0.073 | **0.561** | 0.195 | 0.146 | 0.024 | | **124** | **124** |

The most frequent allele at each locus is indicated in bold.

## 14.5 Conclusion

Microsatellites have been extremely useful for studies of human evolution, particularly on a world-wide scale. At this level they have provided so far considerable support for the recent out-of-Africa model of human origins. The collection of additional data for more populations and markers could confirm the inferences made so far, and allow finer estimates based on statistics incorporating the stepwise mutational model such as the $(\delta\mu)^2$ genetic distance. These statistics extract more information from microsatellite data but suffer from the drawback of a large variance, thus requiring the typing of a considerable number of loci to produce reliable estimates (Goldstein *et al.* 1995a,b). Preliminary analyses of additional microsatellite data have produced estimates of the African–non-African split consistent with the figure of 156 000 years ago calculated by Goldstein *et al.* (1995a) (Cavalli-Sforza 1997).

Additional microsatellite data should provide deeper insights into various questions regarding modern human origins. Of particular interest is the estimation of the coalescent time for different genomic regions based on microsatellite information. Under the recent human origins hypothesis, these coalescent times should revolve quite tightly around a fairly recent date. If there was a genetic contribution from ancient *Homo* to modern humans, coalescent times are likely to vary greatly between different genomic regions, and some could be of considerable age (Goldstein *et al.* 1996; Ruvolo 1996). Another area that is just starting to be explored is the use microsatellite data to infer some aspects of demographic history (e.g. population expansions: Shriver *et al.* 1997; Reich and Goldstein, in preparation).

Obviously, more needs to be known about microsatellite mutation rates and mechanisms in order to select loci that are particularly informative for specific population studies. Considering the large number of microsatellites already available in humans, there should be no trouble in finding adequate ones for most situations. At a practical level, and considering initiatives such as the Human Genome Diversity Project, care needs to be taken in defining optimal sets of microsatellite loci that would allow multilateral comparisons to be made. Standardization is also needed at the level of allele nomenclature, optimally referring to number of repeats, as recommended by the International Society for Forensic Haemogenetics (ISFH 1994), instead of allele sizes.

Microsatellites seem specially informative for regional human population studies. Their high level of diversity makes them particularly informative, perhaps even more in the case of the Y chromosome, and it is going to be of great interest to compare results obtained using autosomal and Y microsatellites with mtDNA data. Furthermore, analysis of linked microsatellite markers should help estimate the age of 'single event' mutations, and to track such mutations in populations (Slatkin and Rannala 1997; Goldstein *et al.*, this volume).

At another level, the role of microsatellite population data in linkage disequilibrium mapping is likely to increase in importance for the identification of genes involved in human disease (Stephens *et al.* 1994; Laan and Pääbo 1997; Thompson and Neel 1997). Microsatellites should also permit a precise estimation of the level of individual and population admixture, thus facilitating an evaluation of the role of ethnicity in disease susceptibility (Shriver 1997; Shriver *et al.* 1997). Through this type of approach, microsatellites should help strengthen ties between human population genetics and disease mapping studies in the near future.

# Acknowledgements

I am very grateful to L.L. Cavalli-Sforza, D.B. Goldstein and C. Schlötterer for helpful discussions and comments on the manuscript and to L.L. Cavalli-Sforza for his support during my involvement with various aspects of the work described in this review.

# 15 Forensic applications of microsatellite markers

David Balding

## Chapter contents

## Abstract

The application of microsatellite (or Short Tandem Repeat, STR) loci to problems in forensic science is reviewed, starting with a brief historical overview pointing out limitations of earlier forensic typing methods. The interpretation of STR profiles in terms of evidential weight is emphasized, leading to match probability calculations, which allow for relatedness and population genetic effects. The situation in which a single defendant is alleged to be the source of a crime scene STR profile is discussed in some detail, including a worked example typical of recent UK casework. In addition, there is a shorter discussion of more complex scenarios, in which paternity, or other relatedness, is investigated, or where the crime scene STR profile reflects the DNA of more than one individual (a so-called 'mixed' profile).

## 15.1 Introduction and overview

Microsatellite loci, generally known in forensic applications as Short Tandem Repeat (STR) loci, are currently widely used for forensic identification and relatedness testing, and seem set to become the predominant genetic marker in this area of application. In forensic identification cases, the goal is typically to link a suspect with a sample of, for example, blood or semen taken from a crime scene and presumed to have originated from the culprit. Alternatively, the goal may be to link, say, blood found on a suspect's clothing with a victim. Relatedness testing in criminal work may involve investigating paternity in order to establish rape or incest. Another application involves linking a DNA sample with the relatives of a missing person. The features which have made microsatellites attractive for use in the various other applications described in this book, such as their relative ease and accuracy of typing and high levels of polymorphism, have also led to their popularity in forensic work. The ability to employ PCR to amplify small samples is particularly valuable in this setting, since in criminal casework only minute samples of DNA may be available.

In this chapter we first present a brief historical overview of the development of forensic DNA profiling systems, highlighting the problems which each system encountered. Some of these problems are overcome by the use of STR markers; others remain. In particular, finding a method of presenting DNA profile evidence in court so that it is comprehensible to jurors and fair to the defendant, yet does not excessively understate evidential strength, poses a challenge to all systems of DNA profiling, and indeed all scientific evidence. We outline methods for assessing evidential weight for STR profiles, and examine their practical implications in the courtroom. As well as straightforward identification problems, we also briefly discuss relatedness testing and identification of one or more contributors to a mixed STR profile.

## 15.2 Background: MLP and SLP profiling systems

The use of DNA markers for forensic applications can be traced to the ground-breaking work described in Jeffreys *et al.* (1985*b*), from which a Multi-Locus Probe (MLP) system was developed, popularly known as a DNA 'fingerprint'. The MLP system records the presence or absence of alleles from diverse, unknown locations on the genome; see Jeffreys *et al.* (1991*b*) for details of the experimental technique. The autoradiogram (or *autorad*) which records an MLP profile typically shows between 10 and 25 bands, each indicating the presence of an allele, and resembles a supermarket bar-code. This resemblance seems to have played an important role in acceptance and understanding by the public of the complex new technology.

The MLP system, introduced into UK casework around 1988, suffers a number of drawbacks which eventually hindered its acceptance by courts. Although experts agree that it is typically very *unlikely* that the defendant, if innocent, would have the same MLP profile as the perpetrator of a crime, such a false match is undeniably *possible*. Moreover, assessing how unlikely a 'chance' match might be in a particular case is difficult, in particular because of the unknown genomic origin of the bands composing the fingerprint. Firstly, there is the problem of possible *linkage*: the matching of several bands originating on the same chromosome might be substantially less incriminating than an apparently similar match of unlinked bands, since linked units may be shared through direct inheritance from a common ancestor. Similarly, *allelism*—two alleles originating from the same locus—affects profile match probabilities. For MLP profiles, however, the pattern of linkage and allelism is unknown. Estimates of average levels of allelism and linkage have been made, but there are reasons to suspect that they may be underestimates (Donnelly 1995).

A practice was established in UK courts of calculating MLP match probabilities using the formula $(0.26)^k$, where $k$ denotes the number of matching bands corresponding to alleles of length at least 4 kb. The constant 0.26 was chosen based on observed frequencies of band sharing, averaged over bands; see Donnelly (1995) for a discussion of the observational studies. Since 0.26 was regarded as high compared with most estimates of band sharing frequencies, it was hoped that its implementation in calculating match probabilities would incorporate a 'conservative' bias (i.e. favouring defendants) which would overcome any non-conservative effects of averaging over bands and ignoring linkage, allelism, and other population genetics factors. There is, however, no satisfactory way of assessing whether or not this aspiration was realized in practice (Donnelly 1995).

Although the unsatisfactory nature of the match probability calculation allowed ample scope for defences to challenge MLP profile evidence, public enthusiasm for DNA evidence, facilitated perhaps by the analogy with bar-codes, combined with a lack of understanding by lawyers and judges of the underlying science, seems to have led to MLP evidence initially meeting little challenge in court. Ironically, perhaps, later and superior methods of DNA profiling received more critical attention from scientists and statisticians, sparked in the USA by the controversy surrounding the *Castro* case (Lander 1989). One consequence has been that some cases which met little challenge at trial have met with greater scrutiny at subsequent appeals.

Because of the weaknesses outlined above, the MLP profiling technology was relatively short-lived in UK casework, and little used elsewhere. It was replaced around 1990 by minisatellite profiles, generally known in the forensic setting as Single-Locus Probe (SLP) profiles (Buffery *et al.* 1991; Gill *et al.* 1991). The name emphasizes its key feature: the profile bands can be distinguished according to their genomic locus. Consequently, allelism and linkage can be accounted for explicitly in the assessment of evidential weight. Initially three, later four to six, loci were typically tested, producing in each case an autorad with two bands (sometimes only one band and, very rarely, no bands). Such an SLP profile is highly informative, possibly less informative than an MLP profile, but the degree of informativeness can be assessed more accurately.

Problems remain, however, with the SLP profiling system. Accuracy of length measurement is one problem: minisatellite allele lengths are measured to an accuracy of, typically, 1–3 per cent, so that two genes of different, but similar, lengths may not be distinguished in practice. Consequently, there is no natural definition of *allele*: allelic classification can be based on either 'fixed' or 'floating' bins, and in either case the bin widths are arbitrary (Roeder 1994). A related problem is false homozygosity: a single-band autorad can result from a genuine homozygote, or from a heterozygote with two DNA fragments of lengths too similar to be distinguished, or from a heterozygote of which one fragment is too short or too long for the gel resolution employed. A third problem is statistical dependence of matches: although the principal causes of dependence affecting MLP profiles, allelism and linkage, can be eliminated or accounted for in the match probability calculations, weaker dependence due to population genetics effects can nevertheless still be important, and becomes relatively more important as the number of loci increases.

## 15.3 STR profiles

Methods for dealing with the problems faced by the SLP profiling system were developed, and the system was widely employed in successful prosecutions. Nevertheless, scientific controversy and legal challenge arose and persisted, particularly in the USA where aspects of the 1992 report of the National Research Council, *The use of DNA evidence in forensic science*, were widely criticized.

Starting around 1994, SLP profiles began to be superseded in the UK by a system based on microsatellite, or STR, loci. The four tetrameric STR loci initially employed, briefly described in Table 15.1, were typed via automated fluorescence, multiplexed in a single gel with loci distinguished by size and dye colour (Kimpton *et al.* 1994). The result of this procedure is an electropherogram, which is a plot recording the strength of the fluorescent signal, with peaks corresponding to allele lengths.

More recently, a second generation of six tetrameric STR loci, including VWA and THO1, have been adopted in the UK for the national intelligence database and other forensic applications (Sparkes *et al.* 1996a,b).

STR markers are less polymorphic than the minisatellite markers employed in the SLP system, so that six to eight STR loci may be required to achieve the same power of discrimination as a four-locus SLP profile. However, this disadvantage is outweighed by the important advantages of STR profiles over previous DNA-profiling technology. Perhaps the most important advantage is

**Table 15.1**
Details of four STR loci introduced into UK casework in 1994. There is some internal variation for some loci, and the number of alleles is approximate

| Locus (short name) | Repeat unit | Number of alleles | Chromosome location | Genbank accession |
|---|---|---|---|---|
| HUMVWFA31/A (VWA) | TCTA | 10 | 12p12-pter | M25858 |
| HUMTHO1 (THO1) | TCAT | 7 | 11p15–15.5 | D00269 |
| HUMF13A1 (F13) | GAAA | 15 | 6p24–25 | M21986 |
| HUMFES/FPS (FES) | ATTT | 8 | 15q25-qter | X06292 |

that STR markers can be employed in conjunction with PCR, allowing the typing of samples containing as little as 100 pg of DNA (Kimpton *et al.* 1994). Other advantages include the fact that the length measurement and false-homozygosity problems are greatly reduced. Typing additional loci brings with it the advantage of added power in discriminating close relatives: systems which are extremely polymorphic but use only one or a few loci, such as the 'digital fingerprints' proposed by Jeffreys *et al.* (1991*a*), are comparatively poor at distinguishing culprits from their close relatives, which reduces their usefulness for forensic applications.

These benefits of STR profiles over their predecessors are substantial but, inevitably, some problems remain, and the new technology brings with it some new potential problems. One of these is contamination: since PCR is effective with very small amounts of DNA, the DNA of bystanders might contaminate a crime sample, perhaps even via breathing, and be amplified by PCR rather than the culprit's DNA. Other potential problems include preferential amplification of one of the two STR alleles, leading to a false homozygote error, as well as 'stutter' and 'shoulder' peaks in the electropherogram. These potential problems have been overcome to a large extent in developmental work (Andersen *et al.* 1996; Sparkes *et al.* 1996*b*), although they can manifest themselves in some circumstances, particularly when very small crime samples are involved.

Another problem faced by STR markers, in common with previous systems, is the role of population genetics effects on assessments of evidential strength. In addition, courts in the UK have begun to be more critical in considering not only the scientific validity of DNA evidence, but the way in which it is presented in court. This additional scrutiny has led to the overturning on appeal of some earlier convictions, the first being that of Andrew Deen, overturned late in 1993 (*The Times*, 10 January 1994).

## 15.4 Weight of evidence for DNA profiles

The problem of conveying evidential strength in a way which is fair, accurate, and readily comprehended by jurors is common to all systems of DNA profiling. It seems intuitively clear that an assessment of the relative frequency of the profile in some appropriate population is relevant to evidential weight, since this can be interpreted as the probability that a person chosen at random in the population has the profile. The problem then arises of how to estimate profile (relative) frequencies.

The direct approach of estimating population profile frequencies via observed frequencies in a random sample of profiles is impractical because of the rarity of each particular profile. Instead, profile frequencies are usually estimated in practice via the *product rule*, which involves multiplying together frequency estimates for each allele represented in the profile. Use of the product rule relies for its validity on an assumption of independence, which is almost certainly false.

The size and practical importance of the error resulting from adopting this false assumption has been the subject of considerable debate (Roeder 1994). A related controversy concerns the choice of population in which the profile frequency should be assessed.

Perhaps more important than the issues of how to estimate profile frequencies and in which population they should be estimated, is the question of how any estimates should be *interpreted*. How should a reasonable juror use an estimate of the profile frequency, however calculated, to assess evidential weight, taking into account factors such as:

- What if close relatives of the accused are among the persons not excluded from being alternative possible culprits?
- How does the possibility of a laboratory or handling error affect evidential weight, as measured by the profile frequency?
- What if the defendant was identified *because* of his DNA profile, perhaps through a search of a large database of DNA profiles?

Because of these and other difficulties, the weight-of-evidence problem for DNA profiles has continued to be the focus of scientific and legal controversy.

The key to resolving these controversies is to focus on the right question, which concerns the truth or otherwise of the hypothesis.

$G$: the defendant is Guilty.

For convenience, we will assume here that 'is guilty' is equivalent to 'is the source of the crime stain', although this is not necessarily the case in practice.

Since the truth or falsity of hypothesis $G$ cannot be definitively established, it is necessary to assess the degree of certainty which can be attached to it on the basis of the evidence. In order to avoid logical inconsistencies, these assessments should conform to the rules of probability. Although jurors cannot be expected to be conversant with probability theory, it is nevertheless helpful for the scientist to consider a formal, probability analysis, at least in skeleton form, in order both to clarify the aspects of the evidence on which a rational juror might reasonably need guidance, and to illuminate the many possibilities for misleading presentation of evidence, some of which have already led to successful appeals (see, for example, Balding and Donnelly 1994).

Common sense, and the elementary rules of probability, concur in requiring that in order to assess hypothesis $G$, it must be compared with all the alternative hypotheses. These may usually be taken to be the hypotheses of the form:

$I_x$: individual $x$ is guilty,

for all alternative possible culprits $x$. Hypothesis $G$ should only be accepted if the *combined* weight of *all* the alternative hypotheses is negligibly small in the light of the evidence presented in court (the definition of 'negligible' is a matter for each juror to decide).

For convenience, we will for the moment assume that the evidence $E$ consists of the STR profile evidence only, which is usually of the form

$E$ : the defendant has STR profile $D$; the crime scene STR profile is $D$.

The information which evidence $E$ conveys in favour of hypothesis $G$ rather than hypothesis $I_x$ is measured by statisticians via the likelihood ratio $R_x$, defined by

$$R_x = \frac{P(E|I_x)}{P(E|G)}. \tag{15.1}$$

It follows that a rational juror might reasonably want assistance from a forensic scientist in assessing $P(E \mid G)$, the probability of the STR evidence if the defendant is guilty, and $P(E \mid I_x)$, its probability if individual $x$ is the culprit.

Notice that $E$ concerns two observed STR profiles, which may or may not have the same source. Under some simplifying assumptions (Balding and Donnelly 1995$a$), the likelihood ratio $R_x$ reduces to a conditional match probability:

$R_x = P$ ($x$ has STR profile $D$ | the defendant has STR profile $D$).

Conditional match probabilities (henceforth just 'match probabilities') form the key to correctly allowing for population genetic effects: the principal effect is that the crime profile may be more common among the possible perpetrators of an offence than among the population for which frequency estimates are available. For example, there may be a geographic effect due to genetic differentiation of the population local to the crime scene, relative to the national population. This effect causes the match probability for alternative possible culprits local to the crime scene to be higher than the product rule estimate of the profile frequency. Similarly, match probabilities for relatives of the defendant are much higher than those applying to unrelated individuals.

More generally, the approach to assessing evidential weight via likelihood ratios illuminates a path to resolving difficulties such as the role of possible laboratory or handling error, the incorporation of the non-DNA evidence, and the effect of searches. See Balding and Donnelly (1995$b$) for a further discussion.

## 15.5 STR match probabilities

If the STR profiles of distinct individuals were independent, then the product rule would provide a reasonable approximation to the match probability. However, STR profiles are not independent. The underlying cause of dependence is shared ancestry, either recently in the case of relatives, or more distantly for partly isolated subpopulations. The effect of shared ancestry is a function both of factors common to all loci, such as migration and mating patterns, and of locus-specific factors such as mutation and, possibly, selection. The product rule provides a good approximation to the match probability only when the alternative possible culprit $x$ is known to have very little ancestry shared with the defendant.

Balding and Nichols (1995) have developed a methodology for modelling shared ancestry in terms of a single parameter $F$, which is essentially the parameter known to population geneticists as $F_{ST}$ (sometimes denoted $\vartheta$). Although not valid completely generally, the likelihood formula for $F$ which they propose is widely applicable, and can be expressed as follows. Suppose that a sample of size $n$ has been drawn from a local population, and we are interested in the allelic type of the next observation (i.e. number $n + 1$). The probability $P_{n+1}(k)$ that it is allele $k$ is given by:

$$P_{n+1}(k) = \frac{n_k F + (1 - F) p_k}{1 + (n - 1) F} \tag{15.2}$$

in which $n_k$ denotes the number of $k$-alleles observed so far, and $p_k$ is the global frequency of $k$-alleles. Initially, when $n = 0$, we have $P_1(k) = p_k$, so that probabilities of allelic types for the first draw correspond to global frequencies. However, as soon as an allele has been observed, possible shared ancestry makes it more likely that further copies of that allele will be observed on the next draw. As the sample size $n$ increases, the ratio in eqn (15.2) becomes dominated by $n_k/n$, the sample relative frequency of allele $k$.

Equation (15.2) has two important applications in connection with STR profile evidence. Firstly, it provides STR match probabilities. Four applications of the formula give the following expression for the probability that a sample of size four consists of two A and two B alleles:

$$P(A^2 B^2) = \frac{(1 - F) p_A p_B (F + (1 - F) p_A)(F + (1 - F) p_B)}{(1 + F)(1 + 2F)}.$$

Dividing by the probability of observing AB in a sample of size two, we obtain the single-locus match probability in the heterozygote case:

$$R_x = \frac{4P(A^2 B^2)}{2P(AB)} = 2\frac{(F + (1 - F) p_A)(F + (1 - F) p_B)}{(1 + F)(1 + 2F)}, \tag{15.3}$$

the four and two arising because of the two possible orderings of a heterozygous genotype. The homozygote match probability is similarly shown to be:

$$R_x = \frac{P(A^4)}{P(A^2)} = \frac{(2F + (1 - F) p_A)(3F + (1 - F) p_A)}{(1 + F)(1 + 2F)}. \tag{15.4}$$

With linkage, allelism, and now shared ancestry allowed for, there seem to be no further genetic effects which would be expected to cause deviations from independence across loci, and so it seems reasonable to calculate multi-locus match probabilities by multiplying together these single-locus match probabilities. Match probabilities based on eqns (15.3) and (15.4) are routinely used for STR profile evidence in UK court cases, and were also recommended for use in the USA by the 1996 report of the National Research Council *The evaluation of forensic DNA evidence.*

The question remains as to what is the appropriate value of $F$ to use in the match probability eqns (15.3) and (15.4). In principle, a different value of $F$ is appropriate for each alternative possible culprit, according to the amount of ancestry which that individual shares with the defendant. Of course, exact levels of shared ancestry are not usually known, but reasonable assessments can be based on estimates of average levels in a range of populations.

Obtaining such estimates brings into play the second important role for eqn (15.2): it allows likelihood-based inference for $F$, relative to global allele frequencies $p_k$. The $p_k$ are usually estimated in forensic work from a convenience sample of individuals connected with previous cases. Unfortunately, even a large sample typically provides only imprecise information about $F$ at a particular locus and population. This imprecision is reflected by likelihood curves which are not sharply peaked. However, Balding *et al.* (1996) developed a hierarchical modelling framework in which information can be shared both across loci and across populations, without restrictive assumptions of constancy, thus allowing much more precise inferences to be drawn. For a range of European populations, they found that values of $F$ around 0.01 are typical at STR loci, relative to global frequencies estimated from a UK mixed Caucasian forensic database, and values as large as 0.03 are supported by the data in some cases.

## 15.6 Identification example

Consider a case in which crime scene and defendant profiles both match the profile shown in Table 15.2(a), and assume the population relative frequencies given in the table. If the alternative possible culprit, $x$, is unrelated to the defendant, and if population genetic effects are ignored (i.e. $F = 0$), then the match probability is given by the product rule, that is, it is the product of the allele frequencies, with an additional factor of 2 for each heterozygote, which gives 1 in 7450.

The product rule is almost always unfair to defendants since most, if not all, alternative possible culprits will in practice have some (usually unknown) level of ancestry shared with the defendant. If, more realistically, $F$ is taken to be 0.01 at each locus, then the match probability for the profile of Table 15.2(a) is increased

**Table 15.2(a)**
Example of a four-locus STR profile with population allele relative frequencies

| STR Locus | VWA | THO1 | F13 | FES |
|---|---|---|---|---|
| Recorded alleles | 14, 16 | 9, 10 | 6 | 10, 11 |
| Relative frequency | 0.11, 0.22 | 0.15, 0.32 | 0.33 | 0.32, 0.40 |

**Table 15.2(b)**
Match probabilities for various relationships for the STR profile of Table 15.2(a)

| Relationship with defendant | Unrelated $F = 0$ | Unrelated $F = 0.01$ | Unrelated $F = 0.02$ | Cousin $F = 0.01$ | Half-sibling $F = 0.01$ | Sibling $F = 0.01$ |
|---|---|---|---|---|---|---|
| Match probability (reciprocal) | 7450 | 5920 | 4780 | 1750 | 712 | 32 |

to 1 in 5920. If, perhaps because the circumstances of the case suggest a relatively high level of shared ancestry among the possible culprits, or because of a desire not to be unfair to the defendant, a larger value of 0.02 is assumed for $F$, then the match probability is further increased to 1 in 4780.

These calculations ignore other factors which affect match probabilities, such as the sampling variability in the allele frequency estimates, and the possibility of non-amplification of an allele, leading to a false homozygote error. For a description of statistical approaches to allow for these effects, see Balding and Nichols (1994) and Balding (1995). Here, we merely report that the match probability of 1 in 5920 increases to 1 in 5650 if the sample size for the allele frequency estimates is taken to be 1000, and increased further to 1 in 5350 if, additionally, 1 allele in 100 is not detected. A thorough assessment of evidential weight should also take into account the possibility of a laboratory or handling error. Accounting for these possibilities is, however, difficult and we do not address this problem here.

Table 15.2(b) also gives match probabilities for various (regular) relationships of alternate possible culprit with defendant, assuming a background level of shared ancestry of $F = 0.01$. For details of the underlying calculations, see Balding and Nichols (1994).

## 15.7 Relatedness testing

Equation (15.2) can also be used to calculate likelihood ratios for investigating relatedness, for example paternity. Suppose that at a particular STR locus a mother is heterozygous with alleles AB, say, while her child is a BC heterozygote. A man can be excluded as a possible father of the child if he has no C allele at the locus. One approach to measuring the strength of evidence against men who do have a C allele at this locus is to calculate the *inclusion probability* $P_{inc}$, which is the population relative frequency of such men. Ignoring shared ancestry and other factors, we have:

$$P_{inc} = 1 - (1 - p_C)^2 = 2p_C - p_C^2. \qquad (15.5)$$

Although the inclusion probability is a meaningful measure of evidential strength, it has a number of drawbacks. One of these is that it is independent of the genotype of an alleged father, and hence ignores the fact that the evidence is stronger for an alleged father who is a CC homozygote than for a CD heterozygote.

To avoid any such logical difficulties, evidential weight should be measured by likelihood ratios. Consider first the case that the alleged father is a CD heterozygote, so that the complete data are:

$E$: mother is AB; child is BC; alleged father is CD.

The likelihood ratio $R_x$ comparing the hypothesis that the alleged father is the true father with the alternative hypothesis that another man $x$ is the true father, based on the evidence $E$, is given by

$$R_x = \frac{P(E \mid x \text{ is true father})}{P(E \mid \text{alleged father is true father})}. \tag{15.6}$$

If the alleged father is the true father, then four independent alleles have been observed: AB in the mother and CD in the father (the child's two alleles are copies of observed parental alleles). Four applications of eqn (15.2) give

$$P(E \mid \text{alleged father}) = \frac{1}{4}P(ABCD) = \frac{(1-F)^3 p_A p_B p_C p_D}{4(1+F)(1+2F)}. \tag{15.7}$$

(The 1/4 corresponds to the probability that the child has genotype BC given parental genotypes AB and CD.) If, however, the true father is $x$, a man not related either to the mother or to the alleged father, but sharing a common level of ancestry with them, then five independent alleles have been observed: AB in the mother, CD in the alleged father, and C in the child. The probability of the evidence $E$ under this scenario is

$$P(E \mid x) = \frac{1}{2}P(ABC^2D) = \frac{(1-F)^3 p_A p_B p_C(F+(1-F)p_C)p_D}{2(1+F)(1+2F)(1+3F)}. \tag{15.8}$$

Dividing eqn (15.8) by eqn (15.7), we obtain the likelihood ratio

$$R_x = 2\frac{F+(1-F)p_C}{(1+3F)}. \tag{15.9}$$

Notice that if C is very common then $R_x$ can exceed one: even though the alleged father is not excluded, the evidence $E$ points away from him, and in favour of $x$ being the true father. This is because only one of his two alleles are C, whereas $x$ may well have two C alleles. This remark highlights another weakness of the inclusion probability, since it counts every inclusion as evidence against the alleged father. When $p_C$ is small, however, $P_{inc}$ is close in value to $R_x$.

If the alleged father is a CC homozygote then

$$R_x = \frac{2F+(1-F)p_C}{(1+3F)}, \tag{15.10}$$

which cannot exceed 1. For further details of paternity and other relatedness calculations, see Balding and Nichols (1994).

## 15.8 Mixed profiles

The analyses discussed so far apply to the simplest setting in which the sample being analysed contains the DNA of only one individual. Crime samples sometimes contain the DNA of two or more individuals, victim and culprit, for example, or each of several culprits. It is impossible to determine from the profiling results how many individuals contributed to the sample. The STR profile of Table 15.2(*a*), for example, could correspond to a mixture of DNA from two individuals, the first homozygous at each locus, with alleles 14, 9, 6, and 10, respectively, while the second individual might be homozygous for alleles 16, 10, 6, and 11. Although this possibility cannot be conclusively excluded, when no more than two alleles are observed at any locus the observed data are much more likely under the hypothesis of one contributor to the sample than under a two-contributors hypothesis, and so the possibility of two contributors can usually be excluded in practice. For example, ignoring shared ancestry and other effects, the probability of observing the STR profile of Table 15.2(a), assuming that the sample contains the DNA of just one individual, is 1 in 7450. Under the assumption of two contributors to the sample, summing over all 729 possibilities for the two genotypes gives a match probability just larger than 1 in 1 000 000. Unless there is reason to suspect that the sample has two contributors, this possibility might be regarded as negligible in practice.

Suppose, however, that the STR profile obtained from the crime sample is the 'mixed profile' of Table 15.3. Since more than two alleles are detected at the VWA and FES loci, there must have been at least two contributors of DNA to the sample. Suppose that it is assumed that one of the contributors to the sample is the victim of the alleged offence. The profiles of the defendant and the victim together account for all the alleles in the mixed profile, and so the STR evidence is consistent with the hypothesis that defendant and victim only are the sources of the DNA in the sample. If the defendant is not a contributor to the sample, there are many possibilities for the genotypes of the unknown contributor(s).

Suppose that it can be assumed that there is only one unknown contributor to the sample (in addition to the victim). Comparing the 'victim' and 'mixed' profiles in Table 15.3, we see that this unknown person must have VWA genotype 14, 16. Further, he or she must have a 6 at F13 and an 11 at FES. The final three rows of Table 15.3 indicate all 18 possible genotypes for the (unique) unknown contributor. Ignoring shared ancestry and other issues, the probability that an unrelated individual has any one of these 18 genotypes, and hence is an alternative possible contributor to the sample, is 1 in 453. Allowing for a common level of ancestry shared among victim, defendant, and alternative possible source corresponding to $F = 0.01$ at each locus, the match probability is increased to 1 in 384.

If two co-defendants of a crime are thought to be the two contributors to a mixed profile then additional complications arise: the evidence may be very strong against the two jointly, but may possibly be relatively weak against either one under the assumption that the other is a contributor; this and related topics

**Table 15.3**
Example of a four-locus mixed STR profile (row 1). Row 3 gives the profile of the victim, assumed to be a contributor of DNA to the sample. The other contributor could be the defendant (row 4), or an individual with any of the genotypes indicated in the final three rows. There may also be more than one contributor, in addition to the victim

| STR Locus | VWA | THO1 | F13 | FES |
|---|---|---|---|---|
| Mixed profile | 14, 16, 17, 18 | 9, 10 | 6, 7 | 10, 11, 12 |
| Allele relative frequencies | 0.27 (17) | | 0.35 (7) | 0.22 (12) |
| (see also Table 15.2(a)) | 0.21 (18) | | | |
| Victim's profile | 17, 18 | 9, 10 | 7 | 10, 12 |
| Defendant's profile | 14, 16 | 9, 10 | 6 | 10, 11 |
| Possible genotypes of | 14, 16 | 9, 9 | 6, 6 | 10, 11 |
| a (assumed unique) | | 9, 10 | 6, 7 | 11, 11 |
| contributor to the sample, | | 10, 10 | | 11, 12 |
| in addition to the victim | | | | |

are beyond the scope of this introduction. For a further discussion of the analysis of mixed profiles, see Weir *et al.* (1997) and Evett and Weir (1998).

The discussion so far has assumed that the absence or presence of alleles is the only useful information derived from an electropherogram. However, it may be possible to draw stronger inferences concerning mixed profiles by interpreting peak height, or area. For example, if three peaks are recorded corresponding to a particular STR locus, and the peak corresponding to allele A is approximately equal in area to the sum of the peak areas for B and C, then it may be possible to infer that the two contributors to the sample had genotypes either AA, BC or AB, AC. If, on the other hand, only the presence of alleles A, B, and C is inferred, then there are six possibilities for the two genotypes. Although peak areas are somewhat variable, it may be that this variability may be controlled sufficiently to allow the interpretation of peak heights in analysing STR mixtures. This is a topic of ongoing research activity.

# Acknowledgement

Helpful discussions with Dr James Walker of University Diagnostics, London, are gratefully acknowledged work supported in part by the UK EPSRC (Grant No. K72599).

# 16 Tracking linkage disequilibrium in admixed populations with MALD using microsatellite loci

J. Claiborne Stephens, Michael W. Smith, Hyoung Doo Shin and Stephen J. O'Brien

## Chapter contents

## Abstract

Microsatellite analyses have already had a profound impact on the construction of a human genetic linkage map, on the mapping of numerous disease gene loci, and on our understanding of human migration, demography, and evolution. In this report we summarize our recent efforts to extend microsatellite analysis to the mapping of loci involved in complex diseases. In particular, we focus on mapping by admixture linkage disequilibrium, dubbed MALD, which makes use both of the existing human microsatellite map and of human population genetic principles and data.

## 16.1 Introduction

Our current level of understanding of the genetic architecture of the human genome was hardly imaginable scarcely twenty years ago. Researchers interested in human polymorphisms currently have over ten thousand molecular markers to choose from, with the prospect of several hundred thousand in the near future (Chee *et al.* 1996). This wealth of resources underpins efforts to map important human disease loci in populations by whole genome screening (Davies *et al.* 1994; Stephens *et al.* 1994; Ewens and Spielman 1995; Hanis *et al.* 1996; McKeigue 1997), and has also proved valuable in addressing candidate genes and genomic regions, such as the human major histocompatibility complex (see Carrington *et al.* this volume). We are particularly interested in identifying that subset of loci that is differentiated among human populations, and in applying analysis of such loci to relevant complex human diseases. The whole class of genetic linkage analyses based on disease association with specific marker alleles is known as 'association analysis', and depends upon linkage disequilibrium (Risch and Merikangas 1996). Our specific interest is in evaluating several hundred microsatellite loci as potential markers for use in MALD studies of the African-American and Hispanic populations (Dean *et al.* 1994; Smith *et al.* unpublished).

The major varieties of association tests in use for human disease gene mapping can be distinguished by whether they deal with familial samples and haplotypes (Ewens and Spielman 1995; McKeigue 1997), or whether the data are purely populational, without family structure, such as MALD (see also Carrington *et al.* this volume). Although both approaches require 'linkage' disequilibrium, several authors (Stephens *et al.* 1994; Ewens and Spielman 1995; Risch and Merikangas 1996; McKeigue 1997) have cautioned about the need to evaluate whether apparent linkage disequilibrium does indeed reflect linkage. The family-based association studies (Ewens and Spielman 1995; McKeigue 1997) explicitly exclude contributions to disequilibrium that are not due to linkage, whereas most population-based approaches rely on other means. For instance, in our MALD applications, we have proposed (Stephens *et al.* 1994) a sampling strategy that will reduce the disequilibrium between unlinked genes.

Although somewhat more prone to such false positive associations, MALD may be the method of choice for a lot of complex diseases, such as infectious disease, that are not very amenable to family studies. The principle underlying MALD is straightforward: any genetic disease (including genetic differences in susceptibility, rate of progression, age of onset, etc.) that has different frequencies between two admixing populations will be in 'linkage' disequilibrium with marker loci that also have different frequencies between the two populations (Chakraborty and Weiss 1988; Stephens *et al.* 1994). The marker loci need not be on the same chromosome as the disease locus to create disequilibrium, but of course, it is desirable to eliminate this source of disequilibrium if it is not minimized naturally by the dynamics of population admixture. Importantly, the disequilibrium produced is numerically proportional to the product of allele frequency differences between the admixing populations, which means that the disequilibrium produced may be orders of magnitude greater than that produced purely by mutation and drift at linked loci, unless the loci are very tightly linked. Therefore, there is a window of time after the initial admixture during which levels of disequilibrium will be detectable even for relatively large recombination distances, yet reduced sufficiently for unlinked loci (Stephens *et al.* 1994). The feasibility depends largely on recombination distance and level of disequilibrium, the latter being dependent on allele frequency differences and the dynamics of admixture.

Ongoing studies (Dean *et al.* 1994; Shriver *et al.* 1997; Smith *et al.* unpublished) have started to identify useful markers for MALD throughout the genome. Both family-based and population-based association studies of admixed populations can be expected to profit from the definition of a sufficient number of such markers. Below, we have begun to explore levels of linkage disequilibrium between a series of microsatellite markers on chromosome 3 and two chemokine receptor loci, *CCR5* and *CCR2*. These two loci differ in one critical aspect with respect to their molecular evolution: the mutant form at *CCR5* (*CCR5-Δ32*) is restricted to Caucasians and appears to be of extremely recent origin (700–2000 years, Libert *et al.* 1998; Stephens *et al.* 1998), whereas the *CCR2* mutation (*CCR2–64I*) appears much older and is found in all major population groups.

Since the product of such studies is genotypes from many loci (loosely, multilocus 'genotypes', but strictly speaking 'phenotypes' since phase is generally unknown), it is important to identify ways in which such data can be analysed. In this paper, we identify three overlapping methods of analysis, and illustrate them by applying them to data collected from our ongoing studies of the *CCR5* and *CCR2* chemokine receptor loci (Dean *et al.* 1996; Smith *et al.* 1997; Stephens *et al.* 1998) on chromosome 3.

## 16.2 Microsatellite marker maps useful for MALD

Earlier studies of diallelic loci, such as RFLPs, (Chakraborty and Weiss 1988; Briscoe *et al.* 1994; Stephens *et al.* 1994) identified the allele frequency difference between the two founding populations ($\partial$) as the key parameter for both the mathematical prescription for linkage disequilibrium due to admixture and for evaluating relevant loci. In subsequent studies, especially those with multiple alleles, it seems likely that alternative measures will need to be explored. We have used one, the 'composite delta' ($\partial_c$), that appears to capture some, although certainly not all, of the relevant information at each locus. The composite delta is defined as a natural extension of the diallelic $\partial$, that is, as the sum of all frequency differences that are in the same direction. For instance, if a microsatellite locus has five alleles that have higher frequencies in population A, and six alleles that have a higher frequency in population B,

$$\partial_c = \sum (f_{iA} - f_{iB}) = \sum (f_{jB} - f_{jA}),$$

where the first sum is over the five alleles that are higher in A and the second sum is over the six alleles that are higher in B. This measure appears to be useful for characterizing loci as to their relevance for MALD gene mapping, although it misses information on the centrality of each allele's frequency, which is also important (Shriver *et al.* 1997; Stephens *et al.* unpublished).

We have used $\partial_c$ in Fig. 16.1 to characterize 23 microsatellite loci on chromosome 3 in a comparison of allele frequencies between Caucasians and African-Americans. These loci were derived from the Applied Biosystems alpha set of markers for development of a fluorescent detection-based microsatellite map. They were further supplemented based on available microsatellites to reach a map with 10 cM resolution and $\partial_c$ of at least 0.30. This schematic represents one of the chromosomes from ongoing efforts to identify a panel of 10 cM spaced markers appropriate for MALD in African-Americans and Hispanics.

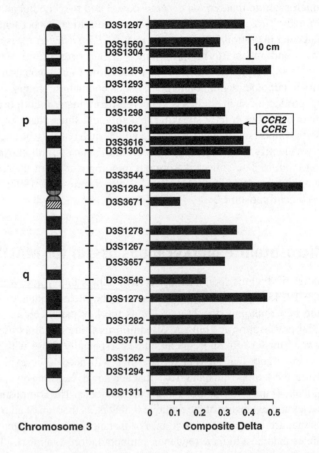

**Fig. 16.1** Characterization of microsatellite loci on human chromosome 3. Loci are drawn at approximate cM positions and are further characterized by the composite delta, $\partial_c$, a measure that reflects the overall differentiation in allele frequency between two populations. In this case, $\partial_c$ is calculated for the comparison between Caucasian and African-American population samples. See text for mathematical definition of $\partial_c$.

# 16.3 Modes of analysis

The best data for evaluating linkage disequilibrium is a sample of unrelated haplotypes, from which phase is known and frequencies are estimated directly. Pragmatically, this generally requires reduction of each phase-unknown multilocus phenotype into its two constituent phase-known haplotypes. Alternatively, one can use the genotypes directly, without inferring phase, in a somewhat cruder test. Since haplotype information may not be available or may be difficult to extract, especially for microsatellite data, we employ three tests of linkage disequilibrium that include these two extremes.

## 16.3.1 Genotype-by-genotype

Our first method of analysis is the comparison between two loci based purely on genotype-by-genotype associations. Although theoretically one could use a standard association test directly, in practice there are generally very large numbers of genotypes for microsatellite loci, which usually violates the assumptions for standard parametric association tests. To overcome this, we compile the matrix of all observed genotypes at locus A versus all observed genotypes at locus B, and compute an association statistic from this matrix. We then use a randomized permutation test to estimate the empirical distribution of this association. More formally, if there are $k$ alleles at locus A and $l$ alleles at locus B, the $k(k + 1)/2$ by $l(l + 1)/2$ matrix is formed, reduced for rows and columns corresponding to unobserved genotypes, and the association statistic is generated from this reduced matrix. Then a simulation is performed, for generally at least 1000 replications, based on the observed frequencies of each genotype, which are the marginal frequencies for this type of association test. The actual statistics we use for this test are the $\chi$-square and the G-test (likelihood ratio). Since both of these statistics would have the $\chi$-square distribution in theory, we divide each by its degrees of freedom to make comparisons across pairs of loci. This convention was used in a related test of genetic differentiation among subspecies (Stephens *et al.* unpublished). While simple to perform and generally applicable, this test is not expected to be very powerful in cases of weak linkage disequilibrium.

## 16.3.2 Tests of allele frequency difference

For disease gene mapping, quite often the phenotype is dichotomous (e.g. affected vs. unaffected), so that a marker locus in linkage disequilibrium with a locus contributing to the affected state might display allele frequency differences between affected and unaffected individuals. This intuitive notion is confirmed more rigorously in proposed transmission disequilibrium tests (TDT, Ewens and Spielman 1995; McKeigue 1997), but can also be extended to admixed population samples (Ewens and Spielman 1995; McKeigue 1997). In fact, whatever the cause of disequilibrium, predictable allele frequency differences arise for marker alleles in disequilibrium with the phenotype (McKeigue 1997; Stephens *et al.* unpublished). Therefore, when investigating linkage disequilibrium between a diallelic locus (e.g. disease states, RFLPs, single nucleotide polymorphisms) and a microsatellite locus, one can often simply tabulate the microsatellite allele frequencies between two groups (e.g. affected vs. unaffected) and then statistically test any allele frequency differences.

The major drawbacks with this approach are that most of the genotypic data is lost for both loci, and that it easily leads to multiple comparisons when many microsatellites are screened for a given phenotype. Since our primary purpose for MALD gene mapping is exploratory, such as the identification of novel genomic regions relevant to a given disease phenotype, we are willing to accept some level of false positives for a whole genome screen, and then further evaluate each positive result empirically (e.g. additional markers in the region, additional patient cohorts, and alternative linkage analyses). On the positive side, the statistical analysis of such data is routine and relatively free of genetic or other assumptions, and it is far more sensitive than the genotype-by-genotype comparisons.

### 16.3.3 Haplotype inference

Although known haplotypes are the data of choice, haplotype estimation, at least for pairs of loci, can come from three general directions. First, good estimates of at least some haplotype frequencies can come from the homozygous segment of the data, as follows. When one of the loci is much less polymorphic than the other, one can estimate haplotype frequencies based on the homozygotes, and use these to estimate, or at least set bounds on, frequencies of all haplotypes involved in double heterozygotes. For instance, if much of the data involves a single homozygous genotype, say 11 at locus A, one can estimate the frequencies of all haplotypes $[1 - x]$, where $x$ is an allele at the second locus, from the 11 segment of the data, and then resolve all double heterozygotes involving allele 1 at locus A as having proportional contributions of the estimated $[1 - x]$ haplotypes. This strategy assumes Hardy–Weinberg equilibrium among haplotypes, but can be very useful in certain situations. However, this strategy is known to introduce a bias in certain situations, such as overestimating frequencies of rare or even non-existent haplotypes.

A second, more conventional method of haplotype estimation is the ascertainment of phase by the transmission of haplotypes through pedigrees. However, this strategy requires the collection of families and especially the need for multiple genotypes from each family, which can be daunting. We have assumed that this information will not be available in general. However, this consideration points to the importance of study designs which collect samples from parents where practical.

Third, it is possible to take a maximum likelihood approach to the resolution of double heterozygotes into their constituent haplotypes, which allows estimation of all haplotype frequencies simultaneously. We do this by introducing a parameter for each class of double heterozygotes. Each parameter reflects the apportionment of that class into its two alternative pairs of haplotypes, and is allowed to take on all values from all of one pair to all of the other. By evaluating the likelihood of each split, one can statistically evaluate the improvement for alternative values of these parameters, ultimately deriving a maximum-likelihood estimate for haplotype frequencies for each data set. Furthermore, there are now several computer programs (Weir 1990; Hawley and Kidd 1995; Long et al. 1995) for estimating these haplotype frequencies from such data, even extending to the situation of more than two loci.

# 16.4 Results

We have focused on assessment of linkage disequilibrium among 15 microsatellite loci and the chemokine receptors CCR5 and CCR2, which are only 18 kilobases apart on chromosome 3 (Dean et al. 1996; Smith et al. 1997). Of particular interest is the contrast in the levels of apparent linkage disequilibrium between Caucasian (CA) and African-American (AA) patient samples. Although the following analyses are provisional, in that much more data is anticipated in the near future, they will illustrate the approaches that are currently being taken. Additionally, we are developing software to facilitate these analyses.

In Table 16.1 we have characterized the 17 loci used in this study—15 microsatellite loci and the two chemokine receptor loci. All microsatellite loci have from 4 to 19 alleles, even though the sample sizes are still somewhat small for some loci. We see that composite delta values ($\partial_c$) can be relatively high (e.g. $\partial_c = 0.52$ for D3S1284), even though this comparison was between Caucasians and African-Americans, not native Africans.

## 16.4.1 Application to CCR5 and CCR2

We will illustrate each of the three methods for assessing linkage disequilibrium by application to our largest data set, that of CCR5/CCR2 genotypes. We do this for both Caucasian and African-American populations in Table 16.2. In part A we show the raw genotypic data, where allele CCR5*2 is the 32-base pair deletion ($\Delta 32$) implicated both in resistance to HIV infection (homozygotes) as well as delay to AIDS onset (heterozygotes) (Liu et al. 1996; Samson et al. 1996; Dean et al. 1996). Allele CCR2*2 is the 64I deletion which also delays onset to AIDS (Smith et al. 1997). We can see immediately from Table 16.2(a) that these mutations are never found together on the same chromosome, that is, as the [2–2] haplotype, except possibly as double heterozygotes (the [1–1]/[2–2] genotype). Thus, even in this extremely simple comparison, there is some ambiguity as to haplotype frequencies.

Our first test of linkage disequilibrium between CCR5 and CCR2 is to calculate the association statistics ($\chi$-square/d.f. and G-test/d.f.) for each data set, and compare our observation to a simulated distribution of each statistic. The results

**Table 16.1**
Loci used in assessing linkage disequilibrium

| | Sample size (ind.) | | Distance (cM) | | |
| | CA | AA | from *CCR5* | $\partial_c$ | Alleles |
| --- | --- | --- | --- | --- | --- |
| *D3S1304** | 45 | 121 | −48 | 0.18 | 9 |
| *D3S1259** | 43 | 120 | −35 | 0.41 | 12 |
| *D3S1293** | 46 | 122 | −28 | 0.25 | 19 |
| *D3S1266** | 0 | 82 | −17 | – | 6 |
| *D3S2354* | 176 | 188 | −2 | 0.39 | 9 |
| *D3S3582* | 172 | 199 | −2 | 0.34 | 10 |
| *D3S3647* | 77 | 0 | −2 | – | 6 |
| *AFMB* | 135 | 155 | −1 | 0.25 | 5 |
| *GAAT* | 180 | 205 | 0 | 0.20 | 6 |
| *STR2* | 353 | 132 | 0 | 0.26 | 4 |
| *CCR2* | 2205 | 994 | 0 | 0.05 | 2 |
| *STR1* | 353 | 132 | 0 | 0.34 | 14 |
| *CCR5* | 2747 | 1188 | 0 | 0.08 | 2 |
| *D3S1621** | 174 | 201 | 1 | 0.32 | 16 |
| *D3S1578* | 169 | 199 | 1 | 0.22 | 18 |
| *D3S1300** | 46 | 122 | 14 | 0.34 | 12 |
| *D3S1284** | 46 | 123 | 38 | 0.52 | 14 |

*Also on Fig. 16.1.

are shown in Table 16.2(b). For the Caucasian data set, both statistics were statistically significant, indicating linkage disequilibrium. This was not the case for the African-American data set.

For microsatellite data sets, we have adopted use of the *t*-test for each allele individually when the other locus or trait is essentially dichotomous. For assessment of linkage disequilibrium between *CCR5* and *CCR2*, a χ-square test would be more conventional, but we retain the *t*-test format for illustrative purposes (Table 16.2(c)). As expected, there is very strong linkage disequilibrium between *CCR5* and *CCR2* in the Caucasian data set, but only moderate association in the African-American data set, presumably due to the paucity of *CCR5-Δ32* positive individuals (37/983).

Our final analysis of linkage disequilibrium between *CCR5* and *CCR2* required resolution of phase of the 53 double heterozygotes in Table 16.2(a). The maximum likelihood estimates (Table 16.2(d)) were used in a permutation test, which shows significant association for both population groups (Table 16.2(e)). We now turn our attention to analyses of both of these loci with the flanking microsatellites on chromosome 3.

## 16.4.2 Genotype-by-genotype

We compiled pairwise genotypes for both chemokine receptor loci and both populations for each of the 15 microsatellite loci, and performed the permutation

**Table 16.2**
Assessment of linkage disequilibrium between *CCR5* and *CCR2*

**(a) Genotypic data**

|  |  | CCR2 |  |  |  |
|---|---|---|---|---|---|
|  |  | 11 | 12 | 22 |  |
| **Caucasians (CA)** |  |  |  |  |  |
| *CCR5* | 11 | 1396 | 338 | 18 | 1752 |
|  | 12 | 363 | 48 | 0 | 411 |
|  | 22 | 15 | 0 | 0 | 15 |
|  |  | 1774 | 386 | 18 | 2178 |
| **African-Americans (AA)** |  |  |  |  |  |
| *CCR5* | 11 | 677 | 249 | 20 | 946 |
|  | 12 | 32 | 5 | 0 | 37 |
|  | 22 | 0 | 0 | 0 | 0 |
|  |  | 709 | 254 | 20 | 983 |

**(b) Genotype-by-genotype comparisions**

|  | $\chi$-square/d.f. |  |  | G-test/d.f. |  |  |
|---|---|---|---|---|---|---|
|  | Obs | Sim | $p \leq$ | Obs | Sim | $p \leq$ |
| CA | 5.41 | 9.06 | 0.005 | 7.19 | 4.78 | 0.001 |
| AA | 2.07 | 7.97 | 0.114 | 2.65 | 5.55 | 0.062 |

**(c) *t*-tests of frequency differences**

|  | CCR5-$\Delta$32 |  |  |  |  |  |
|---|---|---|---|---|---|---|
|  | + | − | $f_1$ | $f_2$ | $t$ | $p \leq$ |
| **Caucasians (CA)** |  |  |  |  |  |  |
| *CCR2\*1* | 804 | 3130 | 0.94 | 0.89 | 4.46 | 0.001 |
| *CCR2\*2* | 48 | 374 | 0.06 | 0.11 | −4.46 | 0.001 |
|  | 852 | 3504 |  |  |  |  |
| **African-Americans (AA)** |  |  |  |  |  |  |
| *CCR2\*1* | 69 | 1603 | 0.93 | 0.85 | 2.02 | 0.022 |
| *CCR2\*2* | 5 | 289 | 0.07 | 0.15 | −2.02 | 0.022 |
|  | 74 | 1892 |  |  |  |  |

**(d) Inferred haplotypes**

|  | [1–1] | [1–2] | [2–1] | [2–2] |  |
|---|---|---|---|---|---|
| Caucasians (CA) | 3493 | 422 | 441 | 0 | 4356 |
| African-Americans (AA) | 1635 | 294 | 37 | 0 | 1966 |

**(e) Permutation tests of haplotype frequencies**

|  | $\chi$-square/d.f. |  |  | G-test/d.f. |  |  |
|---|---|---|---|---|---|---|
|  | Obs | Sim | $p \leq$ | Obs | Sim | $p \leq$ |
| CA | 52.63 | 17.63 | 0.001 | 95.03 | 21.58 | 0.001 |
| AA | 6.63 | 12.08 | 0.005 | 12.11 | 12.11 | 0.003 |

tests described above to evaluate the observed level of association (Table 16.3). In general, only loci very close to *CCR5/2* showed strong associations (e.g. *AFMB*, *STR1*, *STR2*), although the African-American population also showed a strong association between *CCR5* and *D3S3582* on the telomeric side, and *CCR2* and *D3S1578* on the centromeric side. The African-American sample also showed some evidence, albeit weak, for association of *CCR2* with *D3S1284*, *GAAT*, and *D3S2354*, as did the Caucasian sample for *CCR5* with *GAAT* and *D3S2354*.

**Table 16.3**
Assessment of linkage disequlibrium by pairwise comparison of genotypes in permutation tests

|  | Distance (cM) from CCR5 | CCR5 CA | AA | CCR2 CA | AA |
|---|---|---|---|---|---|
| *D3S1304* | −48 | ns (45, 22) | ns (129, 31) | ns (40, 22) | ns (121, 28) |
| *D3S1259* | −35 | ns (43, 22) | ns (127, 35) | ns (39, 20) | ns (120, 35) |
| *D3S1293* | −28 | ns (46, 26) | ns (130, 49) | ns (41, 24) | ns (122, 47) |
| *D3S1266* | −17 | – – | ns (87, 14) | – – | ns (82, 14) |
| *D3S2354* | −2 | 0.245, 0.041 (175, 13) | ns (190, 16) | ns (167, 14) | 0.145, 0.029 (181, 16) |
| *D3S3582* | −2 | ns (172, 22) | 0.008, 0.002 (195, 31) | ns (164, 22) | ns (187, 31) |
| *D3S3647* | −2 | ns (77, 14) | – – | ns (76, 14) | – – |
| *AFMB* | −1 | 0.008, 0.001 (135, 5) | 0.044, 0.017 (147, 7) | ns (132, 5) | 0.001, 0.001 (143, 7) |
| *GAAT* | 0 | 0.079, 0.006 (179, 6) | ns (197, 9) | ns (171, 6) | 0.101, 0.043 (187, 9) |
| *STR2* | 0 | 0.001, 0.001 (352, 4) | 0.119, 0.023 (132, 7) | 0.019, 0.001 (304, 4) | 0.003, 0.001 (124, 7) |
| *STR1* | 0 | 0.001, 0.001 (352, 45) | 0.001, 0.001 (132, 31) | 0.008, 0.001 (304, 43) | 0.001, 0.001 (124, 31) |
| *D3S1621* | 1 | ns (173, 22) | ns (193, 40) | ns (166, 22) | ns (185, 41) |
| *D3S1578* | 1 | ns (169, 66) | ns (192, 69) | ns (161, 64) | 0.002, 0.010 (183, 68) |
| *D3S1300* | 14 | ns (46, 20) | ns (131, 39) | ns (40, 18) | ns (122, 37) |
| *D3S1284* | 38 | ns (45, 16) | ns (132, 54) | ns (41, 16) | 0.019, 0.284 (123, 53) |

Note: ns if not significant by either the $\chi$-square test or G-test; if significant by either test, significance under $\chi$-square test and G-test are both given. Second row values are (sample size, number of microsatellite genotypes observed).

### 16.4.3 $t$-tests

The genotype-by-genotype contingency tables used above were reduced into tables appropriate for testing allele frequency differences between $CCR5$-$\Delta32$ positive ($CCR2$-$641$ positive) and negative individuals within each population group. We observed 160 different microsatellite alleles among the 15 loci in our study. In Caucasians, 23 alleles were in either positive or negative linkage disequilibrium with $CCR5$, and nine with $CCR2$ (Table 16.4). In African-Americans, 12 alleles were in linkage disequilibrium with $CCR5$, and 17 with $CCR2$.

There are two primary sources of linkage disequilibrium that we expected to detect with this analysis. First, $CCR5$-$\Delta32$ has been shown to be of relatively recent origin in Caucasians (Martinson *et al.* 1997; Libert *et al.* 1998; Stephens *et al.* 1998) and in strong linkage disequilibrium with $GAAT$ and $AFMB$ (Stephens *et al.* 1998). This work extends that association to $STR1$, $STR2$, $D3S3582$, $D3S2354$, $D3S1304$, and more weakly, to $D3S1621$ and $D3S1284$.

The second source of linkage disequilibrium we expected to see was the replacement of African haplotypes with Caucasian haplotypes in the African-American population. We used $CCR5$-$\Delta32$ as a paradigm for this. Indeed, most of the African-American associations seen with $CCR5$-$\Delta32$ reflect this precisely: in the $CCR5$-$\Delta32$ positive African-American sample an allele or alleles generally higher in the African-Americans has been replaced by an allele or alleles that are generally more frequent in Caucasians. A counter-example of note is $AFMB*217$. This allele is apparently more frequent in Caucasians than in African-Americans. Yet, among the $CCR5$-$\Delta32$ positive African-Americans, this allele frequency is reduced relative to the $CCR5$-$\Delta32$ negative African-American sample. However, this departure is easily explained by the strong, almost exclusive, association of $CCR5$-$\Delta32$ with $AFMB*215$ in Caucasians. That is, the $AFMB*217$ frequency in African-Americans is reduced even further among $CCR5$-$\Delta32$ positive individuals by replacement with $AFMB*215$. This shows quite clearly that it is not enough to look at $\partial$ values for each allele; one must also ask whether there is linkage disequilibrium in the founding population(s).

We have also modified the $t$-test approach by categorizing all alleles at each locus as either CA or AA, depending on which population has a higher frequency, and then doing a standard $\chi$-square association test on the resulting $2 \times 2$ contingency table. For instance, in testing associations with $CCR5$-$\Delta32$ in African-Americans, these results confirmed the strong association of $D3S2354$, $D3S3582$, $GAAT$, and $STR2$, but failed for other loci such as $STR1$ and $AFMB$.

### 16.4.4 Haplotype inference

The reliability of haplotype inference depends strongly on the number of double heterozygotes and number of classes of double heterozygotes. Accordingly, in this preliminary analysis, we have only attempted haplotype inference for a few select cases of immediate importance: ($CCR5$, $GAAT$) and ($CCR5$, $AFMB$) for both population groups. Many other pairs of potential interest have double

**Table 16.4**

Assessment of linkage disequilibrium by *t*-tests of microsatellite allele frequencies in *CCR5*2* positive (or *CCR2*2+*) versus *CCR5*2* negative (*CCR2*2−*) groups

| | Distance (cM) from CCR5 | w/CCR5*2 in CA | w/CCR5*2 in AA | w/CCR2*2 in CA | w/CCR2*2 in AA |
|---|---|---|---|---|---|
| *D3S1304* | −48 | *258, −2.08, 0.019<br>*264, 2.35, 0.009 | none | none | *254, 2.34, 0.001 |
| *D3S1259* | −35 | none | *194, 2.19, 0.014 | none | none |
| *D3S1293* | −28 | none | none | *134, 2.91, 0.002<br>*136, 2.06, 0.020 | *112, 1.92, 0.027 |
| *D3S1266* | −17 | — | *297, −2.06, 0.020 | — | *289, −2.41, 0.008 |
| *D3S2354* | −2 | *119, −2.93, 0.002<br>*135, 3.65, 0.0002<br>*151, 2.79, 0.003 | none | *151, 2.53, 0.006 | *141, 1.94, 0.026 |
| *D3S3582* | −2 | *224, −2.35, 0.009<br>*232, 2.05, 0.020 | *234, −2.40, 0.008<br>*236, 6.23, ≤0.0001 | none | *230, 2.52, 0.006 |
| *D3S3647* | −2 | none | — | *287, 2.01, 0.22 | — |
| *AFMB* | −1 | *215, 3.37, 0.0004<br>*217, −3.27, 0.0005 | *217, −2.09, 0.018 | none | *215, −3.21, 0.0007<br>*217, 7.08, ≤0.0001<br>*219, −4.10, ≤0.0001 |
| *GAAT* | −1 | *193, −2.27, 0.012<br>*197, 2.30, 0.011<br>*201, 2.00, 0.023 | *193, −2.42, 0.008<br>*197, 2.49, 0.006 | *191, 2.90, 0.002 | *193, 3.56, 0.0002<br>*197, −3.01, 0.001 |
| *STR2* | 0 | *143, −7.65, ≤0.0001<br>*146, 7.37, ≤0.0001 | *146, 3.07, 0.001 | *143, 5.22, ≤0.0001<br>*146, −5.14, ≤0.0001<br>*148, −4.04, ≤0.0001 | *143, 4.40, ≤0.0001<br>*149, −4.42, ≤0.0001<br>*154, 3.34, 0.0004 |
| *STR1* | 0 | *146, 3.28, 0.0005<br>*148, 16.37, ≤0.0001<br>*154, −2.39, 0.008<br>*156, −3.18, 0.0007<br>*158, −3.74, 0.0002<br>*160, −5.00, ≤0.0001<br>*162, −2.55, 0.005 | *148, 9.50, ≤0.0001 | *156, 7.55, ≤0.0001 | *156, 3.48, 0.0003<br>*160, −2.53, 0.006<br>*164, −2.99, 0.001 |
| *D3S1621* | 1 | *117, −2.12, 0.017 | none | none | none |
| *D3S1578* | 1 | none | *155, 1.97, 0.024,<br>*251, −2.04, 0.021 | none | *157, 3.47, 0.0003 |
| *D3S1300* | 14 | none | *150, 2.02, 0.022 | none | none |
| *D3S1284* | 38 | *176, 2.03, 0.021 | *150, 2.02, 0.022 | | |

heterozygotes distributed into far too many double heterozygote classes to be of current value. For instance, to investigate the apparently strong association of *D3S3582* with *CCR5* in African-Americans, we would need to enumerate and evaluate the 2880 alternative apportionments of the 15 double heterozygotes. A computer program is being written to facilitate such analyses. As expected from previous work (Stephens *et al.* 1998), both *GAAT* and *AFMB* show very strong linkage disequilibrium with *CCR5* in Caucasians. A newer result is the strong association of these loci with *CCR5* in African-Americans, shown by the permutation test of haplotypes as well as by the *t*-tests above.

## 16.5 Discussion

In this paper, we have focused on the assessment of whether any existing linkage disequilibrium in a data set was statistically significant. Methods presented herein are subject to the usual sorts of problems for ascertainment from genetic data, such as bias in frequency due to over-sampling, or the confounding of other factors (both genetic and non-genetic) with linkage disequilibrium. For these reasons, we sought to develop an approach to such data that will be both robust and verifiable. As mentioned earlier, our primary goals are exploratory, with the idea being that all key findings can be scrutinized further with additional data or analyses.

Both the genotype-by-genotype phase of analysis and the tests of microsatellite allele frequency differences between dichotomized samples seem to be generally applicable to the data sets we envision being available in the future. In particular, mapping genes associated with a disease or other phenotype by their disequilibrium, and hence allele frequency distortion, with marker microsatellite alleles appears promising. Our third proposed test of linkage disequilibrium uses inferred haplotype frequencies. Although considerably more work is involved, either experimentally or computationally, the payoff is in having a more sensitive test for disequilibrium.

As shown here, the pattern of linkage disequilibrium is complicated, being the consequence of multiple processes including population history, recombination distance, genetic drift, and possibly natural selection. Even so, some attributes of these processes are still decipherable, such as the general tendency of *CCR5-Δ32* positive African-Americans to have higher frequencies of other predominantly Caucasian alleles in the vicinity of *CCR5*. Future analyses will profit greatly by improvements in both the theory and the data addressing linkage disequilibrium in human populations. For instance, much of our current analysis lacks proper controls for levels of admixture or potentially different ethnicities in our population groups. Additional insights into microsatellite evolution, as addressed in other chapters of this book, will guide us toward a better theoretical framework for understanding the evolutionary and demographic dynamics of microsatellite haplotypes.

## Acknowledgements

We thank Sadeep Shrestha, Keith Byrd, Raleigh Boaze, Mike Malasky, Bernard Gerrard, Mary McNally, and Mike Weedon. We are grateful to our colleagues Drs Michael Dean, Mary Carrington, George Nelson, and Cheryl Winkler for helpful discussions.

# 17 Microsatellite markers in complex disease: mapping disease-associated regions within the human major histocompatibility complex

Mary Carrington, Darlene Marti, Judy Wade, William Klitz, Lisa Barcellos, Glenys Thomson, John Chen, Lennart Truedsson, Gunnar Sturfelt, Chester Alper, Zuheir Awdeh and Gavin Huttley

**Chapter contents**

## Abstract

Microsatellite markers have facilitated identification of a number of disease genes, most of which have a monogenic inheritance. Genes of the human major histocompatibility complex (*HLA*) have been shown to be associated with diseases of various aetiologies, and microsatellite loci mapping within the *MHC* may provide a means to identify accurately disease-associated regions. Allele frequencies of microsatellite loci located throughout the *HLA* complex were determined in patients with diseases known to be associated with a particular gene or region within the complex. Four of these diseases (IDDM, multiple sclerosis, narcolepsy, and uveitis) are associated with genes encoding molecules involved in antigen presentation, and one, C2 deficiency, is caused by a single mutation in the class III C2 gene. Significant associations ($p < 0.01$) between disease and microsatellites located in closest proximity to the known 'disease gene' were observed in all of the five diseases studied. This demonstrates the power of microsatellite markers in disease studies where no previous *HLA* association has been reported. The strong linkage disequilibrium that exists between pairs of *HLA* genes has often caused difficulty in distinguishing the primary gene(s) associated with a particular disease. Stratification analysis was used to determine whether linkage disequilibrium can explain all of the statistical association at a given marker locus relative to a known disease locus for multiple sclerosis, narcolepsy, and IDDM. The analysis suggested few, if any, additional components of disease susceptibility to these three diseases, despite the fact that additional *HLA* region variation (as yet unidentified) contributes to IDDM.

## 17.1 Introduction

Genetic mapping has been greatly facilitated by the use of polymorphic microsatellite markers over the past several years (Hearne *et al.* 1992; Bruford and Wayne 1993*b*). The microsatellite repeats of $CA_n$ are interspersed throughout the eucaryotic genome, occurring about every 30 kb in DNA from euchromatic regions (Stallings *et al.* 1991). Their potential function is covered by Soller *et al.* in this volume. The utility of microsatellites has promoted the localization of disease loci in monogenic Mendelian disorders, but has not been thoroughly examined in studies of complex diseases having environmental and multiple genetic components.

Disease gene mapping may be accomplished using family-based or association-based methods. Linkage analysis using data derived from families depends on short term recombination rates (i.e. large recombination distances) transmitted in pedigrees. Linkage analysis is particularly advantageous if data from large pedigrees containing several members with the disease can be analysed. Association-based mapping using a case-control experimental design is dependent on linkage disequilibrium between a marker locus and the disease gene locus at the population level. Association-based mapping has been performed using both whole genome scans (discussed in detail by Stephens *et al.* this volume), where no preference in genomic location is applied, and a candidate gene approach, where available information allows selection of genes (or markers near those genes) having high potential for causing the disease. If strong candidate genes are identifiable, then this approach can be more direct and rapid.

Genes of the human major histocompatibility complex (*MHC*), *HLA*, encode cell surface molecules which present antigenic peptides to T cells. These genes are characterized by extraordinary levels of polymorphism, presumably to enhance immune responsiveness to a large array of antigens. Given their role in antigen presentation, *HLA* can be considered as candidate genes for a large number of diseases having a variety of aetiologies. Microsatellite markers have been mapped among the inter-related genes of the *MHC*, and used for purposes such as determination of recombination rates across the human *MHC* (Martin *et al.* 1995) and localization of the relative position of the haemochromatosis gene (e.g. Raha-Chowdhury *et al.* 1995). These loci may serve as useful markers when in linkage disequilibrium with other histocompatibility loci that are associated with disease.

Among the large number of diseases shown to be associated with the human *MHC*, four of the most thoroughly characterized diseases in terms of their association with *HLA* loci are insulin-dependent diabetes mellitus (IDDM), multiple sclerosis (MS), narcolepsy, and the spondyloarthropathies, including uveitis. Evidence indicates a polygenic inheritance for IDDM (Davies *et al.* 1994), and this is likely to be the case for MS, narcolepsy, and uveitis as well. Although haplotypes of *DRB1* and *DQB1* show the strongest HLA association with IDDM, with haplotypes encoding DR3/DR4 having the strongest predispositional effects in Caucasians (reviewed in Tisch and McDevitt 1996), a number of additional studies indicate that *DPB1* and loci in the class I region may contribute to a lesser extent (e.g. Erlich *et al.* 1996). Both MS and narcolepsy are associated with the DR2 haplotype *DRB1*1501-DQA1*0102-DQB1*0602* (e.g. Haegert and Marrosu 1994). The strongest disease association with loci in the class I region observed to date involves the spondyloarthropathies. These inflammatory disorders, including the ocular manifestation uveitis, have been shown to be associated with HLA-B27 (Tiwari and Terasaki 1985). A large number of studies have convincingly shown that associations between these four diseases and *HLA* exist, but in some cases identification of all of the possible *HLA* loci involved remains questionable as does the basis of their relationships.

The ability to detect a locus influencing predisposition to disease, based on a nearby microsatellite locus, depends on the extent of linkage disequilibrium occurring between the disease and marker loci. A number of variables affect the strength of linkage disequilibrium, including the recombination rate ($r$) between the two loci, the mutation rate ($u$), the selection coefficient ($s$) at each of the two loci, and the demographic history of the population. Considerable information has been published regarding $r$, $u$, and $s$ for *HLA* region loci. Recombination fractions have been measured between *HLA* loci using the CEPH families (Martin *et al.* 1995), and mutation rates have been estimated using phylogenetic analyses. For example, there are ten major allelic lineages for *DRB1* which are separated by an average of six million years, but specific subtypes are generated at a much higher rate, with several dozen arising over the last 200 000 years (Erlich *et al.* 1996). Microsatellite mutation rates are at least an order of magnitude higher, ranging from $10^{-3}$ to $10^{-4}$ (Banchs *et al.* 1994). Selection on microsatellite loci has been assumed to be absent, except when the locus occurs in a region affecting gene regulation. The mean selection coefficient for the *HLA* loci has been estimated to be in the neighbourhood of one per cent per generation (Satta *et al.* 1995). Selection is thought to be much more important than mutation and recombination on disequilibrium when considering all pairs of histocompatibility loci, except for those separated by more than a few per cent recombination frequency. Pertinent to the utility of *HLA* region microsatellite loci as useful markers of disease susceptibility, the high mutation rate of microsatellite loci is still at least an order of magnitude less than positive selection on *HLA* alleles. On this basis, we predict that *HLA* region microsatellites alleles should maintain high levels of disequilibrium through hitchhiking with the selected *MHC* alleles, and therefore be effective markers for *HLA* disease loci.

We have tested the ability of microsatellite markers in the *MHC* to identify regions of disease association defined previously for IDDM, MS, narcolepsy, and uveitis using classical *HLA* loci. The markers were also typed in a group of patients with *C2* deficiency, which is caused by a 28-base pair deletion in the class III *C2* gene (Johnson *et al.* 1992). As *C2* deficiency is considered a monogenic disease, it serves as a control for association with microsatellites in the region, and can be used as a reference point when analysing diseases with a more complex genetic aetiology in the *MHC*.

The strong linkage disequilibrium between loci in the *HLA* region, particularly between the classical *HLA* class I and II loci, complicates the ability to identify a specific locus involved in disease resistance and/or susceptibility. Therefore, the ability of microsatellites to identify *HLA* involvement in disease additional to the known 'disease region' of *DR-DQ* for IDDM, MS, and narcolepsy was also tested using five microsatellite markers located within the *MHC*. One strategy to identify multiple sites of genetic influence within a defined genomic region is to stratify the data according to alleles known to be associated with the disease. Stratification analysis did not reveal strong evidence for additional disease components in the regions near the five markers tested.

## 17.2 Materials and methods

### 17.2.1 DNA samples

DNA prepared from the Tenth International Histocompatibility Workshop B lymphoblastoid cell lines (HTCs; Yang *et al.* 1989) was used as size controls for alleles at each microsatellite locus. DNA from 59 CEPH pedigrees was used to determine the non-disease distribution of allele frequencies at each locus (number of independent chromosomes $= 400$). DNA samples were prepared from Caucasian individuals with IDDM ($n = 60$), MS ($n = 108$), uveitis ($n = 21$), narcolepsy ($n = 28$), or *C2* deficiency ($n = 11$).

### 17.2.2 Microsatellites

The list of citations describing identification and, in some cases, characterization of the microsatellites used in this study are provided in Martin *et al.* (1998). The sequences of the oligonucleotide primers used for amplification will be provided upon request. Amplification and typing of each locus was performed as described previously (Martin *et al.* 1995).

### 17.2.3 HLA typing

*HLA-A*, *-B*, and *-C* were typed using the standard National Institutes of Health microcytoxicity techniques on peripheral blood leukocytes. High resolution *DRB1*, *DQA1*, *DQB1*, and *DPB1* types were determined either by amplification with sequence-specific primers, or by sequence-specific oligonucleotide probe hybridization after PCR amplification of genomic DNA (Kimura and Sasazuki 1991).

### 17.2.4 Statistics

Microsatellite and *HLA* loci were evaluated for allele associations with diseases (with one exception, see below) using Fisher's exact tests. The $p$ value for the exact tests were estimated by adapting a Monte-Carlo resampling strategy (Guo and Thompson 1992). A statistic was estimated from the table of observed allele frequencies in the disease and non-disease (independent CEPH chromosomes) groups, and also from each of 17 000 additional randomly generated tables that had the same marginal totals. The proportion of randomly generated tables that had a statistic greater than or equal to that calculated from the observed table was taken as an estimate of the $p$ value. Uveitis patients were determined to be either B27+ or B27−, as opposed to having complete class I typing, and therefore a G-test for association between uveitis and B27 was performed (Zar 1984).

Stratification analysis between the known disease allele and the microsatellite locus were run in an attempt to identify novel associations independent of the disease allele. Microsatellite alleles were combined for $\chi$-square tests under two circumstances: (1) when both patient and control frequencies were less than 5 per cent in each group and (2) when the per cent difference was less than 5 per cent and the per cent in one population was less than 5 per cent. When a table was relatively sparse, $2 \times k_a$ Fisher's exact tests were run instead of $\chi$-square, where $k_a$ is the adjusted number of microsatellite alleles after combining rare alleles. The combined alleles class never exceeded 20 per cent of the total sample.

## 17.3 Results

### 17.3.1 Disease association with microsatellites neighbouring the known 'disease gene'

DNA from patients with IDDM, MS, narcolepsy, or uveitis was typed at the disease-associated class I or class II gene(s) which had been characterized pre-viously. Significant *DRB1* and *DQB1* associations ($p < 10^{-6}$) were observed for both the IDDM and MS cohorts. *DRB1* was also significantly associated with narcolepsy ($p < 10^{-6}$; *DQB1* was not tested), as was *HLA-B* with uveitis ($p < 10^{-6}$). Thus, the disease cohorts used herein exhibited *HLA* associations

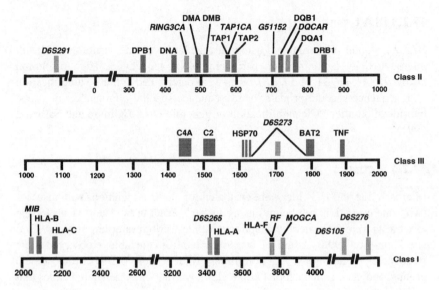

**Fig. 17.1** Map of the human major histocompatibility complex. Microsatellite loci (shown in red and named in italics) are shown relative to defined loci within the *MHC* genomic region.

similar to those described previously. DNA from individuals with these four diseases, as well as from a group of *C2* deficient individuals, were typed at seven to ten microsatellite loci located within or near the *MHC* (see Fig. 17.1 for a map of the *MHC* illustrating the locations of microsatellites used). If levels of linkage disequilibrium between the disease susceptibility alleles of *HLA* genes and a neighbouring microsatellite locus are strong enough, an association between the microsatellite locus and each disease should be observed.

Initially, the ten microsatellite loci were characterized in terms of allele frequencies, heterozygosity, and mutation frequency using DNA from cell lines representing 59 CEPH families (Table 17.1). Allele numbers ranged from 7–16, with an average heterozygosity of 75 per cent. Four mutations observed for locus *D6S276* occurred in a single family, as did the two *MOGCA* mutations. The ten loci tested exhibited a mean mutation rate of $1.1 \times 10^{-3}$ mutations per meiosis. This rate was not significantly different ($p = 0.15$) from that estimated from similar data of Banchs *et al.* (1994), but was significantly higher ($p = 0.004$) than a rate estimate corrected for transformation related mutations (Banchs *et al.* 1994). These results imply that some of the observed mutations probably occurred during the transformation process.

Figure 17.2 illustrates the relationship between physical distance and significance of associations between microsatellite loci and each disease analysed. Significant associations were observed between the recessive *C2* deficiency, and five microsatellite loci flanking the *C2* gene (Fig. 17.2(a)), the furthest of which was 2.3 Mb telomeric to the *C2* gene (*MOGCA*; see Fig. 17.1). However, absolute association was only observed between *C2* deficiency and a single allele at

**Table 17.1**
Characterization of microsatellite loci in the *HLA* complex based on data derived from
59 CEPH families

| Locus | Number of alleles | Heterozygosity (per cent) | Number of mutations | Informative meioses | Mutation frequency |
|---|---|---|---|---|---|
| *D6S276* | 16 | 79 | 4[a] | 1115 | 0.36 |
| *MOGCA* | 15 | 77 | 2[a] | 609 | 0.33 |
| *D6S265* | 14 | 76 | 0 | 630 | – |
| *MIB* | 15 | 82 | 0 | 603 | – |
| *D6S273* | 8 | 78 | 2 | 1113 | 0.18 |
| *DQCAR* | 13 | 82 | 0 | 1106 | – |
| *G51152* | 11 | 81 | 0 | 1200 | – |
| *TAP1CA* | 9 | 58 | 1 | 1093 | 0.09 |
| *RING3CA* | 9 | 73 | 1 | 1105 | 0.09 |
| *D6S291* | 7 | 72 | 0 | 597 | – |

[a] All mutations listed were observed in a single family.

locus *D6S273*, the microsatellite marker closest to the *C2* gene (150–300 kb telomeric to the *C2* gene) (data not shown). Association between *C2* deficiency and
the closest marker centromeric to the *C2* gene, *DQCAR* (a distance of 750 kb),
was also highly significant ($p < 10^{-6}$). However, no association was seen with
either the *TAP1CA* or *RING3CA* markers, which are only an additional 175 kb
and 300 kb, respectively, further centromeric to *C2* than *DQCAR*.

Four microsatellite loci analysed herein were significantly associated with
IDDM (Fig. 17.2(b)); two of these map on opposite sides of the *DQB1* gene
(*DQCAR* and *G51152*) and the other two map near the *HLA-B* and *HSP70-BAT2*
genes (*MIB* and *D6S273*, respectively). No association was observed between
IDDM and *D6S265*, a microsatellite mapping approximately 100 kb centromeric
to *HLA-A*. The *TAP1CA* and *RING3CA* markers were weakly associated with
IDDM, perhaps indicating a functionally significant association between IDDM
and a gene(s) in the vicinity of *TAP1CA* and *RING3CA*, such as *DPB1*.

The clear association between MS and the microsatellite markers closely flanking the *DQB1* locus (*DQCAR* and *G51152*) (Fig. 17.2(c)) supported the large body
of data demonstrating a relationship between the *DR-DQ* loci and MS (Haegert
and Marrosu 1994). A similar situation was observed in the case of uveitis, where
the single microsatellite marker *MIB*, which maps close to the *HLA-B* locus, was
significantly associated with this *HLA-B* related disease (Fig. 17.2(d)). Patterns
of association between microsatellite markers and MS as well as uveitis indicated
that single, defined regions were responsible for these *HLA* associations.

The significance of association between narcolepsy and the microsatellites
*DQCAR* and *G51152* was less pronounced than with the other two *DR-DQ* associated diseases, IDDM and MS, though *DQCAR* was still significant (1 per cent
level) after applying the Bonferroni correction for multiple tests (Fig. 17.2(e)).

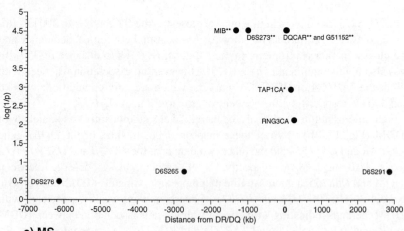
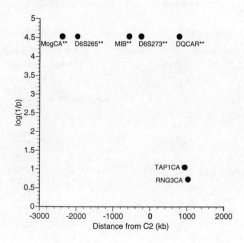
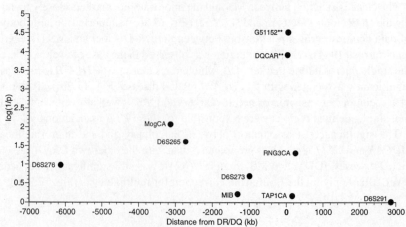

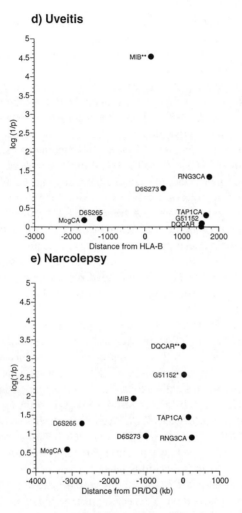

**Fig. 17.2** Relationship between physical distance and significance of association. *p*-values were derived from Fisher exact tests. 5 per cent (*) and 1 per cent (**) significance levels for each disease were determined by adjusting for number of loci using the Bonferroni correction. Numbers of individuals used are: (a) *C2*, *n* = 11; (b) IDDM, *n* = 60; (c) MS, *n* = 108; (d) Uveitis, *n* = 21; (e) Narcolepsy, *n* = 28.

## 17.3.2 Use of microsatellites to detect additional, novel disease loci

Having established that microsatellite markers can identify regions of primary disease association in the *MHC*, it was of interest to determine whether these markers could detect additional regions of disease association that have not been previously defined. Determining the influence of a second locus upon disease predisposition in a region having extensive disequilibrium presents particular challenges. Confounding effects include: (1) the possibility that the predisposing allele at the second locus may be in positive or negative disequilibrium with the primary predisposing allele; (2) differential magnitude of effects between the two loci; and (3) different modes of inheritance, including the interaction of the two loci in causing the disease state.

MS, IDDM, and narcolepsy all have well established associations with variation at the class II *DR/DQ* loci. Other loci within the *MHC* contribute to susceptibility to IDDM (Robinson *et al.* 1993), although the precise number and identification of these genes remain speculative, due to the strong linkage disequilibrium between pairs of loci within the complex. Disease gene candidates located telomeric to the *DR-DQ* loci include *HLA-B*, *TNF*, *MHC*-linked *heat shock protein-70* genes, and/or unidentified genes mapping near the boundary of the class I and III regions of *HLA*. Additional regions of disease association for MS and narcolepsy apart from the *DR-DQ* region have not been reported.

If variation at a microsatellite is associated with disease, then three explanations are possible: (1) the association is due to linkage disequilibrium with a known disease susceptibility allele; (2) the association is due to disequilibrium with a known disease locus and/or an additional, novel disease locus; or (3) there is Type I statistical error. Three association tests were used to differentiate explanations (1) and (2) as follows.

Test I determines the level of linkage disequilibrium between the known disease allele and a microsatellite marker using a control population (independent CEPH haplotypes) in a $2 \times k$ contingency table consisting of individuals with or without the known disease allele in one dimension, stratified on the k alleles of the microsatellite locus in the other dimension. For example, *DQCAR*, *D6S273*, and *MIB* are in strong disequilibrium ($p < 0.001$) with *DRB1*1501*; and *RING3CA*, *DQCAR*, *D6S273*, and *MIB* are in disequilibrium with *DRB1*03/DRB1*04* (Table 17.2, Test I). *DQCAR* is located between *DRB1* and *DQB1* (Fig. 17.1), and shows the anticipated high disequilibrium with these class II loci. If all *HLA* associations can be accounted for by *DRB1*04* and *DRB1*03* in IDDM and *DRB1*1501* in the case of narcolepsy and MS, then no additional *DQCAR* association is expected after the removal of the influence of *DRB1* through disequilibrium.

Two separate tests were employed to determine the level of association, independent of the disease allele, between the microsatellites and the diseases. Test II (Table 17.2) compares the microsatellite allele distribution in patient and control groups with only individuals positive for the disease allele, and Test III employs only individuals lacking the disease allele. If disequilibrium is present (Test I significant), and no independent disease effect exists at the microsatellite locus, then both Tests II and III will be negative. For MS and narcolepsy, no significant differences in microsatellite allele distribution between control and patient population with genotypes *1501/1501* plus *1501/X* (Table 17.2, Test II) or non-*1501* genotypes (Table 17.2, Test III), where *X* represents non-*1501* alleles, were observed. Thus, all association between the microsatellites and these two diseases can be accounted for by their disequilibrium with *DRB1/DQB1*. If disequilibrium

**Table 17.2**
Tests of disequilibrium and independent genetic effects among microsatellites for three *HLA* diseases. *P* values for disease association tests

| Disease | Disease allele | Microsatellite | I disequilibrium (total control counts) | II 'disease allele' genotypes only (control: patient counts) | | | III 'not-disease allele' genotypes only (control: patient counts) |
|---|---|---|---|---|---|---|---|
| Multiple sclerosis | DRB1*1501 | | (314) | *1501/1501, 1501/X* (86:124) | | | *XX* (228:92) |
| | | RING3CA | 0.078 | 0.187 | | | 0.309 |
| | | DQCAR | 0.001 | 0.106 | | | 0.103 |
| | | D6S273 | 0.001 | 0.184 | | | 0.417 |
| | | MIB | 0.001 | 0.089 | | | 0.148 |
| | | D6S265 | 0.611 | 0.345 | | | 0.220 |
| | | MOGCA | 0.460 | 0.609 | | | 0.058 |
| Narcolepsy | DRB1*1501 | | (314) | *1501/1501, 1501/X* (86:44) | | | *XX* (228:12) |
| | | TAP1CA | 0.007 | 0.214 | | | 1.000 |
| | | RING3CA | 0.078 | 0.109 | | | 0.464 |
| | | DQCAR | 0.001 | 0.418 | | | 0.161 |
| | | MIB | 0.001 | 0.975 | | | 0.316 |
| | | D6S265 | 0.611 | 0.092 | | | 0.471 |
| | | MOGCA | 0.461 | 0.602 | | | 0.555 |
| IDDM | DR3, DR4 DRB1*03 DRB1*04 | | (314) | *DRB1*03/04* (6:38) | *DRB1*03/Y* (42:24) | *DRB1*04/Z* (96:28) | *DRB1*X/Y* (136:10) |
| | | RING3CA | 0.001 | 0.407 | 0.064 | 0.019 | 0.443 |
| | | DQCAR | 0.001 | 0.127 | 0.161 | 0.137 | 0.798 |
| | | D6S273 | 0.001 | 0.637 | 0.367 | 0.481 | 0.178 |
| | | MIB | 0.008 | 0.515 | 0.133 | 0.012 | 0.889 |
| | | D6S265 | 0.709 | 0.885 | 0.286 | 0.296 | 0.062 |

For multiple sclerosis and narcolepsy, Tests II and III contrast patient and control genotype frequencies with and without the known class II disease susceptibility allele (*DRB1*1501*). For IDDM four genotypes are tested. '*X*' is any allele at the disease susceptibility locus except for the disease allele(s). *Y* is '*not DRB1*04*', *Z* is '*not DRB1*03*'.

is present (Test I significant), and an independent disease effect is also present at the microsatellite, then given sufficient power, Tests II or III will be significant. An independent *RING3CA* signal is present for IDDM with genotypes *DRB1\*03/Y* and *DRB1\*04/Z* (Test II), as is the case for *MIB* in the *DRB1\*04/Z* comparison. These results hint at possible *HLA* associations with IDDM apart from *DR/DQ* and warrant further investigation.

## 17.4 Discussion

Microsatellite markers were capable of identifying regions of disease association in the *MHC* for all five diseases studied, indicating their usefulness in identifying potential *HLA* associations with diseases not already known to have an *HLA* association. Identification of appropriately located microsatellite loci near (<100 kb) most of the polymorphic *HLA* genes functioning in antigen presentation, as well as other genes in the *MHC*, allows a thorough and rapid analysis of disease association across the complex.

Data obtained from C2-deficient individuals showed that markers telomeric to the *C2* gene by distances as great as 2.3 Mb were in strong disequilibrium with the mutant *C2* allele. Low frequency of recombination across the class I region (Martin *et al.* 1995) might have contributed to the strength of association between this disease and markers flanking the telomeric end of class I. The abrupt decline in significance of association between *C2* deficiency and some point between *DQCAR* and *TAP1CA* (Fig. 17.2(a)) is consistent with reports that a hotspot(s) for recombination exists within this segment of DNA (Cullen *et al.* 1995). Given that all C2-deficient patients tested are homozygous for a single allele at *D6S273*, an estimate for the age of the *C2* deficiency mutation can be calculated with 95 per cent confidence using the equation $p = 0.05 = (l - p_{cs})^{ml}$, where $m$ is the number of meioses, and $l$ is the number of independent chromosomal lineages. $p_{cs}$ is the probability of a change in allelic state at *D6S273* and is estimated from $p_{cs} = m + \theta f_0$, where $m$ is the mutation rate at *D6S273*, $\theta$ is the recombination fraction between *C2* and *D6S273*, and $f_0$ is the population frequency of all other *D6S273* alleles (those not observed in C2-deficient patients). Given 1 per cent recombination per megabase per meiosis and published dinucleotide mutation rates (Weber and Wong 1993), we solve the first equation for $m$ and estimate that the *C2* deficiency mutation occurred within the last 242 generations.

Associations between narcolepsy and microsatellite loci located on either side of *DQB1* were less pronounced than those seen with the two other *DR/DQ*-associated diseases studied. Assuming that the *DR/DQ* region contains a gene responsible for an association between *HLA* and narcolepsy, numerous mutations at the *DQCAR* and *G51152* repeat loci could account for the relatively weak association between the disease and these markers if the disease is sufficiently old. Alternatively, another novel gene telomeric to *DRB1* in a region relatively uncharacterized may contribute to susceptibility to narcolepsy, though data suggests that this is unlikely to be the case (Mignot *et al.* 1997).

Many haplotypes composed of two or more *MHC* loci have been observed to be in strong linkage disequilibrium (Imanishi *et al.* 1992), complicating the identification of specific disease-associated loci within the *MHC*. A second issue we addressed using microsatellite typing data was the identification of disease-associated loci in addition to and independent of the known disease locus. This analysis requires large sample sizes due to the relative absence of disease genotypes in controls (e.g. *DRB1\*03/DRB1\*04* heterozygotes in controls for IDDM), and non-disease genotypes in patients (e.g. the *DRB1\*X/X* class in narcolepsy and IDDM). Power may have been insufficient in this study to detect novel associations with loci having very weak effects, but the exploratory data analysis indicated a lack of additional components of disease susceptibility for MS, narcolepsy, and IDDM near the six microsatellite loci tested. Haplotypes derived from family data may be more informative in this analysis, and fewer samples would be required to obtain adequate power to detect weak associations.

It is unlikely that microsatellite markers are themselves conferring susceptibility to disease nor, accordingly, are they constrained by direct selective pressure. On the other hand, levels of linkage disequilibrium between microsatellite markers and closely linked genes (depending on recombination fractions in the region) appear to be very high, as implicated by the data reported herein. Therefore, microsatellites serve as useful markers for identification of regions within the *MHC* which are associated with specific diseases. Our rapidly increasing knowledge of linkage disequilibrium and recombination fractions across the *MHC* strengthens proper interpretation of disease gene mapping in *HLA*. We can apply this information in determining which loci are truly associated with a particular disease, as opposed to those that are simply hitchhiking with a linked disease gene.

# Acknowledgements

We wish to thank Michael Cullen and George Nelson for helpful discussions. This work was supported in part by grants from the Swedish Medical Research Council and the Swedish National Association against Rheumatism (L.T. and G.S.), and NIH grants GM35326 (L.B. and W.K.) and GM56688 (G.T. and J.C.).

# 18 Microsatellites, a neutral marker to infer selective sweeps

Christian Schlötterer and Thomas Wiehe

## Chapter contents

## Abstract

In recent years microsatellites have become popular as a highly variable neutral marker. In natural populations, microsatellite variability is determined by new mutations, genetic drift, and selection at linked chromosomal regions. In this chapter, we focus on the consequences of directional selection on observed microsatellite variability in natural populations. A test statistic is introduced that permits the identification of such 'selective sweeps'. The test is designed to detect reduced genetic variation in a specific locus–population combination from a data set of several microsatellite loci in multiple populations. Theoretical considerations, however, predict that selective sweeps may only be detected where microsatellites have a mutation rate which is below a certain threshold.

## 18.1 Introduction

Until now, microsatellites have been mainly used for genetic mapping (Dib *et al.* 1996; Dietrich *et al.* 1996), paternity testing (Queller *et al.* 1993; Schlötterer and Pemberton 1998) and estimating genetic variation in natural populations (Bruford and Wayne 1993*b*). Especially for the latter application, one major assumption is the effective neutrality of microsatellite variability. A common strategy is to select highly variable microsatellite loci and to type them in a set of populations. The average variation at these loci is used as an estimator of genetic variability within and between populations. Based on such data sets, inferences on population structure and demographic processes have been made for a wide range of species. Furthermore, in conservation genetics this information is now widely used for the implementation of strategies to conserve biodiversity. In this chapter we focus on how directional selection at linked loci may cause microsatellite variation in natural populations to deviate from their neutral expectations.

## 18.2 The concept of selective sweeps

Consider a gene which acquires a novel mutation that is selectively favoured over the ancestral allele. Natural selection will spread the novel mutation in the population until it eventually becomes fixed. The timespan until an advantageous mutation becomes fixed depends on its selective advantage and on the effective population size. Fixation times of strongly selected mutations may range from less than one hundred to several thousand generations. While the new allele is being fixed, polymorphism in the flanking region may be wiped out. Therefore, such a 'selective sweep' leads to a reduction in genetic variation in this region.

At the molecular level, the advantageous effect of an allele may be caused by a single point mutation. In this case, directional selection is acting on a single nucleotide change. While nucleotide polymorphism in the flanking region may be selectively neutral, close linkage to the selected site causes the fixation of the flanking region. This phenomenon has been termed hitchhiking (Maynard Smith and Haigh 1974). The size of the flanking region which is subject to this hitchhiking effect depends on the selection coefficient of the favourable allele and the recombination rate. When the recombination rate is high, or when the newly introduced allele has a low selective advantage, only a small portion of the flanking region will be affected. Low recombination rates or large selection coefficients increase the affected genomic region. Immediately after a selective sweep, levels of variation will be significantly reduced in the genomic region linked to the substitution. This footprint of a selective sweep will gradually disappear by the accumulation of novel mutations. However, until the mutation–drift equilibrium is restored, there is a surplus of rare variants.

When the genomic region over which hitchhiking occurs is large, a subsequent selective sweep is likely to happen before natural variation has recovered in this region, and possibly even before the previous sweep is completed. In regions of very low recombination, a permanent reduction of natural variation can be caused by the occurrence of multiple selective sweeps. It should be noted that the same pattern of reduced variation in regions of low recombination is expected for reproductively isolated populations. In these populations, multiple independent selective sweeps may fix different alleles, but a large region of the chromosome surrounding the selected site will be affected. Thus, even though there may be sequence differences between populations, the level of variation will be low.

On the other hand, the loss of variation in regions of intermediate and high recombination will be less severe and more transient. Because selective sweeps are rare, it is expected that only a few populations have reduced variation at a given chromosomal segment, while the majority of populations maintain their average level of variation. Below, we explain how this difference in variability could be used to infer selective sweeps in a population-based assay.

## 18.3 Microsatellites and selective sweeps

Microsatellites can be regarded as neutral markers, which are randomly distributed over the euchromatic part of the genome. Models of microsatellite evolution which incorporate the stepwise mutation process of microsatellites permit a good prediction of the microsatellite variation at mutation–drift equilibrium. Hence, test statistics can be designed which allow the detection of deviations from this equilibrium.

Given the random distribution of microsatellites and the relative ease of their analysis, it is possible to use microsatellite markers from the entire genome (Dib *et al.* 1996; Dietrich *et al.* 1996), covering a large spectrum of recombination frequencies and chromosomal environments. In the wake of the various genome projects, improved genetic maps have already become available for a wide range of organisms, providing the basis for reliable estimators of recombination rates in the genome. Screening the levels of genetic variation at different chromosomal positions, and comparison of multiple, reproductively isolated populations is a promising way to identify selective sweeps.

## 18.4 Detecting selective sweeps with microsatellite analysis

### 18.4.1 The effect of a selective sweep on the variance of the microsatellite allele distribution

A deterministic model of hitchhiking was first developed by Maynard Smith and Haigh (1974). It predicts that a selective sweep reduces heterozygosity at a neutral locus by an amount which is inversely proportional to $r/s$, the ratio of the recombination rate ($r$) between a neutral marker and the selected locus and the selective advantage ($s$). Since the selection coefficient determines the time for fixation of an advantageous substitution, the reduction in heterozygosity may be written equivalently as a product of the recombination rate $r$ and the fixation time $t$. Let $t_0$ and $t_1$ be the times of beginning and completion of the selective sweep, respectively. Then, $t = t_1 - t_0$. Let $H$ be heterozygosity at the neutral adjacent locus. Then

$$\frac{H(t_1)}{H(t_0)} = 1 - \exp(-rt) \approx \min\{rt, 1\}. \tag{18.1}$$

If the ratio is zero, then variation has been eliminated completely; if it is one, then it remained constant during the selective sweep. It is possible to generalize eqn (18.1) and to determine the expected long-term equilibrium level of heterozygosity under recurrent selective sweeps. Predictions of this model can be applied to data of nucleotide variability, and yield estimates of various population genetic quantities.

A new problem arises when the simple hitchhiking model is applied to cases where microsatellites are used as markers. The fast mutation process at these loci (mutation rates of $10^{-5}$ to $10^{-3}$ per generation and locus have been documented for mammals; Dallas 1992; Ellegren 1995; Weber and Wong 1993) quickly erases the footprints which a selective substitution would otherwise leave in its chromosomal neighbourhood. In particular, the mutation process needs to be incorporated into the theoretical analysis and may not be neglected. We have studied a generalized version of Maynard Smith and Haigh's model (Wiehe 1998) which accounts for this effect. Instead of heterozygosity, the variance ($V$) in the number of repeats is used as another measure of variability at a microsatellite locus.

$$V = \sum_i (i - \bar{i})^2 p_i, \tag{18.2}$$

where $p_i$ indicates the frequency of alleles which have $i$ motif repeats, and $\bar{i}$ denotes the mean number of repeats. Reduction of variability due to a selective sweep can then be described by a formula which is very similar to Eqn (18.1), but which includes the mutation rate $\mu$ and the number of neutral alleles $n$ as additional parameters:

$$\frac{V(t_1)}{V(t_0)} = 1 - \exp\left(-\left(r + 4\frac{\mu}{n}\right)t\right) \approx \min\left\{\left(r + 4\frac{\mu}{n}\right)t, 1\right\} \quad \text{(Wiehe 1998)}. \tag{18.3}$$

Equation (18.3) shows that not only recombination but also mutation reduces the hitchhiking effect. In particular, if $\mu$ is large compared to $r$, then mutation may contribute more than recombination to this dilution. Note that even in regions of no recombination, traces of hitchhiking will eventually be wiped out due to mutation. The grey shaded areas in Fig. 18.1 show for which combination of mutation (abscissa) and recombination rates (left ordinate) a selective sweep produces a reduction of variability by at least 10 per cent. There is a critical mutation rate

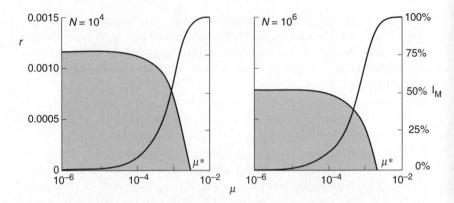

**Fig. 18.1** Effect of a single hitchhiking event with strong selective advantage ($s = 0.01$). If recombination rate ($r$, left ordinate) and mutation rate ($\mu$, abscissa) are within the shaded area, then the variance of the allele distribution at the microsatellite locus is reduced to 90 per cent or less of its neutral value as a result of hitchhiking. The right ordinate ($I_M$) shows to what percentage the variance is reduced in the case of complete linkage ($r = 0$). Other parameters: Population size $N = 10^4$ (left panel), $N = 10^6$ (right panel), number of alleles at the ms locus $n = 10$.

$\mu^*$ beyond which there is virtually no hitchhiking, independent of the recombination rate. Similarly, there is a critical recombination rate $r^*$, beyond which no hitchhiking can be observed. The same figure shows (black line and right ordinate) the relative influence of mutation ($I_M$) upon the hitchhiking effect: a value of zero means that there is no difference to the simple hitchhiking model, a value of one means that—as a result of mutation—there is no hitchhiking. Even when selection is strong, as we assume here ($s = 0.01$), then the critical mutation rate is close to what is typically observed in humans. Therefore, the chances of tracing hitchhiking events should be much better for organisms where the microsatellite mutation rate is lower, as in *Drosophila melanogaster*. Only two mutations were observed for 45 microsatellite loci in 30 mutation accumulation lines maintained for 200 and 230 generations (Schug *et al.* 1997a,b). This estimated mutation rate of $6.5 \times 10^{-6}$, which is at least one order of magnitude lower than in mammals, has recently been confirmed in a completely different set of lines and microsatellites (Schlötterer *et al.* 1998). With such low mutation rates, selective sweeps in regions of low recombination should be detectable. It should be noted, however, that variation in microsatellite mutation rates depending on repeat motif (Chakraborty *et al.* 1997), repeat length (Jin *et al.* 1996; Schlötterer *et al.* 1998), and chromosomal position (Harr *et al.* 1998) could obscure any patterns generated by hitchhiking effects (see below).

## 18.4.2 Reduced genetic variation in regions of low recombination

Significant variation in recombination rates has been described for *Drosophila melanogaster*, with the tips of the chromosomes and the centromeric regions exhibiting extremely low levels of recombination (Lindsley and Sandler 1977).

Genes located in chromosomal regions covering a wide spectrum of recombination rates show a significant correlation between nucleotide variation and recombination rate (Begun and Aquadro 1992). Interestingly, cross-species comparisons do not show significant differences between genes located in regions of high and low recombination rates. This indicates that the differences in nucleotide variability are not caused by variation in mutation rates. The best explanation is that selection has reduced variation in regions of low recombination below the neutral expectation. Two different selection processes have been suggested, hitchhiking (Aguadé *et al.* 1989; Begun and Aquadro 1992) and background selection (Charlesworth *et al.* 1993). Hitchhiking reduces variation in regions of low recombination through recurrent fixation of new advantageous alleles. Background selection operates through the recurrent removal of chromosomes with deleterious mutations from the population. In principle, the impact of these processes can be distinguished by comparing two different estimators of neutral variation, the number of segregating sites and the average number of pairwise differences. The two processes make different predictions for the distribution of the alleles in these regions (Table 18.1). The effect of background selection on the gene genealogy is analogous to that of a reduced effective population size. The effect of a selective sweep, on the other hand, is analogous to a population expansion resulting in a pattern of a star-like phylogeny, with an over representation of rare alleles. Tajima's D (Tajima 1989), a test statistic which is based on the normalized difference between the two estimators, should be negative under the hitchhiking model, as selective sweeps generate an excess of rare sites.

However, despite clear predictions, this issue has not yet been solved by DNA sequence analysis. One of the reasons is the lack of sufficient mutations in regions of low recombination to favour one hypothesis over the other. With more

**Table 18.1**

| | Mechanism | Affect on observed variability in regions of low recombination rates |
|---|---|---|
| **Background selection** | removal of deleterious alleles *could be compared to a permanent reduction of effective population size* | nucleotide polymorphism: low microsatellite polymorphism: reduced |
| **Selective sweeps** | fixation of an advantageous allele by directional selection *could be compared to a population expansion resulting in a star-like phylogeny* | nucleotide polymorphism: low, excess of rare alleles, microsatellite polymorphism: no or very little difference to loci in high recombination regions |

polymorphisms available, this question should be solved. As microsatellites have a higher mutation rate than base substitutions, analysis of microsatellite polymorphism in regions of low recombination is a very promising tool. Microsatellites with high mutation rates allow for microsatellite-specific predictions under the background selection and hitchhiking model. While in regions of low recombination the frequency of selective sweeps is too high to allow the recovery of nucleotide polymorphism between two subsequent sweeps, high microsatellite mutation rates ($10^{-5}$–$10^{-3}$) result in a restoration of microsatellite variability. Background selection, on the other hand, is an equilibrium process, and has the same effect as a reduced effective population size, and will therefore result in a permanent reduction of microsatellite variability in low recombining regions.

Michalakis and Veuille (1996) were the first to compare microsatellites located in genomic regions with different recombination rates. Eleven coding trinucleotide loci were studied in four natural *Drosophila melanogaster* populations, and no correlation between recombination rate and microsatellite variation was observed in individual populations or in pooled data sets. The authors attribute their result to a high mutation rate of microsatellites (Michalakis and Veuille 1996), which would be consistent with the hitchhiking model. Given the recent result of low mutation rates of *Drosophila melanogaster* microsatellites (Schug *et al.* 1997*a*; Schlötterer *et al.* 1998), the original explanation needs further justification. An alternative explanation might be that the loci analysed are not behaving uniformly, which could obscure a reduction in variability. If one or more highly mutable loci are located in a low recombining region, a reduction in variability would not have been detected. Such differences could be caused by either differential selective constraints on microsatellite variation due to the protein coding of the loci analyzed, or general variation in mutation rates between loci. The latter hypothesis is supported by reported differences in mutation rates between microsatellite loci (Weber and Wong 1993; Chakraborty *et al.* 1997; Harr *et al.* 1998).

Interestingly, a later study which used eighteen dinucleotide repeats in a *Drosophila melanogaster* population from Maryland, USA, found a significant correlation between recombination rate and microsatellite variability (Schug *et al.* 1997*b*). More data on microsatellite mutation rates and mutation patterns are needed to resolve the impact of background selection and hitchhiking in *Drosophila melanogaster*. Alternatively, microsatellite variability could be studied in other organisms with a higher microsatellite mutation rate than in *Drosophila melanogaster*, such as humans or mice.

# 18.5 Identification of local selective sweeps

## 18.5.1 Local selective sweeps in *Drosophila melanogaster*

The phenomenon of reduced natural variability in regions of low recombination is quite general, and may be caused by the succession of independent selective sweeps during evolution. It can be studied relatively easily in natural populations, provided such selective sweeps occur at a high enough rate and with a sufficient selective advantage. However, it is much more difficult to identify single selective sweeps in regions of intermediate to high recombination rates. Knowledge about the chromosomal region and the affected population is required for a detailed analysis. Currently, only a few instances are known where this information is available. Taylor *et al.* (1995) recently used pyrethroid insecticide resistance to compare patterns of variability at the locus causing resistance with those patterns in a genomic region which is not involved in the resistance pathway. The authors show that directional selection has reduced allelic diversity at the resistance locus by removing uncommon alleles. The control locus did not follow this pattern, ruling out other population-specific phenomena. A similar example is the pattern of variability at the *Superoxide dismutase* (*SOD*) locus, which suggest that a recently arisen polymorphism has been rapidly driven to intermediate frequency (Hudson *et al.* 1997). As expected when strong directional selection is acting, Hudson *et al.* (1997) observe that a region of more than 20 kb has been included in the affected region. Both studies indicate that local selective sweeps exist in natural populations. Without *a priori* information on candidate regions, however, these sweeps may have remained undetected. Recently, a novel screening strategy has been suggested, which could lead to the discovery of genes involved in local adaptation (Schlötterer *et al.* 1997).

Natural populations are constantly challenged by selection pressures required for adaptation to the local environment. Many of these genes are very unlikely to be detected in a traditional screening approach, because their effects are too subtle to be seen in the laboratory, or the appropriate selection scheme is not yet known. This problem is best illustrated by the example of yeast, where the knockout of the identified genes sometimes fails to uncover a specific phenotype (Oliver *et al.* 1992). The proposed screening strategy is not dependent on phenotypes which could be scored in the laboratory, but makes use of the pattern which selection imposes on genetic variability in natural populations. Levels of variation at many neutral markers, such as microsatellites, which are distributed all over the genome, can be analyzed in several natural populations. Deviations from neutral expectations can be detected with the test statistic described below. Given the fact that the footprints of a selective sweep will eventually be wiped out by the accumulation of new mutations, only recent selective sweeps can be detected.

## 18.5.2 Test statistics to infer local selective sweeps

The mutational behaviour of a locus consisting of tandemly repeated elements, such as microsatellites, is well described by the stepwise mutation model, which rests on earlier work by Moran (1975). One interesting observation is that in populations in mutation–drift equilibrium, the mean number of repeats at a given locus never reaches an equilibrium value. Variances in repeat number, however, converge with time to an equilibrium value. An appropriate test statistic for selective sweeps would be the comparison of the expected equilibrium variance in repeat number to the observed value. One potential complication arises from the fact that the neutral variance $V$ of a given microsatellite is dependent on the effective population size $N_e$ and the mutation rate $\mu$ of the microsatellite locus:

$$V = 2N_e\mu \quad \text{(Moran 1975)}. \tag{18.4}$$

Both parameters could vary considerably. Hence, to calculate the expected variance of a given locus in one population, one needs good estimates of the mutation rate and the population size. When data are available for $L$ microsatellite loci in $P$ populations, the following test statistic can be employed: Calculate $v_{pl}$, the logarithms of the variances in repeat number for each combination of a population and a locus. Obtain the means of the observed log variances across populations $(v_{p.})$ and across loci $(v_{.l})$ and the grand mean $(v_{..})$. With the stepwise mutation model and under mutation–drift equilibrium the expected variance for a given combination of a locus and a population is:

$$u_{pl} = v_{p.} + v_{.l} - v_{..}. \tag{18.5}$$

Therefore, the deviation from predicted values can be calculated as:

$$(v_{pl} - u_{pl}) \frac{PL}{(P-1)(L-1)\sqrt{\hat{V}[v_{pl}]}}, \tag{18.6}$$

where $L$ is the number of loci, $P$ the number of populations, and $\hat{V}[v_{pl}] \approx 0.85$ (Schlötterer *et al.* 1997). The level of significance $\alpha$ can be looked up from a table for the normal distribution. To account for multiple testing, the specified error rate $\alpha_E$ can be corrected by the Bonferroni method: $\alpha_E = \alpha/((K-1)(L-1))$ for a one sided test and $\alpha_E = \alpha/(2(K-1)(L-1))$ for a two-sided test.

### 18.5.3 Local selective sweeps in *Drosophila melanogaster* inferred by microsatellite analysis

*Drosophila melanogaster* is especially appropriate for the study of selective sweeps in natural populations. The reported large effective size of natural *Drosophila melanogaster* populations ensures that the fate of a significant fraction of advantageous mutations will be determined by natural selection, rather than by genetic drift. Furthermore, *Drosophila melanogaster* spread from Africa over the rest of the world less than 10 000 years ago (David and Capy 1988). During this colonization, local *Drosophila melanogaster* populations are expected to have adapted to a broad variety of local environments. Hence, adaptation should also have involved a number of selective sweeps. Apart from regions of low recombination, natural variation will be reduced at different loci in different populations. Recently, Schlötterer *et al.* (1997) screened ten microsatellite loci in natural *Drosophila melanogaster* populations from six different geographic regions (Schlötterer *et al.* 1997). For several loci they observed a low heterozygosity in a given population. As predicted for local selective sweeps, levels of polymorphism were not affected at other loci or populations. By applying the test statistic outlined above, one locus, DS01001, had a significantly reduced variation in an Indian population but nowhere else. This reduction is best explained by a selective sweep in a chromosomal region adjacent to the microsatellite DS01001. For three further locus/population combinations the authors found a single allele which was fixed in the particular population. As the logarithm of zero is not defined, an arbitrary value of 0.1 was added to all variances in order to apply the above test. With this procedure, only DS01001 had a significant reduction in variability (Schlötterer *et al.* 1997).

With the availability of many mapped *Drosophila melanogaster* microsatellites, a large scale survey of microsatellite loci in *Drosophila melanogaster* populations from different habitats will provide an efficient strategy to infer selective sweeps and to characterize genomic regions involved in local adaptation.

## 18.6 Perspective

### 18.6.1 Molecular evolution

Microsatellites provide a powerful tool to study the selective forces acting on the genome in greater detail. While we have concentrated on genetic variability as an indicator of selective events, this procedure may have its limits under certain circumstances. In reproductively isolated populations, the effective population sizes may become quite small, and therefore it is harder to disentangle systematic forces, such as directional selection, from random forces, such as genetic drift. Estimates of parameters become less accurate in this case. However, reproductive isolation provides additional means, other than overall variability, to detect the presence of hitchhiking: it can be shown (Slatkin and Wiehe 1998) that, under some conditions, selective sweeps raise Wright's $F_{st}$ statistic, indicating an excess of differentiation with respect to what is expected under neutrality. It is evident that such elevated $F_{st}$ values would be restricted to the chromosomal neighbourhood of a putative selective substitution, and would also depend on the recombination rate in the region.

### 18.6.2 Conservation genetics

The conservation of genetic variability and the identification of independent genetic units are among the primary goals in conservation genetics. Genetic distances are based on the idea that random mutations and genetic drift are mainly responsible for the differentiation of populations over time. Neutral markers should thus be able to provide a good measure of differentiation between populations. That this assumption may not always be fulfilled was demonstrated by a recent study measuring gene flow and morphological divergence in twelve populations of a common rain forest bird, *Andropadus virens*. The authors showed that a lack of genetic divergence measured by microsatellite variability coincides with significant morphological divergence (Smith *et al.* 1997). This result is particularly important, because it implies that the absence of both morphological and molecular divergence cannot be taken as a guarantee that the populations have not diverged. By searching for reduced genetic variability in specific locus/population combinations, selective sweeps, which are important for local adaptation, may be detected. In combination with the traditional genetic distances, this approach may be very useful in conservation biology as long as population sizes are not very small (see above).

### 18.6.3 Breeding

Many of the interesting genetic traits in animal and plant breeding are quantitative trait loci. Recently, it has been shown in *Drosophila melanogaster* that much of the variation affecting bristle number, a classic quantitative trait in *Drosophila*, is attributable to alleles with large phenotypic effects at a small number of candidate loci (Mackay 1995). To explain the strong selective response at quantitative trait loci (QTLs), a high mutation rate at a small number of loci must be assumed. Selection will increase the frequency of new favourable alleles, resulting in hitch-hiking of flanking markers, such as microsatellites. If QTLs in economically important species follow the same pattern, then reduced microsatellite variability could also serve as a mapping strategy for QTLs. Most importantly, the identification of QTLs would not be dependent on segregation in the study populations, as homozygous loci will be highly informative.

## Acknowledgements

Many thanks to M. Schug and M. Nachman for helpful comments on the manuscript. The laboratory of C.S. is supported by grants from the European Community (Biotechnology) and the Fonds zur Förderung der Wissenschaften (FWF).

# 19 Y chromosome microsatellite haplotypes and the history of Samoyed-speaking populations in north-west Siberia

T. Karafet, L. P. Osipova, O. L. Posukh, V. Wiebe and M. F. Hammer

## Chapter contents

## Abstract

We employed a set of five microsatellites and five biallelic polymorphic sites mapping to the non-recombining portion of the Y chromosome to (1) investigate the paternal relationships among linguistically related populations from western Siberia, and (2) assess the relative utility of these two marker systems for inferring the structure of recently diverged populations. The surveyed populations included three Samoyed-speaking groups (Selkups, Forest Nentsi, and Tundra Nentsi) from nine villages along the

Pur, Taz, and Yenisey River basins, three non-Samoyed-speaking populations (Komi, Kets, and Evenks) inhabiting the same villages, and a sample of Altais from south-western Siberia (the hypothesized territory of ancient Samoyeds). We found that genetic and linguistic patterns were not entirely concordant. In particular, the Selkups were very different from the Tundra Nentsi and Forest Nentsi, and contained a considerable paternal component from southern populations. Separate analyses of the different marker systems allowed us to infer that a small set of microsatellites was more useful than the same number of biallelic sites for resolving relationships among closely related populations. By combining information from both marker systems we obtained higher resolution haplotype networks and more consistent clustering of Samoyed sub-populations.

## 19.1 Introduction

Over the last three decades genetic approaches have greatly aided efforts to address questions concerning the genetic structure, evolution, migration, and adaptation of populations in northern Asia (Rychkov 1969; Sukernik *et al.* 1978, 1980; Rychkov and Sheremet'eva 1980; Szathmary 1981; Crawford and Enciso 1982; Karafet *et al.* 1994). Earlier studies based on blood groups and classical markers showed the existence of a relationship among geography, linguistic, and genetic variation (Karafet 1986; Cavalli-Sforza *et al.* 1994; Crawford *et al.* 1997). More recently, mtDNA has been used to trace matrilineal genetic links between aboriginal populations in Siberia and the Americas (Torroni *et al.* 1993; Shields *et al.* 1993; Petrishchev *et al.* 1993; Sukernik *et al.* 1996). Studies of polymorphisms on the non-recombining portion of the human Y chromosome (NRPY) provide the opportunity to compare maternally- and paternally-inherited patterns of variation, and to complement studies of biparental nuclear polymorphisms. Indeed, investigations of variation on the NRPY are beginning to offer insights into population history that were not apparent from studies of other kinds of markers (Hammer and Horai 1995; Underhill *et al.* 1996; Cavalli-Sforza and Minch 1997; Poloni *et al.* 1997; Zerjal *et al.* 1997; Hammer *et al.* 1998).

Genetic variation in the form of point mutations and insertion/deletions (indels) has been useful for investigating Y chromosome evolution during the past $\sim 200\,000$ years (Jobling and Tyler-Smith 1995; Hammer *et al.* 1997; Underhill *et al.* 1997). On the other hand, more highly mutable markers such as microsatellites are most suitable for examining recent events and the relationships among closely related populations. Currently, 23 microsatellites have been mapped to the NRPY, several of which have now been examined in human populations (Jobling and Tyler-Smith 1995; Hammer and Zegura 1996). These markers, including di-, tri-, tetra- and pentanucleotide repeats, show high levels of heterogeneity within and between populations, and appear to be as variable as their autosomal counterparts (Ciminelli *et al.* 1995; Roewer *et al.* 1996; de Knijff *et al.* 1997; Zerjal *et al.* 1997). The inability of the NRPY to undergo recombination means that the combination of variants at several Y-linked markers can be considered as haplotypes. We have taken advantage of this property of the NRPY to compare the usefulness of haplotypes based on diallelic markers, microsatellites, and the combination of both kinds of markers.

## 19.1.1 Samoyed populations

The peoples of north-west Siberia speak different languages belonging to the Uralic linguistic family. The Uralic family includes the Finno-Ugric and Samoyedic languages, which are spoken in a vast territory of Northern Eurasia. Within the Samoyeds two main linguistic branches are distinguished: Northern Samoyedic (Nentsi, Entsi, and Nganasans) and Southern Samoyedic (Selkups). In Siberia, Samoyed peoples lead a nomadic or semi-settled existence in the forest and tundra zone between the Ural Mountains and the Yenisey River. It is generally accepted by Russian ethnographers that Nentsi, Entsi, Nganasans, and Selkups are descendants of Palaeoasiatic (autochthonous) tribes who inhabited the arctic tundra zone for a long time, and who were then assimilated by ancient Samoyeds during their northward movements (Prokof'ev 1940; Vasilyev 1985). Who were these autochthonous tribes? According to Simchenko (1968) all Arctic populations from the Kola Peninsula up to the Bering Straight represented an ethnically homogeneous population. Other ethnographers suggested that different autochthonous components contributed to extant populations. For example, Prokof'ev (1940) suggested that the Forest and Tundra Nentsi, Entsi, and Nganasans are descendants of one aboriginal tribe, while Selkups, Ugric-speaking Khants, and Mansi shared a separate Palaeoasiatic component.

In order to address hypotheses on the origin and affinities of the Samoyed peoples, several populations were studied on a number of field expeditions from 1991–95 led by the Institute of Cytology and Genetics, Novosibirsk. In this chapter, we describe patterns of Y chromosome variation in Samoyed populations (Selkups, Tundra and Forest Nentsi) from villages in western Siberia, non-Samoyed-speaking populations (Komi, Kets, and North Evenks) living in the same villages, and a Turkic-speaking group (Altais) inhabiting the putative ancestral homeland of the Samoyeds. In addition, we describe patterns of variation in several Selkup and Forest Nentsi sub-populations that were hypothesized to be closely affiliated, based on historical and demographic information. Analyses of these sub-populations helped us to assess the relative utility of different Y-specific marker systems for inferring population relationships.

## 19.2 Subject and methods

### 19.2.1 Population samples

Blood samples were collected with informed consent of volunteers from ten Samoyed-speaking populations (three Forest Nentsi, one Tundra Nentsi, and six Selkup populations) and four non-Samoyed-speaking populations: the Komi, Kets, North Evenks, and Altais. The geographic locations of the sampling sites are shown in Fig. 19.1 and the linguistic characteristics, geographic settings, and collection sites of the populations are given in Table 19.1. From demographic data and genealogical information we were able to identify 257 paternally-unrelated males (for at least three to six generations) as the subjects of our study: 191 unrelated Samoyed-speaking males and 66 unrelated males from the other four population groups.

### 19.2.2 Genotyping Y chromosome markers

The polymorphic sites in our survey included five diallelic polymorphisms and five microsatellite markers. The *DYS287* (YAP) locus was scored according to the method of Hammer and Horai (1995). Polymorphic nucleotide sites in the YAP region were genotyped by SSO hybridization (Hammer *et al.* 1997). An $G \to A$ transition within the *DYS257* sequence-tagged-site (Hammer *et al.* 1998) and an $A \to G$ transition ($SRY_{10831}$) within the SRY region (Whitfield *et al.* 1995) were genotyped according to methods reported by Hammer *et al.* (1998). We also genotyped a $T \to C$ transition within the single-copy Tat (RBF5) locus using the procedure of Zerjal *et al.* (1997), and a $C \to T$ transition in the *RPS4Y* gene (Bergen and Goldman, in press).

Five Y-chromosomal microsatellite markers were genotyped: *DYS19*, *DYS390*, *DYS391*, *DYS393*, and *DXYS156*. The primer sequences and amplification conditions are reported in Jobling and Tyler-Smith (1995) and Kayser *et al.* (1997). The amplified products were separated on 6–8 per cent polyacrylamide gels in TBE buffer for 8–16 hours at 200–350 V. The gels were stained with ethidium bromide and the fragments were visualized by u.v. light. Identification of alleles was verified by comparing PCR product sizes with allelic ladders kindly provided by P. de Knijff.

### 19.2.3 Statistical analyses

All analyses were performed on Y chromosome haplotypes (a haplotype is the combination of variants at each polymorphic site for each Y chromosome). We analyzed three sets of Y haplotypes: those based only on five diallelic polymorphisms (unique haplotypes), those based only on five microsatellites (microsatellite haplotypes), and those based on both diallelic and microsatellite haplotypes (combination haplotypes). Population genetic structure indices (molecular variance and $\Phi$ statistics) were estimated by use of the AMOVA software package (Excoffier *et al.* 1992). Both haplotype frequencies and molecular differences between haplotypes are taken into account with this approach. Distance matrices containing pairwise comparisons among all unique, microsatellite, and combination haplotypes were constructed in following manner: (1) the number of mutational differences (both point and indel) were counted for unique haplotypes (linear distances); (2) the sum of squared differences in repeat number over all loci was determined for microsatellite haplotypes (squared Euclidean distances);

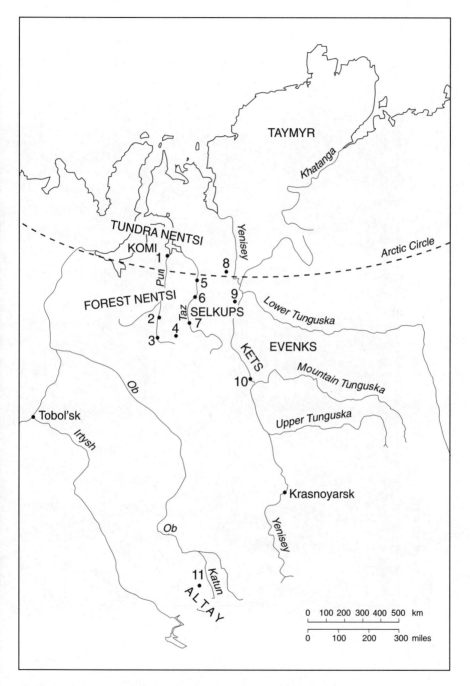

**Fig. 19.1** A map of western Siberia showing the approximate geographic locations of the study populations. The numerical codes of the collection sites are defined in Table 19.1.

**Table 19.1**

Characteristics of Siberian populations under investigation

| Population (n) | Linguistic affiliation | Population size | Geographic setting | Collection sites (map position*) | Major occupation |
|---|---|---|---|---|---|
| Forest Nentsi (27) | Samoyedic | 1500 | between Lower Ob and Taz Rivers | Samburg (1) Harampur (2) Holyasavay (3) | hunting, fishing, reindeer breeding |
| Tundra Nentsi (51) | Samoyedic | 32 000 | between Yamal Peninsula and Lower Yenisey River | Samburg Village (1) and Tundra Camps | reindeer breeding |
| North Selkups (113) | Samoyedic | 2000 | Taz and Turukhan Rivers | Tolka Purovskaya (4) Krasnoselkup (5) Tolka Selkupskaya (6) Ratta (7) Sov. Rechka (8) Farkovo (9) | hunting, fishing |
| Komi (8) | Finno-Ugric | 1000 | Lower Ob River | Samburg village (1) and Tundra Camps | reindeer breeding |
| Kets (12) | Language isolate | 500 | Tributaries of middle Yenisey River | Sov. Rechka (8) Farkovo (9) Sulomay (10) | hunting, fishing |
| Evenks (17) | Tungus | 29 000 | Central Siberia | Krasnoselkup (5) Sov. Rechka (8) Farkovo (9) | hunting, fishing, reindeer breeding |
| Altais | Turkic | 5600 | Altai Republic | Mendur-Sokkon (11) | cattle herding |

and (3) linear distances and squared Euclidean distances were summed for each combination haplotype. When used with AMOVA, the squared Euclidean matrix yields an analogue of Slatkin's (1995) $R_{st}$ statistic (Michalakis and Excoffier 1996). Variance components due to different sources of variation (between individuals within demes, between demes within populations, and between populations) were estimated, and their significances tested by use of a nonparametric permutation procedure (Excoffier et al. 1992). Genetic distances between populations were also obtained based on all pairwise $\Phi_{st}$ values.

We also used the GENDIST Program in PHYLIP 3.4 (Felsenstein 1993) to compute genetic distances according to the methods of Cavalli-Sforza and Edwards (1967) ($D_c$), Reynolds et al. (1983) ($F_{st}$), and Nei (1972) ($D_s$). For each of the four genetic distance statistics employed ($\Phi_{st}$, $D_c$, $F_{st}$, and $D_s$), comparisons of geographic and genetic distance were made for the eighteen possible pairs of sub-populations (within six Selkup and within three Forest Nentsi sub-populations). The linear increase in genetic distance as a function of geographic distance was evaluated using linear regression.

Clustering diagrams based on genetic distances were generated using neighbour-joining (NJ) (Saitou and Nei 1987) and the method of Fitch and Margoliash (FM) (1967). The PHYLIP package was also used for constructing maximum-likelihood (ML) phylogeny estimates based on Y chromosome haplotype frequencies. Majority-rule consensus trees based on the eight possible combinations of two tree-building (NJ, FM) and four distance measures ($\Phi_{st}$, $D_c$, $F_{st}$, and $D_s$), as well as maximum likelihood, were constructed for the seven Siberian populations and sub-populations using the CONSENSE program in PHYLIP. We noticed that dendrograms based on $D_s$ were inconsistent and dissimilar to those based on other distance measures, and that correlation values involving $D_s$ were significantly lower than pairwise correlations between the other three distance measures (see above). Therefore, we do not include population trees based on $D_s$. Haplotype diversities and their variances were determined as described by Nei (1978, 1987).

## 19.3 Results

### 19.3.1 Unique, microsatellite, and combination haplotype variation

Variation at four polymorphic transitions and the YAP indel in the 257 unrelated males in this survey produced six unique haplotypes, which varied in frequency among the different populations surveyed (Table 19.2). While the allelic distribution for each microsatellite was essentially unimodal, the most frequent allele differed from population to population. The combination of variation at the five microsatellite markers gave rise to 26 microsatellite haplotypes, none of which was shared by all groups. By combining microsatellite and unique haplotype variation, a total of 31 'combination' haplotypes were produced (Table 19.2). This slight increase in number of haplotypes from 26 indicates a fairly strong

**Table 19.2**
Y chromosome combination haplotype frequencies

| | DXYS 156 | DYS 19 | DYS 390 | DYS 391 | DYS 393 | Forest Nentsi | Tundra Nentsi | Selkups | Komi | Kets | Evenks | Altai |
|---|---|---|---|---|---|---|---|---|---|---|---|---|
| **Unique haplotype: 1B** $(+, -, -, -, -)$ | | | | | | | | | | | | |
| 1. | 12 | 13 | 23 | 10 | 13 | | | | | | | 1 |
| 2. | 12 | 14 | 22 | 10 | 13 | | 3 | | | | | |
| 3. | 12 | 14 | 22 | 11 | 14 | | 3 | | | | | |
| 4. | 12 | 14 | 23 | 10 | 13 | 3 | 16 | 9 | 2 | | 2 | |
| 5. | 12 | 14 | 23 | 11 | 13 | 12 | 7 | | 1 | | | |
| 6. | 12 | 14 | 23 | 11 | 14 | | 6 | | | | | |
| 7. | 12 | 14 | 24 | 11 | 13 | | | | | | | 1 |
| 8. | 12 | 15 | 23 | 10 | 13 | | 2 | | | | 1 | |
| 9. | 12 | 15 | 23 | 11 | 13 | | | | | | | 1 |
| **Unique haplotype: 1C** $(+, +, -, -, -)$ | | | | | | | | | | | | |
| 10. | 12 | 12 | 23 | 9 | 13 | | | 2 | | | | |
| 11. | 12 | 12 | 23 | 10 | 13 | | | 4 | | | | |
| 12. | 12 | 13 | 23 | 10 | 13 | 1 | 1 | 66 | | 5 | 4 | 2 |
| 13. | 12 | 13 | 23 | 10 | 14 | | | | | 5 | | |
| 14. | 12 | 13 | 24 | 10 | 13 | | | 2 | | | | |
| 15. | 12 | 13 | 25 | 10 | 13 | | | 1 | | | | |
| 16. | 12 | 14 | 22 | 11 | 13 | | | 3 | | | | |
| 17. | 12 | 14 | 23 | 10 | 13 | | | 4 | | | | |
| 18. | 12 | 14 | 24 | 11 | 12 | | | 3 | | | | |
| **Unique haplotype: 1D** $(-, +, -, -, -)$ | | | | | | | | | | | | |
| 19. | 12 | 16 | 25 | 10 | 13 | | | 11 | | | | 14 |
| 20. | 12 | 16 | 25 | 11 | 13 | | | 5 | | | | 1 |
| 21. | 12 | 16 | 26 | 10 | 13 | | | | | | | 1 |
| 22. | 12 | 16 | 26 | 11 | 13 | | | 1 | | | | |

**Unique haplotype: 1F** $(+, -, +, -, -)$

| # | | | | | | 27 | 51 | 113 | 8 | 12 | 17 | 29 |
|---|---|---|---|---|---|---|---|---|---|---|---|---|
| 23. | 11 | 15 | 25 | 10 | 13 | | | 2 | | | | 2 |
| 24. | 11 | 16 | 24 | 9 | 13 | | | | | 2 | 10 | 4 |
| 25. | 11 | 15+17 | 24 | 9 | 13 | | | | | | | 1 |

**Unique haplotype: 1I** $(+, -, -, +, -)$

| # | | | | | | 27 | 51 | 113 | 8 | 12 | 17 | 29 |
|---|---|---|---|---|---|---|---|---|---|---|---|---|
| 26. | 12 | 14 | 23 | 10 | 13 | 1 | | 6 | | | | |
| 27. | 12 | 14 | 23 | 11 | 13 | | | 3 | | | | |
| 28. | 12 | 14 | 23 | 11 | 14 | | 10 | 3 | 3 | | | |
| 29. | 12 | 14 | 24 | 11 | 14 | | 1 | | | | | |
| 30. | 12 | 15 | 23 | 11 | 14 | | | | 2 | | | |

**Unique haplotype: 3G** $(+, -, -, -, +)$

| # | | | | | | 27 | 51 | 113 | 8 | 12 | 17 | 29 |
|---|---|---|---|---|---|---|---|---|---|---|---|---|
| 31. | 11 | 15 | 26 | 10 | 13 | | | | | | | 1 |
| $N$ | | | | | | 27 | 51 | 113 | 8 | 12 | 17 | 29 |
| $h$ | | | | | | 0.675 | 0.856 | 0.641 | 0.821 | 0.682 | 0.618 | 0.756 |
| $\pm$SEM | | | | | | 0.013 | 0.012 | 0.005 | 0.047 | 0.030 | 0.029 | 0.015 |

Note: Unique haplotypes were defined by allelic states at each of the following diallelic markers (shown in parentheses): $SRY_{10831}$, $DYS257$, $RPS4Y$, Tat, and $DYS287$.

**Table 19.3**
Analysis of variance components (AMOVA) for three data sets and seven
Siberian populations

| Haplotypes | Unique | Microsatellite | Combination |
|---|---|---|---|
| Among groups (per cent) | 50.6 | 38.0 | 41.4 |
| Within populations (per cent) | 49.4 | 62.0 | 58.6 |
| $\Phi_{st}$ | 0.506 | 0.380 | 0.414 |

relationship between unique and microsatellite haplotype variation, and is interesting in light of the fact that identical microsatellite haplotypes can arise solely
by chance through recurrent mutation (Cooper *et al.* 1996; de Knijff *et al.* 1997;
Jobling *et al.* 1997). The five instances where identical microsatellite haplotypes
occurred on different unique haplotype backgrounds (see Table 19.2 and below)
probably resulted from convergence rather than from recent common ancestry.

With respect to microsatellite haplotype variation, the seven Siberian populations surveyed here were characterized by low levels of within-group variation
and high levels of among-group variation. For example, the microsatellite haplotype diversity value ($0.847 \pm 0.014$) was much lower than typically observed in
surveys of closely related populations (Cooper *et al.* 1996; Roewer *et al.* 1996;
de Knijff *et al.* 1997). On the other hand, AMOVA analyses (Table 19.3) indicated levels of population differentiation that were much higher than observed
between closely related European populations (de Knijff *et al.* 1997). This pattern
of reduced haplotype diversity within populations and elevated levels of differentiation among populations may be a feature of indigenous groups that have been
isolated for long periods of time (Kayser *et al.* 1997).

### 19.3.2 Correlations between genetic and geographic distances

We compared genetic and geographic distances in order to assess the influence
of geography on the distribution of genetic variation in Samoyed populations.
Although human populations generally do not exhibit high correlations between
geographic and genetic distances (Jorde 1980), some of the highest known correlations have been observed in Siberian populations that existed in relative isolation for centuries (Crawford and Enciso 1982; Karafet 1986). Four different
genetic distance statistics ($\Phi_{st}$, $D_c$, $F_{st}$, and $D_s$) were used to compare geographic and genetic distances for pairs of subpopulations within the Selkup and
the Forest Nentsi groups (Table 19.1). All genetic distances showed significant
linear regressions on geographic distances, and three of the four distance statistics exhibited a trend toward higher regression values for genetic distances based
on combination haplotypes versus microsatellite haplotypes versus unique haplotypes (Table 19.4). Correlations became lower when we compared geographic
and genetic distances at the population level (data not shown). Interestingly, some

**Table 19.4**
Correlation ($R^2$) between genetic and geographic distances*

|  | Unique haplotypes | Microsatellite haplotypes | Combination haplotypes |
|---|---|---|---|
| $\Phi_{st}$ | 0.571 | 0.581 | 0.637 |
| $F_{st}$ | 0.661 | 0.622 | 0.620 |
| $D_c$ | 0.493 | 0.684 | 0.694 |
| $D_s$ | 0.561 | 0.661 | 0.663 |

*Note: Correlations based on comparisons within 6 Selkup and within three Forest Nentsi sub-populations.

of the traditional genetic distance statistics gave higher regression values than $\Phi_{st}$, a statistic with analogous properties to $R_{st}$ that was developed specifically for microsatellite data (Michalakis and Excoffier 1996). Similar results observed in brown bear and human population studies led to the conclusions that genetic drift plays the main role in generating the distributions of microsatellite alleles among populations, and that the variance of the genetic distance statistic used is a more important consideration than accurate mutational models (Paetkau *et al.* 1997; Pérez-Lezaun *et al.* 1997).

## 19.3.3 Population trees

Majority-rule consensus trees based on the three different data sets (unique, microsatellite, and combination haplotype) were constructed for the nine Selkup and Forest Nentsi sub-populations and the four other population groups (Fig. 19.2). There was a similarity among the consensus trees based on the different data sets in that all three were composed of two major clusters: one cluster (top of Fig. 19.2(a)–(c) contained the Samoyed-speaking Selkups and the non-Samoyed-speaking Evenks, Kets and Altais, and the second cluster (bottom of Fig. 19.2(a)–(c)) contained the other two Samoyed-speaking populations (Forest and Tundra Nentsi) and the Komi. Also, all three data sets consistently showed a close affinity of one of the Selkup sub-populations, the Tolka Purovskaya (Se4), with the Altais. In contrast to the agreement regarding the Tolka Purovskaya, there were major differences among the three data sets in their ability to resolve the relationships among the other eight sub-populations.

We suspect that some of the inconsistency encountered in the analysis of the microsatellite haplotypes was due to the aforementioned convergence of microsatellite haplotypes, especially involving the sharing of microsatellite haplotypes among the 1B, 1C, and 1I unique haplotypes. These convergent microsatellite haplotypes (i.e. 4, 17, and 26 in Table 19.2) were present in the Selkup and Nentsi populations at fairly high frequencies. This illustrates one of the potential disadvantages of using haplotypes based solely on microsatellite alleles to resolve relationships among taxonomic groups above the sub-population level

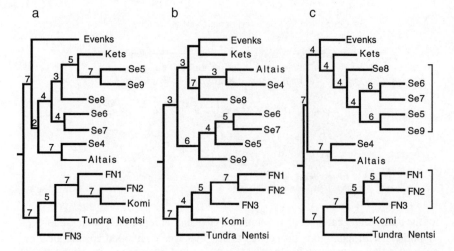

**Fig. 19.2** Majority-rule consensus trees based on (a) unique haplotypes, (b) microsatellite haplotypes, and (c) combination haplotypes for six Selkup (Se) and three Forest Nentsi (FN) subpopulations, and four other population groups. The number associated with each subpopulation corresponds to collection sites listed in Table 19.1. Seven trees were generated using combinations of tree-building and distance methods (see Subjects and methods). The numbers on the branches indicate the number of methods in which the group consisting of the composite populations to the right of that node occurred. Brackets to the right of 3C represent the two clusters of Selkup (top) and Forest Nentsi (bottom) subpopulations.

(or among populations carrying convergent Y lineages). Potentially this can be overcome by including more stable binary markers in the analysis, as evidenced by the more consistent placement of the Selkup sub-populations into their own cluster in the tree based on combination haplotypes (Fig. 19.2(c)).

### 19.3.4 Haplotype networks

A haplotype network based on the 26 microsatellite haplotypes was characterized by ten reticulations and complex, sub-optimal tree solutions (Fig. 19.3(a)). There was a general clustering of haplotypes from Forest Nentsi, Tundra Nentsi, and Komi populations, and a lack of clustering of haplotypes from Selkup, Evenk, Ket, and Altai populations. Some of the factors that have been identified to explain similar complex patterns observed in networks constructed for other populations include small sample sizes, small numbers of loci studied, and high rates of homoplasy (Cooper *et al.* 1996; Deka *et al.* 1996). Additionally, male migration among populations and strong Y chromosome bottlenecks could also lead to complex networks and the absence of clustering within populations.

A more resolved combination haplotype network resulted after including information from the binary markers (Fig. 19.3(b)). The combination haplotype network differed from the microsatellite haplotype network in several important respects: (1) for most of the links only one single-step mutation was required, especially for microsatellite haplotypes within the 1B and 1I haplotypes, (2) only two reticulations were observed (within the 1C and 1D haplotypes), and (3) there was a general clustering of microsatellite haplotypes within a population.

## 19.4 Discussion

### 19.4.1 Utility of Y-linked microsatellite markers for comparing closely related populations

Because of their high mutation rates and the possibility that any given microsatellite haplotype has evolved more than once (convergence), these markers are probably not very useful for constructing haplotype trees for diverse Y chromosomes, or for inferring the relationships of divergent human populations (Cooper *et al.* 1996; Deka *et al.* 1996; Jobling *et al.* 1997; de Knijff *et al.* 1997). Convergence of microsatellite haplotypes is an expected consequence of the stepwise mutation model, which has been shown to describe best the mode of Y microsatellite evolution (Cooper *et al.* 1996; Paetkau *et al.* 1997). Given that the average mutation rate of Y microsatellites may be as high as 0.23 per cent per generation (Heyer *et al.* 1997), it is not surprising that we found five cases of convergent microsatellite haplotypes in relatively closely related populations. Thus, single

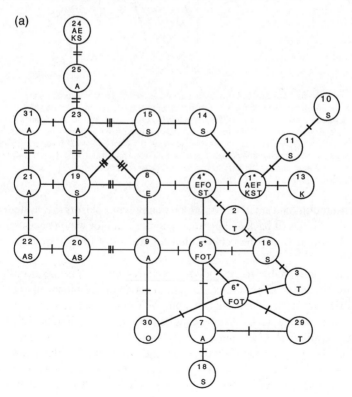

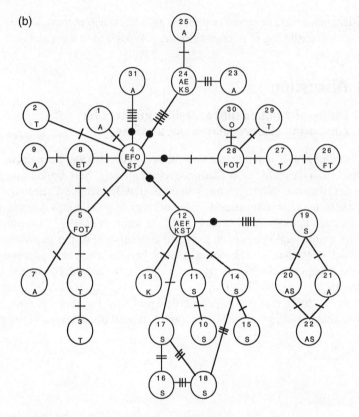

**Fig. 19.3** Networks based on (a) 26 microsatellite haplotypes and (b) 31 combination haplotypes. The network was constructed by starting with the most frequent haplotype and then adding adjacent haplotypes. All possible adjoining associations are shown by connecting lines. Each cross-hatch represents a one-step mutation at one of the five microsatellites. Black nodes indicate a point mutation or YAP insertion. The numbers at the top of each circle indicate the haplotypes shown in Table 19.2. An asterisk refers to microsatellite haplotypes that are shared between different unique haplotypes. One letter codes at the bottom of circle refer to populations with the following abbreviations: A, Altais; E, Evenks; F, Forest Nentsi; K, Kets; O, Komi; S, Selkups; T, Tundra Nentsi.

base substitutions and small insertion/deletion polymorphisms (i.e. systems with lower probabilities of back and parallel mutation, and for which ancestral states can be determined) may be better suited for studies of deeper splits in human evolution.

However, these stable binary polymorphic systems are not necessarily best suited for studies of closely related populations. Y-linked binary markers have generally been ascertained in small samples of Y chromosomes from individuals representing geographically diverse populations (Hammer 1995; Whitfield et al. 1995; Underhill et al. 1997) and may not 'mark' Y lineages that are informative in regional studies. This is primarily a consequence of the fact that many binary Y polymorphisms are rare and only a few are shared among major geographic regions (Jobling and Tyler-Smith 1995; Underhill et al. 1997). The ages of most of the known binary polymorphisms predate the time of separation of closely related populations, and differences in unique haplotype frequencies among populations are mainly the result of genetic drift and migration after divergence from a common ancestral population. On the other hand, differences in frequencies of microsatellite haplotypes among regional populations can also be the result of new mutations that randomly 'mark' Y chromosome lineages. Therefore, microsatellite markers do not suffer from the same degree of ascertainment bias that is associated with the use of unique binary markers.

This study was undertaken, in part, to assess the relative utility of these two kinds of markers in determining the relationships of male lineages in the Samoyed speakers of western Siberia. We found that microsatellite haplotypes were generally more informative than unique haplotypes: our earlier analysis of unique haplotypes based on one less polymorphic site (Tat) were much less informative (data not shown). This illustrates the need to select carefully a set of binary polymorphic sites that are known to be informative for the populations of interest. In contrast, a small set of 'randomly' chosen Y-linked microsatellites is likely to be equally or more informative for resolving the relationships among closely related populations. Whenever possible it is recommended that both kinds of markers are employed together because the combination of multiallelic variation at microsatellites and diallelic variation at a few sites on the NRPY can result in haplotype networks with less reticulation and greater resolution of sub-population affinities.

### 19.4.2 Origin of samoyeds

Our results demonstrate that Samoyed populations do not cluster genetically strictly according to linguistic affiliations. Close relationships were found between the Samoyed-speaking Forest and Tundra Nentsi, but neither was affiliated with a third Samoyed-speaking group, the Selkups. The Forest and Tundra Nentsi were also affiliated with the Komi, a group that speaks a language belonging to the Finno-Ugric branch of the Uralic linguistic family. Haplotypes 1B and 1I and associated microsatellite alleles were the two major unique haplotype groups in these populations. In contrast, the Selkups share the 1C haplotype and associated microsatellite alleles with the Kets, and the 1D haplotype and associated microsatellite alleles with the Altais. This was reflected in the consistent clustering of the Selkups with these two non-Samoyed-speaking groups. Our data support the hypothesis of Prokof'ev (1940) that the Palaeoasiatic male ancestors were different for Nentsi (Forest and Tundra) and for Selkup populations. Moreover, it is very probable that autochthonous ancestors of present Selkups and Nentsi have been influenced by different waves of ancient Samoyeds. The data are also consistent with the following hypotheses: (1) the Ugric-speaking Komi and the Samoyed-speaking Nentsi share common male ancestors; (2) the Kets and Selkups share a different set of common male ancestors; and (3) the contribution of Turkic-speaking people to contemporary Samoyed populations is higher among Selkups than in Tundra or Forest Nentsi. The evidence for the latter hypothesis is that Selkups and Turkic-speaking Altais share four unique and many microsatellite haplotypes. The Selkup group that we have studied settled in the Taz and Turukhan River basins only recently, in the seventeenth century. Prior to this time, all Selkups lived along the Middle Ob and its tributaries (Fig. 19.1). Due to their traditional geographical location, the Selkups had more opportunities for contact with Turkic-speaking tribes in the course of their history. In fact, the Selkup language possesses many Turkic loan-words (i.e. the names of relatives, heroes, certain words connected with the raising of animals, and cultural terms) whereas many fewer Nentsi words were borrowed from Turkic languages (Hajdú 1962). More western Siberian populations need to be examined to clarify the origin of the Samoyeds and their relations with ancient and modern populations. Comparative studies on genetic variation of classical and mitochondrial markers in the same Samoyed populations is expected to shed further light on the factors that have been involved in the origin of Samoyeds.

## 19.5 Conclusions

(1) The cluster and $\Phi_{st}$ analyses indicate that even a limited number of microsatellite haplotypes (five in our study) is sufficient to resolve the structure and relationships of recently diverged populations. With the information now available for global Y-linked microsatellite variation (Deka *et al.* 1996; Scozzari *et al.* 1997; de Knijff *et al.* 1997; Kayser *et al.* 1997; Karafet *et al.* 1998), informative microsatellites can be chosen for any geographical region of interest.

(2) Combining variation at Y-linked microsatellites and diallelic sites represents a powerful tool for human evolutionary studies, especially at the regional level. Combination haplotypes are a valuable source of information for identifying the effects of migrations, founder effects, and mating patterns.

(3) Our results did not show a simple correspondence between linguistic affiliation and Y chromosome-based genetic structure. The results also indicated a moderate relationship between genetic and geographic distances between closely related populations.

(4) Our data are consistent with the hypothesis that the Palaeoasiatic male ancestors were different for Nentsi (Forest and Tundra) and for Selkup populations. Male ancestors could be the same for Ugric-speaking Komi and Samoyedic-speaking Nentsi on the one hand, and for Selkups and Kets on the other. The contribution of ancient Samoyeds and/or Turkic people in modern Samoyeds is higher among Selkups than in Tundra or Forest Nentsi.

# Acknowledgements

This publication was made possible by grant OPP-9423429 from the National Science Foundation (to M.F.H.). Its contents are solely the responsibility of the authors and do not necessarily represent the official views of the NSF.

# 20 Microsatellite analysis of human tumours

Darryl Shibata

## Chapter contents

## Abstract

Human tumours arise after the accumulation of a number of somatic mutations. Many aspects of tumour evolution or progression are unknown, since serial observations are rare. Since tumours consist of populations of tumour cells, phylogenetic approaches may help to unravel their evolution. The recent discovery of human tumours with high frequencies of mutations at microsatellite loci (up to 0.01 per division), due to losses in DNA mismatch repair, provide practical opportunities to follow tumour evolution, since detectable frameshifts accumulate after relatively few divisions. In addition, the availability of cell lines and transgenic mice lacking DNA mismatch repair provide practical experimental models of microsatellite mutational processes.

## 20.1 Introduction

This chapter will concentrate on using microsatellite loci to analyse the behaviour of individual cells within single multicellular organisms. It will focus on the unique properties, problems, and opportunities of such tissue populations.

## 20.2 Microsatellite loci as molecular tumour clocks

Metazoan organisms have finite lifetimes, which limit their potential number of somatic mutations. Some cell types divide infrequently, whereas others divide constantly. For a cell dividing daily, only about 25 500 divisions will have occurred by 70 years of age. Mutation rates in most tumours are less than $10^{-6}$ per locus per division (Loeb 1991). Consequently, detection of somatic mutations in most normal and tumour tissues is extremely difficult.

In 1993, tumours with greatly increased mutation frequencies were uncovered (Aaltonen *et al.* 1993; Ionov *et al.* 1993; Thibodeau *et al.* 1993). These *mutator phenotype tumours* (also called RER+ or USM+ or MI+) have mutations in DNA *mismatch repair* (*MMR*) genes and lack the ability to correct DNA replication errors (Kinzler and Vogelstein 1996). Both hereditary (*hereditary nonpolyposis colorectal cancer* or HNPCC) and sporadic tumours are seen. Approximately 15 per cent of all human colorectal, gastric, and endometrial carcinomas have this mutator phenotype. The increased mutation rate is thought to accelerate tumour evolution.

Mutation rates are elevated about 100–1000-fold compared to non-mutator type tumours. Studies in mutator cell lines indicate mutation frequencies between $10^{-3}$ and $10^{-2}$ mutations per division per microsatellite locus (Bhattacharyya *et al.* 1994; Shibata *et al.* 1994). Mutation rates are also elevated 100–1000-fold in other loci, but are relatively low ($<10^{-4}$) compared to microsatellite loci. Microsatellite loci should rapidly become polymorphic in mutator phenotype tumours (Shibata *et al.* 1996; Shibata 1997).

## 20.3 Tumour progression: a great mystery and problem

Despite the intense focus on cancer, relatively little is known about how cancers cells actually divide, die, evolve, and spread. Direct serial observations are rare and unethical. Tumour growth is variable, and most macroscopic tumours do not exhibit exponential growth. Division rates are much greater than growth rates, leading to the conclusion that more than 90 per cent of all tumour cells die (Steele 1977). Genetic *tumour progression* has been deduced by analysing large numbers of tumours at different stages of evolution from different patients. Tumours appear to arise from a single progenitor cell (Nowell 1976). Progression is a multistep process, as multiple somatic mutations are probably necessary for complete malignant transformation. Current models of tumour progression (Kinzler and Vogelstein 1996) are exemplified by the *adenoma–carcinoma sequence*—a linear process with sequential replacements by populations with greater numbers of somatic mutations (Fig. 20.1). Each new expansion is thought to occur when a single cell in the current population acquires a new somatic mutation which provides a selective advantage and allows clonal dominance and replacement.

Evidence for this linear or phyletic model of tumour progression is strong but circumstantial. Comparisons between progression 'phases' are limited in scope as most tumours are experimentally characterized at only one to three loci, and the absolute frequency of mutation per locus is typically 50 per cent or less. One weakness of this approach is that most early stage tumours do not advance further. For example, adenomas or polyps are small neoplastic growths in the colon. Although adenomas are thought to be direct precursors of colon cancer, the ratio of adenomas to cancers is greater than 30 : 1 (Koretz 1993). Although adenomas are considered 'premalignant', most are genetic 'dead ends' and never advance to cancer within the lifetimes of their hosts.

The 'phase' of tumour progression cannot be reliably deduced by the analysis of specific oncogenic loci, since mutations are not specific and can be present or absent in all phases. Cancers of the same 'type' have heterogeneous combinations of mutations. Identical somatic mutations in both adenomas and adjacent carcinomas are considered strong evidence of direct progression. However, definite conclusions are somewhat limited, since these loci acquire mutations infrequently. Mutation rates in most tumours are less than $10^{-6}$ (Loeb 1991).

The 'ages' of tumours are also unknown. Tumour initiation may occur early or late, and evolution to cancer may require decades. Of note, Fig. 20.1 suggests relatively 'compressed' evolution since many critical mutations are acquired in polyps which typically become evident after 50 years of age. The intervals

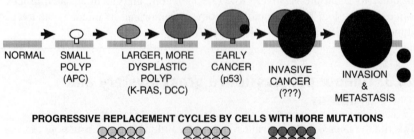

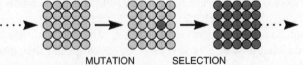

**Fig. 20.1** Diagram of multistep adenoma–carcinoma progression (modelled after Kinzler and Vogelstein 1996). The current model implies sequential replacement by cells which acquire selective advantages with greater numbers of mutations (such as in APC, K-ras, DCC and p53). This direct or phyletic replacement suggests earlier lesions such as polyps are direct precursors to more advanced lesions.

between tumour progression phases are also relatively undefined. A clinically fatal phase of tumour progression is its spread or metastasis to remote parts of the body. It is unclear whether metastasis is an early or late event and if multiple metastases are related or independent.

## 20.4 Experimental strategies: extracting phylogenetic information from tissues

Tumours are large and complex populations. A 1 cm$^3$ tumour (usually the smallest detectable lesion) contains approximately one billion cells. DNA can be extracted in bulk from entire tumours, but some information may be lost if tumour populations are heterogeneous. Tumours are mixtures of normal and tumour cells, and special care must be taken to ensure that the majority (>70 per cent) of the extracted DNA originates from primary tumour cells.

Our current approach relies heavily on tissue microdissection, and the ability to amplify single alleles. Microdissection may involve separation of relatively large (1 cm$^2$) but distinct tumour regions or multiple distinct smaller (200–400 cell) populations. Tumours are physical manifestations of large phylogenetic trees. Neighbouring cells within small microdissected tumour regions are likely to be more related to each other than to cells present in other tumour regions, since physical constraints often present barriers to widespread migration. Tumours fixed in formalin and embedded in paraffin are used. Such processing occurs routinely with human specimens, and the paraffin blocks are stored for decades. The DNA present in these fixed tissues is degraded, but PCR targets less than 180 base pairs can be routinely amplified. A single thin tissue slice is placed on a microscope slide. By analogy, tumours are continents and cells are individuals (Fig. 20.2). Microscop*ical* analysis reveals a rich diversity of tumour cell morphologies topographically defined in two dimensions. Typically, many morphological phases of tumour evolution (non-invasive, invasive, and metastatic populations) are present within a single tumour. Small populations (200–400 cells) with defined and relatively homogeneous morphological characteristics can be dissected from a single microscope slide.

The distribution of alleles can be compared within and between microdissected regions. The extracted DNA is diluted to essentially single molecules with PCR products from 50–70 per cent of amplifications. Radioactive labelling and optimization is required to attain the single molecule sensitivity. DNA extracted from normal tissue defines germline alleles from which the tumour alleles originated. Germline alleles from contaminating normal tissues should be eliminated from the distributions, since they are not an integral part of the tumour population. They can be eliminated by truncation when tumour alleles are clearly different from germline. Otherwise, they can be subtracted by estimating the mix of tumour versus normal cells in the microdissected tissue.

The distribution frequencies can be analysed by many of the phylogenetic techniques outlined in other chapters of this book. Some of the specific questions

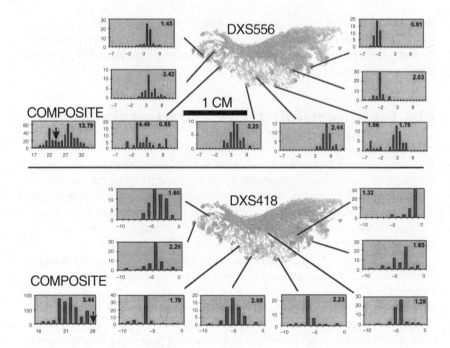

**Fig. 20.2** A cancer and its microsatellite frequency distributions. Lines indicate the locations of different tumour population (200–400 cells) microdissected from the outlined cancer. The outline is a photocopy of the actual microscope slide, magnified twice its actual size. The cancer is from a male patient and the microsatellite loci are X chromosome dinucleotide repeats. The variance of each population is given. In some cases, bimodal populations are present. Germline alleles (indicated by '0' and the arrows) are $(CA)_{24}$ for DXS556 and $(CA)_{26}$ for DXS418. Current tumour approaches presume evolution as in Fig. 20.1, but cannot tell the age of this cancer or its pattern of spread. The microsatellite distributions potentially reveal much more.

which can be analysed phylogenetically include relative ages of tumour populations (How long has this tumour been here?), and the timing of critical progression events (When did distant migration or metastasis to the brain occur?). Multiple microsatellite loci can be analysed from the same tumour sections. Microsatellite mutation rates appear to be constant, since homozygous loss of a single repair component greatly reduces MMR (Marsischky *et al.* 1996). However, exceptions to constant microsatellite mutation rates have been observed (Richards *et al.* 1997), and this assumption remains hypothetical. Microsatellite tumour mutation rates also appear related to the number of repeat units (Fig. 20.3). Microsatellite mutations in MMR-deficient cells arise predominately due to slippage during replication (Strand *et al.* 1993). Therefore, the number of microsatellite mutations can be related directly to the number of cell divisions. Of interest, most tumour cells appear to lose the ability to control cell division (Strauss *et al.* 1995), which may contribute to the relentless cycles of cell division and death which occurs in human tumours (Steele 1977).

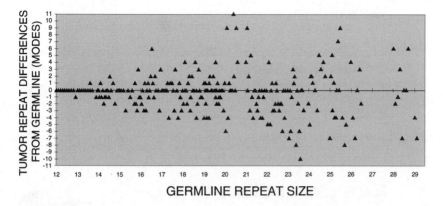

**Fig. 20.3** Tumour microsatellite loci mutation rates appear to vary with respect to the number of germline repeat units. The figure represents unpublished experimental data from multiple mutator phenotype tumours from three different patients. The patients are male and the microsatellite loci (approximately 30 different CA-dinucleotide loci) are on the X-chromosome and therefore hemizygous. Both additions and deletions are observed. The differences between tumour and germline repeat sizes are calculated from the modes of the tumour microsatellite distributions, since there is a distribution of microsatellite alleles present in the mutator phenotype tumour populations (see Fig. 20.2). Germline microsatellite loci with less than 16 dinucleotide repeat units appear to have lower mutation rates, since their tumour microsatellite allele modes are seldom different from germline. For clarity, the loci have been distributed around their absolute numbers of germline dinucleotide repeat units.

Cell division is binary, and mutation in tumour cell lines appears to be step-wise with predominantly single repeat unit additions or deletions (Table 20.1). Since tumours do not grow exponentially and some appear to be stable for years (Otchy *et al.* 1996), one simplification is that small tumour population sizes are constant. Another simplification which minimizes the typical complexity of het-erozygous germline autosomal loci is the use of male tissues and hemizygous X-chromosome microsatellite loci, as mutator phenotype tumours are usually diploid (Kinzler and Vogelstein 1996). Cell division may be asymmetrical with one daughter surviving and one dying. *Asymmetrical division* is characteristic of normal intestinal epithelium (Potten *et al.* 1990). Division may also be *sym-metrical* with either both or no surviving daughter cells. These two patterns of cell division can be distinguished, since lineages terminate frequently with symmetrical division (reducing microsatellite variance) whereas lineages persist indefinitely with asymmetrical division. It appears that malignant cells undergo random proliferation with both asymmetrical and symmetrical divisions (unpub-lished observations).

An example of the microsatellite distributions present in a cancer is illustrated in Fig. 20.2. Multiple distinct populations are present in this cancer. Interpretation of the information recorded by the microsatellite loci should reveal details of tumour evolution not incorporated in current tumour progression models (Fig. 20.1). The data in this cancer may be more consistent with multiple origins and branching evolution rather than direct phyletic replacement.

**Table 20.1**
Frequencies of repeat unit size changes observed in human tumor cell lines

| Repeat unit shifts observed | | | |
|---|---|---|---|
| One | Two | Three | Four |
| 81 | 16 | 3 | 1 |

Data are from Shibata *et al*. 1994. Both CA- and CT-dinucleotide microsatellite loci were examined after limited numbers of divisions (about 30) in tissue culture. Predominately single unit frameshifts (additions and deletions) were observed with PCR size analysis.

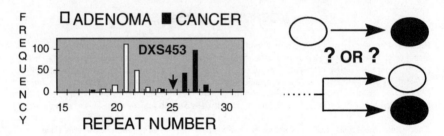

**Fig. 20.4** The microsatellite distributions of a cancer and its immediately adjacent adenoma. The germline allele, $(CA)_{25}$, is indicated by the arrow. According to Fig. 20.1, the adenoma is the immediate precursor to the cancer. The microsatellite distributions suggest a more distant relationship.

Another example is illustrated in Fig. 20.4. In this tumour, a cancer was immediately adjacent to an adenoma. Such morphological proximity is usually interpreted as evidence that the cancer arose directly from the adenoma. However, the microsatellite distributions were virtually distinct between the cancer and adjacent adenoma. Such divergent distributions suggest that tumour evolution is more deeply rooted than implied by current models (Fig. 20.1).

## 20.5 Experimental systems: observing microsatellite changes in tissues

Below are brief summaries of experimental models composed of cells lacking DNA mismatch repair.

### 20.5.1 Xenografts

Human cell lines derived from cancers deficient in mismatch repair are widely available. Cultures originating from a single cell can be injected subcutaneously into immune deficient mice to form tumours or xenografts. These xenografts are relatively stable in size (after an initial growth phase of several weeks), and can be maintained for several months before compromising the health of the host. The xenografts are experimental models of human cancers, and offer opportunities for serial observations with defined initiation times and genotypes.

## 20.5.2 Transgenic mice

The genotype of mice can be manipulated experimentally. Genes can be inserted or inactivated ('knocked out'). Mice deficient in DNA mismatch repair have been constructed with homozygous inactivation of repair genes. The mice develop normally despite having elevated mutation rates in all cells since conception. As expected of molecular clocks, microsatellite mutations accumulate with age in these mice (Tsao *et al.* 1997).

Transgenic mice provide unique experimental opportunities to investigate how microsatellite loci mutate in different genetic and environmental backgrounds. Similarly, microsatellite loci can also unravel unexpected phylogenetic relationships between cells, since current technology cannot readily follow or distinguish between morphologically similar cells. It may be possible to define better the developmental branches involved in the division and differentiation of the different cell types which originate from a single zygote. Mice typically live one to two years.

## 20.5.3 Human tumours

It is difficult to compare tumours arising in heterogeneous human populations. However, about 5 per cent of colon cancers arise in HNPCC families, and their individuals will have more similar genetic backgrounds. HNPCC individuals are heterozygous, with one normal and one mutated repair allele. The heterozygous cells are repair proficient, and microsatellite somatic mutations are extremely rare in their normal tissues. However, the normal repair allele is lost early in tumour formation (Shibata *et al.* 1994), and theoretically the majority of microsatellite mutations in a tumour can be traced to this initiating event in a single MMR deficient progenitor cell. Multiple independent tumours often arise in a single HNPCC individual, thereby allowing analysis of independent populations arising from identical genetic backgrounds and similar environmental exposures. The long lifetime of humans potentially allows large differences to arise between initial germline microsatellite genotypes and their tumour populations (Figs 20.2–4).

# 20.6 Summary

Cancers are populations of tumour cells. Mutator phenotype tumours provide unique opportunities to translate relatively sophisticated phylogenetic approaches to the mysteries of cancer. The legacies recorded in tumour microsatellite loci should provide key insights into their origins and patterns of division and migration. Such missing information is critical for rational strategies to prevent and cure cancer.

## Acknowledgements

The author thanks Drs. Reijo Salovaara and Lauri A. Aaltonen for their collaborative efforts, and support by grants from the Public Health Service (CA58704 and CA70858).

# Appendix

## Coalescent modelling and numerical methods

In this section we describe in more detail some of the technical aspects of the likelihood-based methods that have been described. The aim is to provide numerically-minded biologists with intuitions behind these methods.

## Drift recursion

The probability, $q(t, \mathbf{n})$, of obtaining a sample configuration $\mathbf{n}$ can be written as (O'Ryan *et al.* 1998):

$$q(t, \mathbf{n}) = \int_0^{T_s - t} \left( \sum_{i=1}^{k} \frac{n_i - 1}{S(\mathbf{n}) - 1} q(s + t, \mathbf{n} - \mathbf{e}_i) \right) a(S(\mathbf{n})) \exp(-a(S(\mathbf{n}))s) \, ds$$
$$+ \left( 1 - \int_0^{T_s - t} a(S(\mathbf{n})) \exp(-a(S(\mathbf{n}))s) \, ds \right) M(\mathbf{n}, \mathbf{x}) \qquad \text{(A.1)}$$

$\mathbf{n} = (n_1, n_2, \ldots, n_k)$ is a vector of length $k$ alleles. It gives the number, $n_i$, of chromosomes of each type in the sample:

$$a(n) = n(n - 1)/2.$$

Time is scaled such that $T_s = T/(2N_e G)$ for a diploid population of effective population size $N_e$ founded $T$ years before the population was sampled, where $G$ is some suitable scaling factor ('generation time').

$n_i$ is the $i$th element of $\mathbf{n}$. $S(\mathbf{n})$ is the sum of the $n_i$.

$\mathbf{e}_i$ is a unit vector with 1 at the $i$th position and 0 elsewhere.

$M(\mathbf{n}, \mathbf{x})$ is the multinomial probability of drawing the sample of lineages $\mathbf{n}$ from the ancestral gene frequencies $\mathbf{x}$.

This has the form:

probability of observing configuration $\mathbf{n}$ (of size $S(\mathbf{n})$) at present time

$=$

    sum (integral) of

        probability density of coalescent event at time $s < T_s$

        $\times$

        sum of

probabilities of prior configurations (of size $S(\mathbf{n})$-1)
that could have given rise to observed configuration

+

probability of no coalescent

×

probability of sampling configuration $\mathbf{n}$ from an ancestral population
with gene frequencies $\mathbf{x}$

where $T_s$ and $\mathbf{x}$ are unknown.

## Numerical method of solution

A method of simulation introduced by Griffiths and Tavaré (1994a) can be used
to solve this recursion. They describe a Markov Chain Monte Carlo (MCMC)
technique, but it is very different in spirit and methodology from the usual methods
such as Metropolis–Hastings sampling. Essentially, the recursion such as given
in eqn A.1, above, has the form:

$$q(\mathbf{x}) = \sum r(\mathbf{x}, \mathbf{y}_k)q(\mathbf{y}_k) + \sum r(\mathbf{x}, \mathbf{z}_k)q(\mathbf{z}_k).$$

In a genealogical setting, $r(\mathbf{x}, \mathbf{y})q(\mathbf{y})$ represents the probability of moving from
some ancestral state $\mathbf{y}$ to the current state $\mathbf{x}$ multiplied by the probability of the
ancestral state occurring. $\mathbf{y}_k$ are values where $q(\mathbf{y}_k)$ cannot be directly evaluated,
and $\mathbf{z}_k$ are values where $q(\mathbf{z}_k)$ can be evaluated. For example, the second term
of eqn (13.1) for $q(t, \mathbf{n})$ is not recursive and can always be solved, and values of
the coefficients of the first term will eventually reach 0. Thus if we follow the
recursion down one branch, it will eventually reach a term that can be directly
evaluated.

For ease of explanation, the following method of subscripting has been used:
immediately ancestral states are subscripted by $k$; states within the sequence of
states down a branch of the recursion are subscripted by $j$; the different sequences
obtained by following different branches of the recursion are subscripted by $i$.
This use of subscripts does not imply that all terms are distinguishable. If the
recursion were written out explicitly it would appear as the sum of products of
the coefficients $r(\mathbf{y}_j, \mathbf{y}_k)$ ending in a known $q(\mathbf{z}_i)$. That is, it would be the sum of
terms such as $r(\mathbf{x}, \mathbf{y}_1)r(\mathbf{y}_1, \mathbf{y}_2)r(\mathbf{y}_2, \mathbf{y}_3) \cdots r(\mathbf{y}_n, \mathbf{z}_i)q(\mathbf{z}_i) \cdots$, which would occur
down one branch of the recursion. The lengths of these products would vary
depending on how rapidly a known $q(\mathbf{z})$ was reached. Thus the recursion could be
written as a potentially infinite sum of terms of the form $a_1q(\mathbf{z}_1) + a_2q(\mathbf{z}_2) \cdots$,
where the $a_i$ are the products. In principle, each $a_i$ could be divided into a product
of two terms, $p_i b_i$, where the $p_i$ are known to sum to 1. One can imagine some
sampling procedure that chooses one of these terms at random in proportion to the
value of the coefficient $p_i$ and records the value $b_i q(\mathbf{z}_i)$. This process is repeated
many times. Since the $p_i$ sum to one, the average value of the sampled $b_i q(\mathbf{z}_i)$
must tend to the required probability $q(\mathbf{x})$ when a very large number of terms are
sampled.

In order to ensure that the $p_i$ have the desired properties, and that the terms are sampled in proportion to the $p_i$, the simulation starts with the sample configuration, $\mathbf{x}$. There are a number of prior configurations with probability $q(\mathbf{y}_k)$, each with a probability $r(\mathbf{x}, \mathbf{y}_k)$ of having given rise to $\mathbf{x}$. The simplest way to split the $r(\mathbf{x}, \mathbf{y}_k)$ into a product of two terms is as $r(\mathbf{x}, \mathbf{y}_k)/ \sum r(\mathbf{x}, \mathbf{y}_k) \times \sum r(\mathbf{x}, \mathbf{y}_k)$. We choose a new $\mathbf{y}$ with probability $r(\mathbf{x}, \mathbf{y}_k)/ \sum r(\mathbf{x}, \mathbf{y}_k)$ and record $\sum r(\mathbf{x}, \mathbf{y}_k)$. The process is repeated down the branch of the recursion, $\mathbf{x}, \mathbf{y}_1, \mathbf{y}_2, \ldots$ until we hit $q(\mathbf{z})$, which can be evaluated. The product of the $r(\mathbf{y}_j, \mathbf{y}_k)/ \sum r(\mathbf{y}_j, \mathbf{y}_k)$ along this branch of the recursion corresponds to the $p_i$ above, and the product of $\sum r(\mathbf{y}_j, \mathbf{y}_k)$ correspond to the $b_i$. Clearly there are other ways to split the $r(\mathbf{y}_j, \mathbf{y}_k)$, which may be useful if we wish to increase the correlation between the $p_i$ and the $b_i$ (otherwise we might sample many unlikely states).

The process is made more complicated by the integral in eqn (A.1), but the principle is the same. Since the only changes of state are coalescences going back in time, an extra coefficient would need to be multiplied to the $a_i q(\mathbf{z}_i)$, described earlier, to represent the probability of obtaining that number of coalescences in the time interval to $T_s$. Clearly, also, coalescences cannot occur when each allelic category has only one member. These considerations can be overcome by adapting the approach of Griffiths and Tavaré (1994a). A coalescence time is simulated from the exponential distribution with rate parameter $S(\mathbf{n})(S(\mathbf{n}) - 1)/2$. With probability $(n_i - 1)/(S(\mathbf{n}) - k)$, where $k$ is the number of alleles, the number in the $i$th category is reduced by 1 (a coalescence) (this corresponds to choosing a new $\mathbf{y}$ with probability $r(\mathbf{y}_j, \mathbf{y}_k)/ \sum r(\mathbf{y}_j, \mathbf{y}_k)$, see eqn (A.1)). The quantity $f_j = (S(\mathbf{n}) - k)/ (S(\mathbf{n}) - 1)$ (i.e. $\sum r(\mathbf{y}_j, \mathbf{y}_k)$) is multiplied across iterations. If a coalescent time is simulated but each allelic category has only one member, $f_j$ is set to 0.0 and the iteration stops. Otherwise the iteration continues until a coalescent event occurs that predates the scaled time of foundation, $T_s$. In this case $f_j$ is multiplied by the multinomial probability of obtaining the remaining sample from the ancestral population. This method of explicitly simulating coalescent times ensures that the $a_i q(\mathbf{z}_i)$ are sampled in proportion to their probability of occurring. The average value of the product of the $f_j$ gives the probability of the sample configuration, given the scaled time of foundation, $T_s$, and the ancestral gene frequencies, $\mathbf{x}$.

## Explicit solution

The recursion can also be solved explicitly (O'Ryan et al. 1998) to give:

$$q(t = 0, n_0) = \sum M(\mathbf{n}_i, \mathbf{x}) \frac{(S(\mathbf{n}_0) - S(\mathbf{n}_i))!(S(\mathbf{n}_i) - 1)! \prod_{j=1}^{k} (n_{0_j} - 1)!}{(S(\mathbf{n}_0) - 1)! \prod_{j=1}^{k} (n_{0_j} - n_{i_j})! \prod_{j=1}^{k} (n_{i_j} - 1)!}$$

$$\times p_c(S(\mathbf{n}_0), S(\mathbf{n}_0) - S(\mathbf{n}_i))$$

$$p_c(n, m) = f(n, m) - f(n, m + 1)$$

$$f(n, m) = 1 + (-1)^m \sum_{i=1}^{m} \left( \prod_{j=1; j \neq i}^{m} \frac{a(n - j + 1)}{a(n - i + 1) - a(n - j + 1)} \right)$$

$$\times \exp(-a(n - i + 1)T_s)$$

$f(n, 0) = 1$

$f(n, m) = 0; \quad n = m.$

(O'Ryan *et al.* 1998), where $\mathbf{n}_0$ is the current configuration, and the summation is over $\mathbf{n}_0$ and all possible subsamples that have $n_{i_j} > 0 \,|\, n_{0_j} > 0$. An equivalent expression for $p_c$ is given in Tavaré (1984)(see also Nielsen *et al.*, 1998).

This has the form:

Probability of observing configuration $\mathbf{n}_0$ with sample size $S(\mathbf{n}_0)$ at present time

=

Sum of (over all possible configurations of the initial lineages present at time $T_s$ before present that could have given rise to $\mathbf{n}_0$)

The probability that the initial configuration was sampled from the ancestral gene frequencies $\mathbf{x}$

×

the probability that the initial configuration gave rise to the sample configuration (rather than some other configuration).

×

the probability that the required number of coalescences occurred.

Although this enables us to calculate the probability exactly, in practice the numerical solution is quicker because the exact solution has potentially a very large number of terms that need to be evaluated.

### Metropolis–Hastings simulation

In Metropolis–Hastings simulation we start with a set of parameter values, $v_1$, and evaluate the likelihood, $l_1$. We then randomly choose a new set of values, $v_2$, with likelihood $l_2$, in such a way that the frequency, $t_{1,2}$, of moving from $v_1$ to $v_2$ is known relative to the frequency $t_{2,1}$ of moving from $v_2$ to $v_1$. We then compute the quantity $p = l_2/l_1 \times t_{2,1}/t_{1,2}$. If $p \geq 1$, we accept the move. Otherwise we accept it with probability $p$. The process is repeated to produce a sequence of $v_i$. The theoretical equilibrium distribution of the $v_i$ is the normalized likelihood function. This can be illustrated by a simple example. Suppose we have two possible states $s_1$ and $s_2$ with likelihoods $l_1$ and $l_2$. The transition function might be $t_{1,2} = t_{2,1} = 1$. Since this is symmetrical, the probability of moving from one state to the other depends on the ratio of $l_1$ and $l_2$. For example, if $l_1 > l_2$, we always accept transitions from $s_2$ to $s_1$, and accept transitions from $s_1$ to $s_2$ with probability $l_2/l_1$. At equilibrium, choosing some point in the sequence at random, the probability of observing a change from $s_1$ to $s_2$ must be the same as observing a change from $s_2$ to $s_1$. If $r_1$, $r_2$ are the long-run probabilities of being in state $s_1$, $s_2$, and $l_1 > l_2$, we have $r_1 l_2/l_1 = r_2$, so that $l_2/l_1 = r_2/r_1$. Since $r_1 + r_2 = 1$ this implies that $r_1$, $r_2 = l_1/(l_1 + l_2), l_2/(l_1 + l_2)$.

# References

Aaltonen, L.A., Peltomaki, P., Leach, F.S., Sistonen, P., Pylkkanen, L., Mecklin, L. *et al.* (1993) Clues to the pathogenesis of familial colorectal cancer. *Science*, **260**, 812–16.

Aguadé, M., Miyashita, N. and Langley, C.H. (1989) Reduced variation in the *yellow-achaete-scute* region in natural populations of *Drosophila melanogaster*. *Genetics*, **122**, 607–15.

Aharoni, A., Baran, N., Manor, H. (1993) Characterization of a multisubunit human protein which selectively binds single stranded $d(GA)_n$ and $d(GT)_n$ sequence repeats in DNA. *Nucleic Acids Research*, **21**, 5221–8.

Ahn, B.Y., Dornfeld, K.J., Fagrelius, T.J. and Livingston, D.M. (1988) Effect of limited homology on gene conversion in a *Saccharomyces cerevisiae* plasmid recombination system. *Molecular Cell Biology*, **8**, 2442–8.

Albon, S.D., Clutton-Brock, T.H. and Guinness, F.E. (1987) Early development and population dynamics in red deer. II. Density-independent effects and cohort variation. *Journal of Animal Ecology*, **56**, 69–81.

Ali, S., Muller, C.R. and Epplen, J.T. (1986) DNA fingerprinting by oligonucleotide probes specific for simple repeats. *Human Genetics*, **74**, 239–43.

Allen, P.J., Amos, W., Pomeroy, P.P. and Twiss, S.D. (1996) Microsatellite variation in grey seals (*Halichoerus grypus*) shows evidence of genetic differentiation between two British breeding colonies. *Molecular Ecology*, **4**, 653–62.

Allendorf, F.W. and Leary, R.F. (1986) Heterozygosity and fitness in natural populations of animals. In *Conservation biology*. (ed. M.E. Soulé), pp. 57–76. Sunderland: Sinauer Press.

Almqvist, E., Spence, N., Nichol, K., Andrew, S.E., Vesa, J., Peltonen, L. *et al.* (1995) Ancestral differences in the distribution of the Δ2642 glutamic acid polymorphism is associated with varying CAG repeat lengths on normal chromosomes: insights into the genetic evolution of Huntington disease. *Human Molecular Genetics*, **4**, 207–14.

Altmann, J., Alberts, S., Coote, T., Dubach, J., Geffen, E., Haines, S.A. *et al.* (1996) Behaviour predicts genetic structure in a wild primate group. *Proceedings of the National Academy of Sciences of the USA*, **93**, 5797–801.

Amos, W. and Rubinsztein, D.C. (1996) Microsatellites are subject to directional evolution. *Nature Genetics*, **12**, 13–14.

Amos, W., Sawcer, S.J., Feakes, R.W. and Rubinsztein, D.C. (1996) Microsatellites show mutational bias and heterozygote instability. *Nature Genetics*, **13**, 390–1.

Andersen, J., Martin, P., Carracedo, A., Dobosz, M., Eriksen, B., Johnnson, V. *et al.* (1996) Report on the third EDNAP collaborative STR exercise. *Forensic Science International*, **78**, 83–93.

Andreassen, R., Egeland, T. and Olaisen, B. (1996) Mutation rate in the hypervariable VNTR g3 (D7S22) is affected by allele length and a flanking DNA sequence polymorphism near the repeat array. *American Journal of Human Genetics*, **59**, 360–7.

Angers, B. and Bernatchez, L. (1997) Complex evolution of a salmonid microsatellite locus and its consequences in inferring allelic divergence from size information. *Molecular Biology and Evolution*, **14**, 230–8.

Anon. (1996) Gene modulates prostate cancer risk. *Science*, **272**, 1271.

Anvret, M., Åhlberg, G., Grandell, U., Hedberg, B., Johnson, K. and Edström, L. (1993) Larger expansions of the CTG repeat in muscle compared to lymphocytes from patients with myotonic dystrophy. *Human Molecular Genetics*, **2**, 1397–400.

Araki, E., Shimada, F., Uzawa, H., Mori, M. and Ebina, Y. (1987) Characterization of the promoter region of the human insulin receptor gene. *Journal of Biological Chemistry*, **262**, 16186–91.

Arcot, S.S., Wang, Z., Weber, J.L., Deininger, P.L. and Batzer, M.A. (1995) Alu repeats: a source for the genesis of primate microsatellites. *Genomics*, **29**, 136–44.

Armour, J.A.L. (1996) Tandemly repeated minisatellites: generating human diversity via recombinational mechanisms. In *Human Genome Evolution* (ed. M. Jackson, T. Strachan and G. Dover) Bios Scientific Publishers, Oxford.

Armour, J.A.L., Povey, S., Jeremiah, S. and Jeffreys, A.J. (1990) Systematic cloning of human minisatellites from ordered array Charomid libraries. *Genomics*, **8**, 501–12.

Armour, J.A.L., Harris, P.C. and Jeffreys, A.J. (1993) Allelic diversity at minisatellite MS205 (D16S309): evidence for polarized variability. *Human Molecular Genetics*, **2**, 1137–45.

Armour, J.A.L., Neumann, R., Gobert, S. and Jeffreys, A.J. (1994) Isolation of human simple repeat loci by hybridization selection. *Human Molecular Genetics*, **3**, 599–605.

Armour, J.A.L., Crosier, M. and Jeffreys, A.J. (1996a) Distribution of tandem repeat polymorphism within minisatellite MS621 (D5S110) *Annals of Human Genetics*, **60**, 11–20.

Armour, J.A.L., Anttinen, T., May, C.A., Vega, E.E., Sajantila, A., Kidd, J.R. *et al.* (1996b) Minisatellite diversity supports a recent African origin for modern humans. *Nature Genetics*, **13**, 154–60.

Asghari, B. *et al.* (1994) Dopamine D4 receptor repeat analysis of different native and mutant forms of the human and rat genes. *Molecular Pharmacology*, **46**, 364–73.

Ashley, C.T. and Warren, S.T. (1995) Trinucleotide repeat expansion and human disease. *Annual Review of Genetics*, **29**, 703–28.

Aslandis, C., Jansen, G., Amemiya, C., Shutler, G., Mahadevan, M., Tsilfidis, C. *et al.* (1992) Cloning of the essential myotonic dystrophy region and mapping of the putative defect. *Nature*, **355**, 548–51.

Ayala, F.J. (1995) The myth of Eve, molecular biology and human origins. *Science*, **270**, 1930–6.

Ayala, F.J., Escalante, A., O'hUigin, C. and Klein, J. (1994) Molecular genetics of speciation and human origins. *Proceedings of the National Academy of Sciences of the USA*, **91**, 6787–94.

Babcock, M., de Silva, D., Oaks, R., Davis-Kaplan, S., Jiralerspong, S., Montermini, L. *et al.* (1997) Regulation of mitochondrial iron accumulation by Yfh1p, a putative homolog of frataxin. *Science*, **276**, 1709–12.

Baker, S.M., Bronner, C.E., Zhang, L. *et al.* (1995) Male mice defective in the DNA mismatch repair gene PMS2 exhibit abnormal chromosome synapsis in meiosis. *Cell*, **82**, 309–19.

Baker, S.M., Plug, A.W. and Prolla, T.A. *et al.* (1996) Involvement of mouse *Mlh1* in DNA mismatch repair and meiotic crossing over. *Nature Genetics*, **13**, 336–42.

Balazs, I., Baird, M., Clyne, M. and Meade, E. (1989) Human population genetic studies of 5 hypervariable DNA loci. *American Journal of Human Genetics*, **44**, 182–90.

Balding, D.J. (1995) Estimating products in forensic identification using DNA profiles. *Journal of the American Statistical Association*, **90**, 839–44.

Balding, D.J. and Donnelly, P. (1994) The prosecutor's fallacy and DNA evidence. *Criminal Law Review* October 1994, 711–21.

Balding, D.J. and Donnelly, P. (1995*a*) Inference in forensic identification. *Journal of the Royal Statistical Society* A, **158**, 21–53.

Balding, D.J. and Donnelly, P. (1995*b*) Inferring identity from DNA profile evidence. *Proceedings of the National Academy of Sciences of the USA*, **92**, 11741–5.

Balding, D.J. and Nichols, R.A. (1994) DNA profile match probability calculation: how to allow for population stratification, relatedness, database selection and single bands. *Forensic Science International*, **64**, 125–40.

Balding, D.J. and Nichols, R.A. (1995) A method for quantifying differentiation between populations at multi-allelic loci and its implications for investigating identity and paternity. *Genetica*, **96**, 3–12.

Balding, D.J., Greenhalgh, M. and Nichols, R.A. (1996) Population genetics of STR loci in Caucasians. *International Journal of Legal Medicine*, **108**, 300–5.

Banchs, I., Bosh, A., Guimera, J., Lazaro, C., Puig, A. and Estivill, X. (1994) New alleles at microsatellite loci in CEPH families mainly arise from somatic mutations in the lymphoblastoid cell line. *Human Mutation*, **3**, 365–72.

Bancroft, D.R., Pemberton, J.M., Albon, S.D., Robertson, A., MacColl, A.D.C., Smith, J.A. *et al.* (1995) Molecular genetic variation and individual survival during population crashes of an unmanaged ungulate population. *Philosophical Transactions of the Royal Society of London*, **347**, 263–73.

Barbujani, G., Magagni, A., Minch, E. and Cavalli-Sforza, L.L. (1997) An apportionment of human DNA diversity. *Proceedings of the National Academy of Sciences of the USA*, **94**, 4516–9.

Barton, N.H., Halliday, R.B. and Hewitt, G.M. (1983) Rare electrophoretic variants in a hybrid zone. *Heredity*, **50**, 139–46.

Beckmann, J.S. and Weber, J.L. (1992) Survey of human and rat microsatellites. *Genomics*, **12**, 627–31.

Begun, D.J. and Aquadro, C.F. (1992) Levels of naturally occurring DNA polymorphism correlate with recombination rates in *D. melanogaster*. *Nature*, **356**, 519–20.

Behe, M.J. (1987) The DNA sequence of the human beta-globin region is strongly biased in favor of long strings of contiguous purine or pyrimidine residues. *Biochemistry*, **26**, 7870–5.

Behe, M.J. (1995) An overabundance of long oligopurine tracts occurs in the genome of simple and complex eukaryotes. *Nucleic Acids Research*, **23**, 689–95.

Bell, G.I. and Jurka, J. (1997) The length distribution of perfect dimer repetitive DNA is consistent with its evolution by an unbiased single-step mutation process. *Journal of Molecular Evolution*, **44**, 414–21.

Bell, G.I., Horita, S. and Karam, J.H. (1984) A polymorphic locus near the human insulin gene is associated with insulin-dependent diabetes mellitus. *Diabetes*, **33**, 176–83.

Benjamin, J., Li, L., Patterson, C., Greenberg, B.D., Murphy, D.L. and Hamer, D.H. (1996) Population and familial association between the D4 dopamine receptor gene and measures of novelty seeking. *Nature Genetics*, **12**, 81–4.

Bennett, L., Minch, E., Feldman, M., Jenkins, T. and A. Bowcock, A. (1998) A set of independent tri- and tetranucleotide repeats for human evolutionary studies; additional evidence for an African origin for modern humans and implications for higher mutation rates in tetranucleotide repeats of the 'GATA' type. In preparation.

Bennett, S.T., Lucassen, A.M., Gough, S.C.L., Powell, E.E., Undlien, D.E., Pritchard, L.E. *et al.* (1995) Susceptibility to human type 1 diabetes at *IDDM2* is determined by tandem repeat variation at the insulin gene minisatellite locus. *Nature Genetics*, **9**, 284–92.

Bergen, A.W. and Goldman, D. (In press) Y chromosome short tandem repeat haplotypes and the origins of Native Americans.

Bhattacharyya, N.P., Skandalis, A., Ganesh, A., Groden, J. and Meuth, M. (1994) Mutator phenotypes in human colorectal carcinoma cell lines. *Proceedings of the National Academy of Sciences of the USA*, **91**, 6319–23.

Bichara, M., Schumacher, S. and Fuchs, R.P.P. (1995) Genetic instability within monotonous runs of CpG sequences in *E. coli*. *Genetics*, **140**, 897–907.

Biggin, M.D. and R. Tjian (1988) Transcription factors that activate the Ultra-bithorax promoter in developmentally staged extracts. *Cell*, **53**, 699–711.

Bingham, P.M., Scott, M.O., Wang, S., McPhaul, M.J., Wilson, E.M., Garbern, J.Y. *et al.* (1995) Stability of an expanded repeat in the androgen receptor gene in transgenic mice. *Nature Genetics*, **9**, 191–6.

Bishop, D.K., Andersen, J. and Kolodner, R.D. (1989) Specificity of mismatch repair following transformation of *Saccharomyces cerevisiae* with heterodu-plex plasmid DNA. *Proceedings of the National Academy of Sciences of the USA*, **86**, 3713–7.

Blanquer-Maumont, A. and Crouau-Roy, B. (1995) Polymorphism, monomor-phism, and sequences in conserved microsatellites in primates species. *Journal of Molecular Evolution*, **41**, 492–7.

Blouin, M.S., Parsons, M., Lacaille, V. and Lotz, S. (1996) Use of microsatellite loci to classify individuals by relatedness. *Molecular Ecology*, **5**, 393–401.

Boness, D.J. and Bowen, W.D. (1996) The evolution of maternal care in pin-nipeds. *Bioscience*, **46**, 645–54.

Bonner, W.N. (1968) The fur seal of South Georgia. *British Antarctic Survey Scientific Reports*, **56**, 1–81.

Borts, R.H. and Haber, J.E. (1989) Length and distribution of meiotic gene conver-sion tracts and crossovers in *Saccharomyces cerevisiae*. *Genetics*, **123**, 69–80.

Borts, R.H., Leung, W.-Y., Kramer, W. *et al.* (1990) Mismatch repair-induced mei-otic recombination requires the *PMS1* gene product. *Genetics*, **124**, 573–84.

Boulva, J. and McLaren, I.A. (1979) Biology of the harbour seal (*Phoca vitulina*), in eastern Canada. *Bulletin of the Fisheries Research Board of Canada*, **200**, 1–24.

Bourke, A.F.G., Green, H.A.A. and Bruford, M.W. (1997) Parentage, reproduc-tive skew and queen turnover in a multiple-queen ant analyzed with microsatel-lites. *Proceedings of the Royal Society of London* B, **264**, 277–83.

Bowcock, A.M., Kidd, J.R., Mountain, J.L., Hebert, J.M., Carotenuto, L., Kidd, K.K. and Cavalli-Sforza, L.L. (1991) Drift, admixture, and selection in human evolution, a study with DNA polymorphisms. *Proceedings of the National Academy of Sciences of the USA*, **88**, 839–43.

Bowcock, A.M., Ruiz-Linares, A., Tomfohrde, J., Minch, E., Kidd, J.R. and Cavalli-Sforza, L.L. (1994) High resolution of human evolutionary trees with polymorphic microsatellites. *Nature*, **368**, 455–7.

Bowen, W.D., Oftedal, O.T., Boness, D.J. and Iverson, S.J. (1994) The effect of maternal age and other factors on birth mass in the harbour seal. *Canadian Journal of Zoology*, **72**, 8–14.

Braaten, D.C., Thomas, J.R., Little, R.D., Dickson, K.R., Goldberg, I., Schlessinger, D. *et al.* (1988) Locations and contexts or sequences that hybridize to poly(dG-dT) (dC-dA) in mammalian ribosomal DNAs and two X-linked genes. *Nucleic Acids Research*, **16**, 865–81.

Brais, B., Bouchard, J.-P., Xie, Y.-G., Rochefort, D.L., Chretien, N. *et al.* (1998) Short GCG expansions in the PABP2 gene cause oculopharyngeal muscular dystrophy. *Nature Genetics*, **18**, 164–7.

Briscoe, D., Stephens, J.C. and O'Brien, S.J. (1994) Linkage disequilibrium in admixed populations: applications in gene mapping. *Journal of Heredity*, **85**, 59–63.

Britten, H.B. (1996) Meta-analyses of the association between multilocus heterozygosity and fitness. *Evolution*, **50**, 2158–64.

Brook, J.D., McCurrach, M.E., Harley, H.G., Buckler, A.J., Church, D., Aburatani, H. *et al.* (1992) Molecular basis of myotonic dystrophy: expansion of a trinucleotide (CTG) repeat at the 3' end of a transcript encoding a protein kinase family member. *Cell*, **68**, 799–808.

Brooker, A.L., Cook, D., Bentzen, P., Wright, J.M. and Doyle, R.W. (1994) Organisation of microsatellites differs between mammals and cold-water teleost fishes. *Canadian Journal of Fisheries and Aquatic Science*, **51**, 1959–66.

Brooker, M.G., Rowley, I., Adams, M. and Baverstock, P.R. (1990) Promiscuity—an inbreeding avoidance mechanism in a socially monogamous species. *Behav. Ecology Socio. Biology*, **26**, 191–9.

Brookes, M. (1997) A clean break. *New Scientist*, **156**, 64.

Broun, P. and Tanksley, S.D. (1996) Characterization and genetic mapping of simple repeat sequences in the tomato genome. *Molecular and General Genetics*, **250**, 39–49.

Bruford, M.W. and Wayne, R.K. (1993*a*) The use of molecular genetic techniques to address conservation questions. In *Molecular techniques in environmental biology* (ed. S.J. Garte), pp. 11–28. CRC Press, USA.

Bruford, M.W. and Wayne, R.K. (1993*b*) Microsatellites and their application to population genetics studies. *Current Opinion in Genetics and Development*, **3**, 939–43.

Brunner, H.G., Brüggenwirth, H.T., Nillesen, W., Jansen, G., Hamel, B.C.J., Hoppe, R.L.E. *et al.* (1993) Influence of sex of the transmitting parent as well as of parental allele size on the CTG expansion in myotonic dystrophy (DM) *American Journal of Human Genetics*, **53**, 1016–23.

Buard, J. and Vergnaud, G. (1994) Complex recombination events at the hypermutable minisatellite CEB1 (D2S90) *EMBO Journal*, **13**, 3203–10.

Buchanan, F.C. and Crawford, A.M. (1993) Ovine microsatellites at the OarFCB111, OarFCB128, OarFCB193, OarFCB266 and OarFCB304 loci. *Animal Genetics*, **24**, 145.

Buffery, C., Burridge, F., Greenhalgh, M., Jones, S. and Willott, G. (1991) Allele frequency distributions of four variable number tandem repeat (VNTR) loci in the London area. *Forensic Science International*, **52**, 53–64.

Burke, J.R., Ikeuchi, T., Koide, R., Tsuji, S., Yamada, M., Pericak-Vance, M.A. *et al.* (1994) Dentatorubral-pallidoluysian atrophy and Haw River syndrome. *Lancet*, **344**, 1711–12.

Burke, J.R., Enghild, J.J., Martin, M.E., Jou, Y.-S., Myers, R.M., Roses, A.D. *et al.* (1996) Huntingtin and DRPLA proteins selectively interact with the enzyme GAPDH. *Nature Medicine*, **2**, 347–50.

Burright, E.N., Clark, H.B., Servadio, A., Matilla, T., Feddersen, R.M., Yunis, W.S. *et al.* (1995) *SCA1* transgenic mice: a model for neurodegeneration caused by an expanded CAG trinucleotide repeat. *Cell*, **82**, 937–48.

Burright, E.N., Davidson, J.D., Durick, L.A., Koshy, B., Zoghbi, H.Y. and Orr, H.T. (1997) Identification of a self-association region within the *SCA1* gene product ataxin-1. *Human Molecular Genetics*, **6**, 513–18.

Buxton, J., Shelbourne, P., Davies, J., Jones, C., Van Tongeren, T., Aslanidis, C. *et al.* (1992) Detection of an unstable fragment of DNA specific to individuals with myotonic dystrophy. *Nature*, **355**, 547–8.

Callen, D.F., Thompson, A.D., Shen, Y., Phillips, H.A., Richards, R.I., Mulley, J.C. and Sutherland, G.R. (1993) Incidence and origin of null alleles in the $(AC)_n$ microsatellite markers. *American Journal of Human Genetics*, **52**, 922–927.

Campuzano, V., Montermini, L., Molto, M.E., Pianese, L., Cossee, M., Cavalcanti, F. *et al.* (1996) Friedreich's Ataxia: autosomal recessive disease caused by an intronic GAA triplet repeat expansion. *Science*, **271**, 1423–7.

Cann, R.L., Stoneking, M. and Wilson, A.C. (1987) Mitochondrial DNA and human evolution. *Nature*, **325**, 31–6.

Carey, N., Johnson, K., Nokelainen, P., Peltonen, L., Savontaus, M.L., Juvonen, V. *et al.* (1994) Meiotic drive at the myotonic dystrophy locus? *Nature Genetics*, **6**, 17–18.

Carlin, B.P. and Chib, S. (1995) Bayesian model choice via Markov chain Monte Carlo methods. *Journal of the Royal Statistical Society* B, **57**, 473–84.

Carpenter, A.T.C. (1994*a*) Chiasmata function. *Cell*, **77**, 959–62.

Carpenter, A.T.C. (1994*b*) The recombination nodule—seeing what you are looking at. *BioEssays*, **16**, 69–74.

Caskey, C.T., Pizzuti, A., Fu, Y.H., Genwick, R.G. Jr. and Nelson, D.L. (1992) Triplet repeat mutations in human disease. *Science*, **256**, 784–9.

Cavalli-Sforza, L. (1997) Genes, peoples, and languages. *Proceedings of the National Academy of Sciences of the USA*, **94**, 7719–24.

Cavalli-Sforza, L.L. and Edwards, A.W.F. (1967) Phylogenetic analysis: models and estimation procedures. *American Journal of Human Genetics*, **19**, 233–57.

Cavalli-Sforza, L.L. and Minch, E. (1997) Paleolithic and Neolithic lineages in the European mitochondrial gene pool. *American Journal of Human Genetics*, **61**, 247–51.

Cavalli-Sforza, L.L., Menozzi, P. and Piazza, A. (1994) *The history and geography of human genes*. Princeton University Press, Princeton.

Cavalli-Sforza, L.L., Piazza, A., Menozzi, P. and Mountain, J. (1988) Reconstruction of human evolution, bringing together genetic, archaeological, and linguistic data. *Proceedings of the National Academy of Sciences of the USA*, **85**, 6002–6.

Cavener, D., Feng, Y., Foster, B., Krasney, P., Murtha, M., Schonbaum, C. and Xiao, X. (1988) The YYGG box (CTGA): a conserved dipyrimidine–dipurine sequence element in *Drosophila* and other eukaryotes. *Nucleic Acids Research*, **16**, 3375–90.

Chakraborty, R. and Jin, L. (1993) A unified approach to study hypervariable polymorphisms: Statistical considerations of determining relatedness and population distances, pp. 153–175 in *DNA Fingerprinting: State of the Science*, edited by S.D.J. Pena, R. Chakraborty, J.T. Epplen and A.J. Jeffreys. Birkhauser, Basel.

Chakraborty, R. and Weiss, K.M. (1988) Admixture as a tool for finding linked genes and detecting that difference from allelic association between loci. *Proceedings of the National Academy of Sciences of the USA*, **85**, 9119–23.

Chakraborty, R., Stivers, D.N., Deka, R., Yu, L.M., Shriver, M.D. and Ferrell, R.E. (1996) Segregation distortion of the CTG repeats at the myotonic dystrophy locus. *American Journal of Human Genetics*, **59**, 109–18.

Chakraborty, R., Kimmel, M., Stivers, D.N., Davison, J. and Deka, R. (1997) Relative mutation rates at di-, tri-, and tetranucleotide microsatellite loci. *Proceedings of the National Academy of Sciences of the USA*, **94**, 1041–6.

Chamberlain, N.L., Driver, E.D. and Miesfeld, R.L. (1994) The length and location of CAG trinucleotide repeats in the androgen receptor N-terminal domain affect transactivation function. *Nucleic Acids Research*, **22**, 3181–6.

Chambers, S.R., Hunter, N., Louis, E.J. and Borts, R.H. (1996) The mismatch repair system reduces meiotic homologous recombination and stimulates recombination-dependent chromosome loss. *Molecular Cell. Biology*, **16**, 6110–20.

Chang, K.-C., Hansen, E., Jaenicke, T., Goldspink, G. and Butterworth, P. (1992) Transformation of a novel direct-repeat repressor element into a promoter and enhancer by multimerisation. *Nucleic Acids Research*, **20**, 1669–74.

Charlesworth, B., Morgan, M.T. and Charlesworth, D. (1993) The effect of deleterious mutations on neutral molecular variation. *Genetics*, **134**, 1289–303.

Charlesworth, D. and Charlesworth, B. (1987) Inbreeding depression and its evolutionary consequences. *Annual Review of Ecology and Systematics*, **18**, 237–68.

Chee, M., Yang, R., Hubbell, E., Berno, A., Huang, X.C., Stern, D. *et al.* (1996) Accessing genetic information with high-density DNA arrays. *Science*, **274**, 610–14.

Chen, S., Supakar, P.C., Vellanoweth, R.L., Song, C.S., Chatterjee, B. and Roy, A.K. (1997) Functional role of a conformationally flexible homopurine/homopyrimidine domain of the androgen receptor gene promoter interacting with SP1 and a pyrimidine single strand DNA-binding protein. *Molecular Endocrinology*, **11**, 3–15.

Chen, Y. and Roxby, R. (1997) Identification of a functional CT-element in the Phytophthora infestans piypt1 gene promoter. *Gene*, **198**, 159–64.

Cheng, J.W., Chou, S.H. and Reid, B.R. (1992) Base pairing geometry in GA mismatches depends entirely on the neighbouring sequence. *Journal of Molecular Biology*, **228**, 1037–41.

Chong, S.S., McCall, A.E., Cota, J., Subramony, S.H., Orr, H.T., Hughes, M.R. and Zoghbi, H.Y. (1995) Gametic and somatic tissue-specific heterogeneity of the expanded SCA1 CAG repeat in spinocerebellar ataxia type 1. *Nature Genetics*, **10**, 344–350.

Chong, S.S., Almqvist, E., Telenius, H., LaTray, L., Nichol, K., Bourdelat-Parks, B. *et al.* (1997) Contribution of DNA sequence and CAG size to mutation frequencies of intermediate alleles for Huntington disease: Evidence from single sperm analyses. *Human Molecular Genetics*, **6**, 301–9.

Chung, M.Y., Ranum, L.P., Duvick, L.A., Servadio, A., Zoghbi, H.Y. and Orr, H.T. (1993) Evidence for a mechanism predisposing to intergenerational CAG repeat instability in spinocerebellar ataxia type I. *Nature Genetics*, **5**, 254–8.

Ciminelli, B., Pompei, F., Malaspina, P., Hammer, M.F., Persichetti, F., Pignatti, P. *et al.* (1995) Recurrent simple tandem repeat mutations during human Y chromosome radiation in Caucasian subpopulations. *Journal of Molecular Evolution*, **41**, 966–76.

Clutton-Brock, T.H. and Albon, S.D. (1989) *Red deer in the Highlands*. Blackwell, Oxford.

Clutton-Brock, T.H., Guinness, F.E. and Albon, S.D. (1982) *Red deer: behaviour and ecology of two sexes*. University of Chicago Press, Chicago.

Collick, A. and Jeffreys, A.J. (1990) Detection of a novel minisatellite-specific DNA-binding protein. *Nucleic Acids Research*, **18**, 625–9.

Collins, I. and Newlon, C.S. (1996) Meiosis-specific formation of joint DNA molecules containing sequences from homologous chromosomes. *Cell*, **76**, 65–75.

Coltman, D.W., Bowen, W.D. and Wright, J.M. (1996) PCR primers for harbour seal (*Phoca vitulina concolour*) microsatellites amplify polymorphic loci in several pinniped species. *Molecular Ecology*, **5**, 161–3.

Coltman, D.W., Bowen, W.D. and Wright, J.M. (1998*a*) Birth weight and neonatal survival of harbour seal pups are positively correlated with genetic variation measured by microsatellites. *Proceedings of the Royal Society of London* B, **265**, 803–809.

Coltman, D.W., Bowen, W.D. and Wright, J.M. (1998*b*) Male mating success in an aquatically mating pinniped, the harbour seal (*Phoca vitulina*), assessed by microsatellite DNA markers. *Molecular Ecology*, **7**, 627–38.

Conway, K., Edmiston, S.N., Hulka, B.S., Garrett, P.A. and Liu, E.T. (1996) Internal sequence variation in the Ha-*ras* variable number tandem repeat rare and common alleles identified by minisatellite variant repeat polymerase chain reaction. *Cancer Research*, **56**, 4773–7.

Cooper, D.N. and Krawczak, M. (1993) *Human Gene Mutation*. Bios, Oxford.

Cooper, G. (1995) *Analysis of genetic variation and sperm competition in dragonflies*. D.Phil. Oxford University.

Cooper, G., Amos, W., Hoffman, D. and Rubinsztein, D.C. (1996) Network analysis of human Y microsatellite haplotypes. *Human Molecular Genetics*, **5**, 1759–66.

Cooper, J.K., Schilling, G., Sharp, A., Kaminsky, Z.A., Masone, J., Gosh, A. *et al.* (1997) Distribution of huntingin *n*-terminus in transiently transfected cells. *American Journal of Human Genetics*, **61** (Supp), A306.

Coote, T. and Bruford, M.W. (1996) A set of human microsatellites amplify polymorphic markers in Old World apes and monkeys. *Journal of Heredity*, **87**, 406–10.

Cornuet, J.-M. and Luikart, G. (1996) Description and power analysis of two tests for detecting recent population bottlenecks from allele frequency data. *Genetics*, **144**, 2001–14.

Cossée, M., Schmitt, M., Campuzano, V., Reutenauer, L., Moutou, C., Mandel, J.-L. *et al.* (1997) Evolution of the Friedreich's ataxia trinucleotide repeat expansion: Founder effect and premutations. *Proceedings of the National Academy of Sciences of the USA*, **94**, 7452–7.

Coulson, T., Pemberton, J., Albon, S., Beaumont, M., Marshall, T., Slate, J. *et al.* (1998) Microsatellites reveal heterosis in red deer. *Proceedings of the Royal Society of London* B, **265**, 489–95.

Coward, P., Nagai, K., Chen, D., Thomas, H.D., Nagamine, C.M. and Yun-Fai, C.L. (1994) Polymorphism of a CAG trinucleotide repeat within *Sry* correlates with B6.YDom sex reversal. *Nature Genetics*, **6**, 245–50.

Crawford, A.M. and Cuthbertson, R.P. (1996) Mutations in sheep microsatellites. *Genome Research*, **6**, 876–9.

Crawford, A., Knappes, S.M., Paterson, K.A. *et al.* (1998) Microsatellite evolution: testing the ascertainment bias hypothesis. *Journal of Molecular Evolution*, **46**, 256–60.

Crawford, M.H. and Enciso, V.B. (1982) Population structure of circumpolar groups of Siberia, Alaska, Canada, and Greenland. In *Current developments in anthropological genetics*. (ed. M.H. Crawford and J.H. Mielke), Vol. 2, pp. 133–206, Plenum Press, New York.

Crawford, M.H., Williams, J.T. and Duggirala, R. (1997) Genetic structure of the indigenous populations of Siberia. *American Journal of Physical Anthropology*, **104**, 177–92.

Crow, J.F. and Kimura, M. (1970) *An introduction to population genetics theory*. Harper and Row, New York, Evanston and London.

Cullen, M., Erlich, H., Klitz, W. and Carrington, M. (1995) Molecular mapping of a recombination hotspot located in the second intron of the human TAP2 locus. *American Journal of Human Genetics*, **56**, 1350–8.

Dallas, J.F. (1992) Estimation of microsatellite mutation rates in recombinant inbred strains of mouse. *Mammalian Genome*, **3**, 452–6.

Datta, A. and Jinks-Robertson, S. (1995) Association of increased spontaneous mutation rates with high levels of transcription in yeast. *Science*, **268**, 1616–19.

David, G., Abbas, N., Stevanin, G., Dürr, A., Yvert, G., Cancel, G. *et al.* (1997) Cloning of the SCA7 gene reveals a highly unstable CAG repeat expansion. *Nature Genetics*, **17**, 65–70.

David, J.R. and Capy, P. (1988) Genetic variation of *Drosophila melanogaster* natural populations. *Trends in Genetics*, **4**, 106–11.

Davies, J.L., Kawaguchi, Y., Bennett, S.T., Copeman, J.B., Cordell, H.J., Pritchard, L.E. *et al.* (1994) A genome-wide search for human type 1 diabetes susceptibility genes. *Nature*, **371**, 130–6.

Davies, S.W., Turmaine, M., Cozens, B.A., DiFiglia, M., Sharp, A.H., Ross, C.A. *et al.* (1997) Formation of neuronal intranuclear inclusions underlies the neurological dysfunction in mice transgenic for the HD mutation. *Cell*, **90**, 537–48.

de Knijff, P., Kayser, M., Caglià, A., Corach, D., Fretwell, N., Gehrig, C. *et al.* (1997) Chromosome Y microsatellites: population genetic and evolutionary aspects. *International Journal of Legal Medicine*, **110**, 134–40.

Dean, M., Stephens, J.C., Winkler, C., Lomb, D.A., Ramsburg, M., Boaze, R. *et al.* (1994) Polymorphic admixture typing in human ethnic populations. *American Journal of Human Genetics*, **55**, 788–808.

Dean, M., Carrington, M., Winkler, C., Huttley, G.A., Smith, M.W., Allikmets, R. *et al.* (1996) Genetic restriction of HIV-1 infection and progression to AIDS by a deletion allele of the CKR5 structural gene. *Science*, **273**, 1856–62.

Deka, R., Chakraborty, R. and Ferrell, R.E. (1991) A population genetic study of six VNTR loci in three ethnically defined populations. *Genomics*, **11**, 83–92.

Deka, R., Shriver, M.D., Yu, L.M., Aston, C., Chakraborty, R. and Ferrel, R. (1994) Conservation of human chromosome 13 polymorphic microsatellite $(CA)_n$ repeats in chimpanzees. *Genomics*, **22**, 226–30.

Deka, R., Shriver, M.D., Yu, L.M., Ferrell, R.E. and Chakraborty, R. (1995*a*) Intra- and inter-population diversity at short tandem repeat loci in diverse populations of the world. *Electrophoresis*, **16**, 1659–64.

Deka, R., Jin, L., Shriver, M.D., Yu, L.M., DeCroo, S., Hundrieser, J. *et al.* (1995*b*) Population genetics of dinucleotide $(dC-dA)_n \cdot (dG-dT)_n$ polymorphisms in world populations. *American Journal of Human Genetics*, **56**, 461–74.

Deka, R., Jin, L., Shriver, M.D., Yu, L.M., Saha, N., Barrantes, R. *et al.* (1996) Dispersion of human Y chromosome haplotypes based on five microsatellites in global populations. *Genome Research*, **6**, 1177–84.

Deka, R., Heidrich-O'Hare, E.M., Wu, L.M., Shriver, M.D., Ferrell, R.E., Zhong, Y. *et al.* (1997) Interlocus variation at triplet-repeat loci suggests higher mutation rates for anonymous and disease-causing loci compared with gene-associated repeats *American Journal of Human Genetics*, **61**, A307.

Deng, H.K., Liu, R., Ellmeier, W., Choe, S., Unutmaz, D., Burkhart, M. *et al.* (1996) Identification of a major coreceptor for primary isolates of HIV-1. *Nature*, **381**, 661–6.

Di Rienzo, A., Peterson, A.C., Garza, J.C., Valdes, A.M., Slatkin, M. and Freimer, N.B. (1994) Mutational processes of simple-sequence repeat loci in human populations. *Proceedings of the National Academy of Sciences of the USA*, **91**, 3166–70.

Di Rienzo, A., Toomajian, C., Sisk, B., Haines, K., Barch, D. and Donnelly, P. (1995) STRP variation in human populations and their patterns of somatic mutations in cancer patients. *American Journal of Human Genetics*, **57** (Suppl.), A41.

Di Rienzo, A., Donnelly, P., Toomajian, C., Sisk, B., Hill, A., Petzl-Erler, M., *et al.* (1998) Heterogeneity of microsatellite mutations within and between loci, and implications for human demographic histories. *Genetics*, **148**, 1269–84.

Dib, C., Fauré, S., Fizames, C., Samson, D., Drouot, N., Vignal, A. *et al.* (1996) A comprehensive genetic map of the human genome based on 5264 microsatellites. *Nature*, **380**, 152–4.

Dietrich, W.F., Miller, J., Steen, R., Merchant, M.A., Damron-Boles, D., Husain, Z. *et al.* (1996) A comprehensive genetic map of the mouse genome. *Nature*, **380**, 149–52.

DiFiglia, M., Sapp, E., Chase, K.O., Davies, S.W., Bates, G.P., Vonsattel, J.P. *et al.* (1997) Aggregation of huntingtin in neuronal intranuclear inclusions and dystrophic neurites in brain. *Science*, **277**, 1990–3.

Dohet, C., Wagner, R. and Radman, M. (1986) Methyl-directed repair of frameshift mutations in heteroduplex DNA. *Proceedings of the National Academy of Sciences of the USA*, **83**, 3395–7.

Donnelly, P. (1995) Match probability calculations for multi-locus DNA profiles. *Genetica*, **96**, 55–67.

Donnelly, P. (1996) Interpreting genetic variability: the effects of shared evolutionary history. In *Variation in the Human Genome*, Ciba Foundation Symposium 197, Wiley, Chichester.

Donnelly, P. and Kurtz, T.G. (1999) Genealogical processes in the Fleming-Viot process with selection and recombination. *Annals of Probability*, in press.

Donnelly, P. and Tavaré, S. (1995) Coalescents and genealogical structure under neutrality. *Annual Review of Genetics*, **29**, 401–21.

Dover, G.A. (1982) Molecular drive: a cohesive mode of species evolution. *Nature*, **299**, 111–17.

Duyao, M., Ambrose, C., Myers, R., Novelletto, A., Persichetti, F., Frontali, M. *et al.* (1993) Trinucleotide repeat length instability and age of onset in Huntington's disease. *Nature Genetics*, **4**, 387–92.

Duyao, M.P., Auerbach, A.B., Ryan, A., Persichetti, F., Barnes, G.T., McNeil, S.M. *et al.* (1995) Inactivation of the mouse Huntington's disease gene homologue *Hdh*. *Science*, **269**, 407–10.

Ebstein, R.P., Novick, O., Umansky, R., Priel, B., Osher, Y., Blaine, D. *et al.* (1996) Dopamine D4 receptor (*D4DR*) exon III polymorphism associated with the human personality trait of Novelty Seeking. *Nature Genetics*, **12**, 78–80.

Echols, H. and Goodman, M.F. (1991) Fidelity mechanisms in DNA replication. *Annual Review of Biochemistry*, **60**, 477–511.

Ede, A.J., Pierson, C.A. and Crawford, A.M. (1995) Ovine microsatellites at the OarCP9, OarCP16, OarCP20, OarCP23, and OarCP26 loci. *Animal Genetics*, **26**, 129–30.

Edwards, A., Civitello, A., Hammond, H.A. and Caskey, C.T. (1991) DNA typing and genetic mapping with trimeric and tetrameric tandem repeats. *American Journal of Human Genetics*, **49**, 746–56.

Edwards, A., Hammond, H.A., Jin, L., Caskey, C.T. and Chakraborty, R. (1992) Genetic variation at five trimeric and tetrameric tandem repeat loci in four human population groups. *Genomics*, **12**, 241–53.

Eichler, E.E., Holden, J.J.A., Popovich, B.W., Reiss, A.L., Snow, K., Thibodeau, S.N. *et al.* (1994) Length of interrupted CGG repeats determines instability in the FMR1 gene. *Nature Genetics*, **8**, 88–94.

Eichler, E.E., Hammond, H.A., Macpherson, J.N., Ward, P.A. and Nelson, D.L. (1995*a*) Population survey of the human FMR1 CGG repeat substructure suggests biased polarity for the loss of AGG interruptions. *Human Molecular Genetics*, **4**, 2199–208.

Eichler, E.E., Kunst, C.B., Lugenbeel, K.A., Ryder, O.A., Davison, D., Warren, S.T. *et al.* (1995*b*) Evolution of the cryptic FMR1 CGG repeat. *Nature Genetics*, **11**, 301–7.

Eisen, J.A. (1998) A phylogenomic study of the Muts family of proteins. *Nucleic Acids Research*, **26**, 4291–300.

Eisen, J.A., Kaiser, D. and Myers, R.M. (1997) Gastrogenomic delights: a movable feast. *Nature Medicine*, **3**, 1076–8.

el Hassan, M.A. and Calladine, C.R. (1996) Propeller-twisting of base-pairs and the conformational mobility of dinucleotide steps in DNA. *Journal of Molecular Biology*, **259**, 95–103.

Ellegren, H. (1995) Mutation rates at porcine microsatellite loci. *Mammalian Genome*, **6**, 376–7.

Ellegren, H., Primmer, C.R. and Sheldon, B.C. (1995) Microsatellite evolution: directionality or bias in locus selection. *Nature Genetics*, **11**, 360–2.

Ellegren, H., Moore, S., Robinson, N., Byrne, K., Ward, W. and Sheldon, B.C. (1997) Microsatellite evolution—A reciprocal study of repeat lengths at homologous loci in cattle and sheep. *Molecular Biology and Evolution*, **14**, 854–60.

Ellsworth, D.L., Shriver, M.D. and Boerwinkle, E. (1995) Nucleotide sequence analysis of the Apolipoprotein B 3′-VNTR. *Human Molecular Genetics*, **4**, 937–44.

Epplen, J.T., Kyas, A. and Mäueler, W. (1996) Genomic simple repetitive DNAs are targets for differential binding of nuclear proteins. *FEBS Letters*, **389**, 92–5.

Erlich, H.A., Bengstom, T.F., Stoneking, M. and Gyllensten, U. (1996) HLA sequence polymorphism and the origin of humans. *Science*, **274**, 1552–4.

Estoup, A., Solignac, M., Harry, M. and Cornuet, J.-M. (1993) Characterization of (GT)$_n$ and (CT)$_n$ microsatellites in two insect species: *Apis mellifera* and *Bombus terrestris*. *Nucleic Acids Research*, **21**, 1427–31.

Estoup, A., Garnery, L., Solignac, M. and Cornuet, J.-M. (1995*a*) Microsatellite variation in honey bee (*Apis mellifera* L.) populations: hierarchical genetic structure and test of the infinite allele and stepwise mutation models. *Genetics*, **140**, 679–95.

Estoup, A., Tailliez, C., Cornuet, J.-M. and Solignac, M. (1995*b*) Size homoplasy and mutational processes of interrupted microsatellites in two bee species, *Apis mellifera* and *Bombus terrestris* (Apidae) *Molecular Biology and Evolution*, **12**, 1074–84.

Estoup, A., Solignac, M., Cornuet, J.-M., Goudet, J. and Scholl, A. (1996) Genetic differentiation of continental and island populations of *Bombus terrestris* (Hymenoptera: Apidae) in Europe. *Molecular Ecology*, **5**, 19–31.

Estoup, A. and Angers, B. (1998) Microsatellites and minisatellites for molecular ecology: theoretical and empirical considerations. In *Advances in molecular ecology*. (ed. G.R. Carvalho), pp. 55–86, Nato Sciences Series, IOS Press.

Evett, I.W. and Weir, B.S. (1998) Interpreting DNA evidence. Sinauer, Sunderland MA.

Ewens, W.J. (1972) The sampling theory of selectively neutral alleles. *Theoretical Population Biology*, **3**, 87–112.

Ewens, W.J. (1979) *Mathematical Population Genetics*. Springer, New York.

Ewens, W.J. and Spielman, R.S. (1995) The transmission/disequilibrium test: history, subdivision, and admixture. *American Journal of Human Genetics*, **57**, 455–64.

Excoffier, L., Smouse, P.E. and Quattro, J.M. (1992) Analysis of molecular variance inferred from metric distances among DNA haplotypes: application to human mitochondrial DNA restriction data. *Genetics*, **136**, 343–59.

Feldman, M.W., Bergman, A., Pollock, D.D. and Goldstein, D.B. (1997) Microsatellite genetic distances with range constraints: analytic description and problems of estimation. *Genetics*, **145**, 207–16.

Felsenstein, J. (1981*a*) Evolutionary trees from gene frequencies and quantitative characters: finding maximum likelihood estimates. *Evolution*, **35**, 1229–42.

Felsenstein, J. (1981*b*) Evolutionary trees from DNA sequences: A maximum likelihood approach. *Journal of Molecular Evolution*, **17**, 368–76.

Felsenstein, J. (1993) PHYLIP: Phylogenetic inference package, Version 3.5p. Joseph Felsenstein and the University of Washington, Seattle.

Fiedel, S.J. (1992) *Prehistory of the Americas*. 2nd edn. Cambridge University Press, Cambridge.

Field, D. and Wills, C. (1996) Long, polymorphic microsatellites in simple organisms. *Proceedings of the Royal Society of London* B, **263**, 209–15.

Filla, A., De Michele, G., Cavalcanti, F., Pianese, L., Monticelli, A., Campanella, G. *et al.* (1996) The relationship between trinucleotide (GAA) repeat length and clinical features in Friedreich ataxia. *American Journal of Human Genetics*, **59**, 554–60.

Fishel, R. and Wilson, T. (1997) MutS homologues in mammalian cells. *Current Opinion in Genetics and Development*, **7**, 105–113.

Fitch, W.M. and Margoliash, E. (1967) Construction of phylogenetic trees. *Science*, **155**, 279–84.

FitzSimmons, N.N., Moritz, C. and Moore, S.S. (1995) Conservation and dynamics of microsatellite loci over 300 million years of marine turtle evolution. *Molecular Biology and Evolution*, **12**, 432–40.

Frankham, R. (1995) Conservation genetics. *Annual Review of Genetics*, **29**, 305–27.

Freimer, N.B. and Slatkin, M. (1996) Microsatellites: evolution and mutational processes. In *Variation in the human genome, Ciba Foundation Symposium* (ed. D. Chadwick and G. Cardew), pp. 51–72. Wiley, Chichester.

Fresco, J.R. and Alberts, B.M. (1960) The accommodation of noncomplementary bases in helical polyribonucleotides and deoxyribonucleic acids. *Proceedings of the National Academy of Sciences of the USA*, **85**, 311–21.

Freudenreich, C.H., Stavenhagen, J.B. and Zakian, V.A. (1997) Stability of a CTG/CAG trinucleotide repeat in yeast is dependent on its orientation in the genome. *Molecular and Cell Biology*, **17**, 2090–8.

Fu, Y.H., Kuhl, D.P., Pizzuti, A., Pieretti, M., Sutcliffe, J.S., Richards, S. *et al.* (1991) Variation of the CGG repeat at the fragile X site results in genetic instability: resolution of the Sherman paradox. *Cell*, **67**, 1047–58.

Fu, Y.H., Pizzuti, A., Fenwick, R.G. Jr., King, J., Rajnarayan, S., Dunne, P.W. *et al.* (1992) An unstable triplet repeat in a gene related to myotonic muscular dystrophy. *Science*, **255**, 1256–8.

Fullerton, S.M. (1995) Allelic sequence diversity at the human $\beta$-globin locus. In *Molecular Biology and Human Diversity* (ed. A.J. Boyce and C.G.N. Mascie-Taylor), pp. 225–41. Cambridge University Press, Cambridge.

Gacy, A.M., Goellner, G., Juranic, N., Macura, S. and McMurray, C.T. (1995) Trinucleotide repeats that expand in human disease form hairpin structures in vitro. *Cell*, **81**, 533–40.

Gagneux, P., Boesch, C. and Woodruff, D.S. (1997) Microsatellite scoring errors associated with noninvasive genotyping based on nuclear DNA amplified from shed hair. *Molecular Ecology*, **6**, 861–8.

Gajic, Z. and Qureshi, M.T.J. (1995) *Lyapunov Matrix Equation in System Stability and Control*. Academic Press, San Diego.

García de León, F.J., Chikhi, L. and Bonhomme, F. (1997) Microsatellite polymorphism and population subdivision in natural populations of European sea bass *Dicentrarchus labrax* (Linnaeus, 1758) *Molecular Ecology*, **6**, 51–62.

Garcia-Moreno, J., Matocq, M.D., Roy, M.S., Geffen, E. and Wayne, R.K. (1996) Relationships and genetic purity of the endangered Mexican wolf based on analysis of microsatellite loci. *Conservation Biology*, **10**, 376–389.

Garreau, H. and Williams, J.G. (1983) Two nuclear DNA binding proteins of Dictyostelium discoideum with a high affinity for poly(dA)-poly(dT) *Nucleic Acids Research*, **11**, 8473–84.

Garza, J.C. and Freimer, N.B. (1996) Homoplasy for size at microsatellite loci in humans and chimpanzees. *Genome Research*, **6**, 211–17.

Garza, J.C., Slatkin, M. and Freimer, N.B. (1995) Microsatellite allele frequencies in humans and chimpanzees, with implications for constraints on allele size. *Molecular Biology and Evolution*, **12**, 594–603.

Gastier, J.M., Pulido, J.C., Sunden, S., Brody, T., Buetow, K.H., Murray, J.C. *et al.* (1995) Survey of trinucleotide repeats in the human genome: assessment of their utility as genetic markers. *Human Molecular Genetics*, **4**, 1829–36.

Gelman, A., Carlin, J.B., Stern, H.S. and Rubin, D.B. (1995) *Bayesian Data Analysis*. Chapman and Hall, London.

Gemmell, N.J., Allen, P.J., Goodman, S.J. and Reed, J.Z. (1997) Interspecific microsatellite markers for the study of pinniped populations. *Molecular Ecology*, **6**, 661–6.

Genstat 5 Committee (1995) *Genstat for Windows Reference Manual Supplement*. Clarendon Press, Oxford.

Georges, M. and Massey, J. (1992) *Polymorphic DNA markers in Bovidae*. World Intellectual Property Organization, Geneva.

Gerber, H.P., Seipel, K., Georgiev, O., Hofferer, M., Hug, M., Rusconi, S. *et al.* (1994) Transcriptional activation modulated by homopolymeric glutamine and proline stretches. *Science*, **263**, 808–11.

Gerloff, U., Schlötterer, C., Rassmann, C., Rambold, I., Hohmann, G., Fruth, B. *et al.* (1995) Amplification of hypervariable simple sequence repeats (microsatellites) from excremental DNA of wild living bonobos (*Pan paniscus*). *Molecular Ecology*, **4**, 515–18.

Gill, P., Evett, I.W., Woodroffe, S., Lygo, J.E., Millican, E.S. and Webster, M. (1991) Databases, quality control and interpretation of DNA profiling in the Home Office Forensic Science Service. *Electrophoresis*, **12**, 204–9.

Gill, P., Sparkes, R. and Kimpton, C. (1997) Development of guidelines to designate alleles using an STR multiplex system. *Forensic Science International*, **89**, 185–97.

Gillies, S.D., Folsom, B. and Tonegawa, S. (1984) Cell-type specific enhancer element associated with a mouse MHC gene, Eb. *Nature*, **310**, 594–7.

Gilmour, D.S., Thomas, G.H. and Elgin, S.C.R. (1989) *Drosophila* nuclear proteins bind to regions of alternating C and T residues in gene promoters. *Science*, **245**, 1487–90.

Giniger, E. and Ptashne, M. (1988) Cooperative DNA binding of the yeast transcriptional activator GAL4. *Proceedings of the National Academy of Sciences of the USA*, **85**, 382–6.

Ginot, F., Bordelais, I., Nguyen, S. and Gyapay, G. (1996) Correction of some genotyping errors in automated fluorescent microsatellite analysis by enzymatic removal of one base overhangs. *Nucleic Acids Research*, **24**, 540–1.

Glenn, T.C., Stephan, W., Dessauer, H.C. and Braun, M.J. (1996) Allelic diversity in Alligator microsatellite loci is negatively correlated with GC content of flanking sequences and evolutionary conservation of PCR amplifiability. *Molecular Biology and Evolution*, **13**, 1151–4.

Goldberg, Y.P., McMurray, C.T., Zeisler, J., Almqvist, E., Sillence, D., Richards, F. *et al.* (1995) Increased instability of intermediate alleles in families with sporadic Huntington disease compared to similar sized intermediate alleles in the general population. *Human Molecular Genetics*, **4**, 1911–18.

Goldberg, Y.P., Kalchman, M.A., Metzler, M., Nasir, J., Zeisler, J., Graham, R. *et al.* (1996) Absence of disease phenotype and intergenerational stability of the CAG repeat in transgenic mice expressing the human Huntington disease transcript. *Human Molecular Genetics*, **5**, 177–85.

Goldman, A., Ramsay, M. and Jenkins, T. (1994) Absence of myotonic dystrophy in Southern African Negroids is associated with a significantly lower number of CTG trinucleotide repeats. *Journal of Medical Genetics*, **31**, 37–40.

Goldstein, D.B. and Clark, A.G. (1995) Microsatellite variation in north American populations of *Drosophila melanogaster*. *Nucleic Acids Research*, **23**, 3882–6.

Goldstein, D.B. and Pollock, D.D. (1994) Least squares estimation of molecular distance—noise abatement in phylogenetic reconstruction. *Theoretical Population Biology*, **45**, 219–26.

Goldstein, D.B. and Pollock, D.D. (1997) Launching microsatellites: a review of mutation processes and methods of phylogenetic inference. *Journal of Heredity*, **88**, 335–42.

Goldstein, D.B., Ruiz Linares, A., Cavalli-Sforza, L.L. and Feldman, M.W. (1995*a*) Genetic absolute dating based on microsatellites and the origin of modern humans. *Proceedings of the National Academy of Sciences of the USA*, **92**, 6723–7.

Goldstein, D.B., Ruiz Linares, A., Cavalli-Sforza, L.L. and Feldman, M.W. (1995*b*) An evaluation of genetic distances for use with microsatellite loci. *Genetics*, **139**, 463–71.

Goldstein, D.B., Zhivotovsky, L.A., Nayar K., Ruiz Linares, A., Cavalli-Sforza, L.L. and Feldman, M.W. (1997) Statistical properties of the variation at linked microsatellite loci: implications for the history of human Y chromosomes. *Molecular Biology and Evolution*, **13**, 1213–18.

Goldstein, D.B., Roemer, G.W., Smith, D., Reich, D.E., Bergman, A. and Wayne, R.K. (1998) The use of microsatellite variation to infer patterns of migration, population structure and demographic history: an evaluation of methods in a natural model system. In press.

Goodman, F.R., Mundlos, S., Muragaki, Y., Donnai, D., Giovannucci-Uzielli, M.L., Lapi, E. *et al.* (1997) Synpolydactyly phenotypes correlate with size of expansions in HOXD13 polyalanine tract. *Proceedings of the National Academy of Sciences of the USA*, **94**, 7458–63.

Goodman, S.J. (1997) Dinucleotide repeat polymorphism at seven anonymous microsatellite loci cloned from the European harbour seal (*Phoca vitulina vitulina*). *Animal Genetics*, **28**, 310–11.

Gordenin, D.A., Kunkel, T.A. and Resnick, M.A. (1997) Repeat expansion—all in a flap. *Nature Genetics*, **16**, 116–18.

Gordon, A.J.E. (1997) Microsatellite birth register. *Journal of Molecular Evolution*, **45**, 337–8.

Gostout, B., Liu, Q. and Sommer, S.S. (1993) 'Cryptic' repeating triplets of purines and pyrimidines (cRRY(i)) are frequent and polymorphic: analysis of coding cRRY(i) in the proopiomelanocortin (POMC) and TATA-binding protein (TBP) genes. *American Journal of Human Genetics*, **52**, 1182–90.

Gottelli, D., Sillero-Zubiri, C., Applebaum, G.D., Roy, M.S., Girman, D.J., Garcia-Moreno, J. *et al.* (1994) Molecular genetics of the most endangered canid—the Ethiopian wolf *Canis simensis*. *Molecular Ecology*, **3**, 301–12.

Gourdon, G., Radvanyi, F., Lia, A.-S., Duros, C., Blanche, M., Abitbol, M. *et al.* (1997) Moderate intergenerational and somatic instability of a 55-CTG repeat in transgenic mice. *Nature Genetics*, **15**, 190–2.

Gray, I.C. and Jeffreys, A.J. (1991) Evolutionary transience of hypervariable minisatellites in man and the primates. *Proceedings of the Royal Society of London* B, **243**, 241–53.

Green, H. (1993) Human genetic diseases due to codon reiteration: relationship to an evolutionary mechanism. *Cell*, **74**, 955–6.

Green, H. and Wang, N. (1994) Codon reiteration and the evolution of proteins. *Proceedings of the National Academy of Sciences of the USA*, **91**, 4298–302.

Green, M. and Krontiris, T. (1993) Allelic variation of reporter gene activation by the *HRAS1* minisatellite. *Genomics*, **17**, 429–34.

Greenwood, P.J., Harvey, P.H. and Perrins, C.J. (1978) Inbreeding and dispersal in the great tit. *Nature*, **271**, 52–4.

Griffiths, R.C. and Tavaré, S. (1994*a*) Simulating probability-distributions in the coalescent. *Theoretical Population Biology*, **46**, 131–59.

Griffiths, R.C. and Tavaré, S. (1994*b*) Sampling theory for neutral alleles in a varying environment. *Philosophical Transactions of the Royal Society of London* B. **344**, 403–10.

Griffiths, R.C. and Tavaré, S. (1994*c*) Ancestral inference in population genetics. *Statistical Science*, **9**, 307–19.

Grimaldi, M.-C. and Crouau-Roy, B. (1997) Microsatellite allelic homoplasy due to variable flanking sequences. *Journal of Molecular Evolution*, **44**, 336–40.

Guo, S.W. and Thompson, E.A. (1992) Performing the exact test of Hardy–Weinberg proportion for multiple alleles. *Biometrics*, **48**, 361–72.

Hackam, A.S., Ellerby, L.M., Wellington, C.L., Bredesen, D.E. and Hayden, M.R. (1997) Development of an in vitro model for Huntington's disease: Evidence that the N-terminal domain of huntingtin is toxic to cells. *American Journal of Human Genetics*, **61** (Supp), A310.

Haegert, D.G. and Marrosu, M.G. (1994) Genetic susceptibility to multiple sclerosis. *Annals of Neurology*, **36**, S204–10.

Hajdú, P. (1962) *The Samoyed peoples and languages*. Indiana University, Bloomington.

Hamada, H., Seidman, M., Howard, B.H. and Gorman, C.M. (1984) Enhanced gene expression by the poly(dT-dG)poly(dC-dA) sequence. *Molecular and Cellular Biology*, **4**, 2622–30.

Hammer, M.F. (1995) A recent common ancestry for human Y chromosomes. *Nature*, **378**, 376–8.

Hammer, M.F. and Horai, S. (1995) Y chromosomal DNA variation and the peopling of Japan. *American Journal of Human Genetics*, **56**, 951–62.

Hammer, M.F. and Zegura, S.L. (1996) The role of the Y chromosome in human evolutionary studies. *Evolutionary Anthropology*, **5**, 116–34.

Hammer, M.F., Spurdle, A.B., Karafet, T., Bonner, M.R., Wood, E.T., Novelletto, A. *et al.* (1997) The geographic distribution of human Y chromosome variation. *Genetics*, **145**, 787–805.

Hammer, M.F., Karafet, T., Rasanayagam, A., Wood, E.T., Altheide, T.K., Jenkins, T. *et al.* (1998) Out of Africa and back again: nested cladistic analysis of human Y chromosome variation. *Molecular Biology and Evolution*, **15**, 427–41.

Hammond, H.A., Jin, L., Zhong, Y., Caskey, C.T. and Chakraborty, R. (1994) Evaluation of 13 short tandem repeat loci for use in personal identification applications. *American Journal of Human Genetics*, **55**, 175–89.

Hammond, R.L., Saccheri, I.J., Ciofi, C., Coote, T., Funk, S.M., McMillan, W.O. *et al.* (1998) Isolation of microsatellite markers in animals. In *Molecular tools for screening biodiversity: plants and animals* (ed. A. Karp, P.G. Isaac and D.S. Ingram), pp. 279-285. Chapman and Hall, London.

Hamshere, M.G., Newman, E.E., Alwazzan, M., Athwal, B.S. and Brook, J.D. (1997) Transcriptional abnormality in myotonic dystrophy affects DMPK but not neighbouring genes. *Proceedings of the National Academy of Sciences of the USA*, **94**, 7394–9.

Hancock, J.M. (1993) Evolution of sequence repetition and gene duplications in the TATA-binding protein TBP (TFIID). *Nucleic Acids Research*, **21**, 2823–30.

Hancock, J.M. (1995*a*) The contribution of slippage-like processes to genome evolution. *Journal of Molecular Evolution*, **41**, 1038–47.

Hancock, J.M. (1995*b*) The contribution of DNA slippage to eukaryotic nuclear 18S rRNA evolution. *Journal of Molecular Evolution*, **40**, 629–39.

Hancock, J.M. (1996*a*) Microsatellites and other simple sequences in the evolution of the human genome. In *Human Genome Evolution* (ed. M.S. Jackson, G. Dover and T. Strachan) pp. 191–210. Bios, Oxford.

Hancock, J.M. (1996*b*) Simple sequences and the expanding genome. *BioEssays*, **18**, 421–5.

Hancock, J.M. (1996*c*) Simple sequences in a 'minimal' genome. *Nature Genetics*, **13**, 14–15.

Hancock, J.M. and Armstrong, J.S. (1994) SIMPLE34: an improved and enhanced implementation for VAX and Sun computers of the SIMPLE algorithm for analysis of clustered repetitive motifs in nucleotide sequences. *Computer Applications in Bioscience*, **10**, 67–70.

Hancock, J.M. and Dover, G.A. (1988) Molecular co-evolution among cryptically simple expansion segments of eukaryotic 26S/28S rRNAs. *Molecular Biology and Evolution*, **5**, 377–91.

Hanis, C.L., Boerwinkle, E., Chakraborty, R., Ellsworth, D.L., Concannon, P., Stirling, B. *et al.* (1996) A genome-wide search for human non-insulin-dependent (type 2) diabetes genes reveals a major susceptibility locus on chromosome 2. *Nature Genetics*, **13**, 161–6.

Harding, R.M., Fullerton, S.M., Griffiths, R.C., Bond, J., Cox, M.J., Schneider, J.A. *et al.* (1997) Archaic African and Asian lineages in the genetic ancestry of modern humans. *American Journal of Human Genetics*, **60**, 772–89.

Hardy, D.O., Sher, H.I., Bogenreider, T., Sabbatini, P., Zhang, A.F., Nanus, D.M. *et al.* (1996) Androgen receptor CAG repeat lengths in prostate cancer: correlation with age at onset. *Journal of Clinical Endocrinology and Metabolism*, **81**, 4400–5.

Harley, H.G., Brook, J.D., Rundle, S.A., Crow, S., Reardon, W., Buckler, A.J. *et al.* (1992) Expansion of an unstable DNA region and phenotypic variation in myotonic dystrophy. *Nature*, **355**, 545–6.

Harley, H.G., Rundle, S.A., MacMillan, J.C., Myring, J., Brook, J.D., Crow, S. *et al.* (1993) Size of the unstable CTG repeat sequence in relation to phenotype and parental transmission in myotonic dystrophy. *American Journal of Human Genetics*, **52**, 1164–74.

Harper, P.S. (1989) *Myotonic dystrophy.* 2nd edn. Saunders, London and Philadelphia.

Harr, B., Zangerl, B., Brem, G. and Schlötterer, C. (1998) Conservation of locus specific microsatellite variability across species: a comparison of two *Drosophila* sibling species *D. melanogaster* and *D. simulans. Molecular Biology and Evolution*, **15**, 176–84.

Harris, A.S., Young, J.S. and Wright, J.M. (1991) DNA fingerprinting of harbour seals (*Phoca vitulina concolour*): male mating behaviour may not be a reliable indicator of reproductive success. *Canadian Journal of Zoology*, **69**, 1862–6.

Harris, R.S., Feng, G., Ross, K.J., Sidhu, R., Thulin, C., Longerich, S. *et al.* (1997) Mismatch repair protein MutL becomes limiting during stationary-phase mutation. *Genes and Development*, **11**, 2426–37.

Hastbacka, J., de la Chapelle, A., Kaitila, A., Sistonen, I., Weaver, A. and Lander, E. (1992) Linkage disequilibrium mapping in isolated founder populations: dystrophic dysplania in Finland. *Nature Genetics*, **2**, 204–11.

Hawley, M.E. and Kidd, K.K. (1995) HAPLO: a program using the EM algorithm to estimate the frequencies of multi-site haplotypes. *Journal of Heredity*, **86**, 409–11.

Hayashi, K. (1991) PCR-SSCP: a simple and sensitive method for detection of mutations in the genomic DNA. *PCR Methods and Applications*, **1**, 34–8.

Hayes, T.E. and Dixon, J.E. (1985) Z-DNA in the rat somatostatin gene. *Journal of Biological Chemistry*, **260**, 8145–56.

Hearne, C.M., Gosh, S. and Todd, J.A. (1992) Microsatellites for linkage analysis of genetic traits. *Trends in Genetics*, **8**, 288–94.

Hedrick, P.W. and Miller, P.S. (1992) Conservation Genetics—techniques and fundamentals. *Ecology Applications*, **2**, 30–46.

Henderson, S.T. and Petes, T.D. (1992) Instability of simple sequence DNA in *Saccharomyces cerevisiae. Molecular and Cell Biology*, **12**, 2749–57.

Heshel, M.S. and Paige, K.N. (1995) Inbreeding depression, environmental stress, and population size variation in scarlet gilia (*Ipomopsis aggregata*) *Conservation Biology*, **9**, 126–33.

Hess, S.T., Blake, J.D. and Blake, R.D. (1994) Wide variation in neighbour-dependent substitution rates. *Journal of Molecular Biology*, **236**, 1022–33.

Hey, J. (1997) Mitochondrial and nuclear genes present conflicting portraits of human origins. *Molecular Biology and Evolution*, **14**, 166–172.

Heyer, E., Puymirat, J., Dieltjes, P., Bakker, E. and de Knijff, P. (1997) Estimating Y chromosome specific microsatellite mutation frequencies using deep rooting pedigrees. *Human Molecular Genetics*, **6**, 799–803.

Hino, O., Testa, J., Buetow, K., Taguchi, T., Zhou, Y., Bremer, M. *et al.* (1993) Universal mapping probes and the origin of human chromosome 3. *Proceedings of the National Academy of Sciences of the USA*, **90**, 730–4.

Hirst, M.C., Grewal, P.K. and Davies, K.E. (1994) Precursor arrays for triplet repeat expansion at the fragile X locus. *Human Molecular Genetics*, **3**, 1553–60.

Hodgson, G., Smith, D., Nichol, K., Bissada, N., McCutcheon, K., LePaine, F. *et al.* (1997) Development of YAC transgenic mice expressing mutant huntingtin containing different sized polyglutamine tracts. *American Journal of Human Genetics*, **61**, A52.

Hoffman, E.K., Trusko, S.P., Murphy, M. and George, D.L. (1990) An S1 nuclease-sensitive homopurine/homopyrimidine domain in the c-Ki-ras promoter interacts with a nuclear factor. *Proceedings of the National Academy of Sciences of the USA*, **87**, 2705–9.

Hoffman, S.M.G. and Brown, W.M. (1995) The molecular mechanism underlying the rare allele phenomenon in a subspecific hybrid zone of the California fieldmouse, *Peromyscus californicus*. *Journal of Molecular Evolution*, **41**, 1165–9.

Hollingsworth, M.A., Closken, C., Harris, A., McDonald, C.D., Pahwa, G.S. and Maher, L.J. III. (1994) A nuclear factor that binds purine-rich single-stranded oligonucleotides derived from S1-sensitivie elements upstream of the *CFTR* gene and the *MUC1* gene. *Nucleic Acids Research*, **22**, 1138–46.

Hoogland, J.L. (1992) Levels of inbreeding among prairie dogs. *American Naturalist*, **139**, 591–602.

Hori, R. and Firtel, R.A. (1994) Identification and characterization of multiple A/T rich cis-acting elements that control expression from *Dictyostelium* actin promoters: the *Dictyostelium* actin upstream activating sequence confers growth phase expression and has enhancer-like properties. *Nucleic Acids Research*, **22**, 5099–111.

Hornstra, I.K., Nelson, D.L., Warren, S.T. and Yang, T.P. (1993) High resolution methylation analysis of the *FMR1* gene in fragile X syndrome. *Human Molecular Genetics*, **2**, 1659–65.

Hudson, R.R. (1990) Gene genealogies and the coalescent process. *Oxford Surveys of Evolutionary Biology*, **7**, 1–44.

Hudson, R.R. (1992) The how and why of generating gene genealogies. In *Mechanisms of Molecular Evolution* (ed. N. Takahata and A.G. Clark) pp. 23–36. Sinauer, Sunderland.

Hudson, R.R., Sáez, A.G. and Ayala, F.J. (1997) DNA variation at the *Sod* locus of *Drosophila melanogaster*: an unfolding story of natural selection. *Proceedings of the National Academy of Sciences of the USA*, **94**, 7725–9.

Hughes, C.R. and Queller, D.C. (1993) Detection of highly polymorphic microsatellite loci in a species with little allozyme polymorphism. *Molecular Ecology*, **2**, 131–7.

Huijser, P., Hennig, W. and Dijkhof, R. (1987) Poly(dC-dA/dG-dT) repeats in the *Drosophila* genome: a key function for dosage compensation and position effect? *Chromosoma*, **95**, 209–15.

Huntington's Disease Collaborative Research Group (1993) A novel gene containing a trinucleotide repeat that is expanded and unstable on Huntington's disease chromosomes. *Cell*, **72**, 971–83.

Hurst, G.D., Hurst, L.D., Barrett, J.A., *et al.* (1995) Meiotic drive and myotonic dystrophy. *Nature Genetics*, **10**, 132–3.

Ikeda, H., Yamaguchi, M., Sugai, S., Aze, Y., Narumiya, S. and Kakizuka, A. (1996) Expanded polyglutamine in the Machado-Joseph disease protein induces cell death *in vitro* and *in vivo*. *Nature Genetics*, **13**, 196–202.

Ikeuchi, T., Igarshi, S., Takiyama, Y., Onodera, O., Oyake, M., Takano, H. *et al.* (1996) Non-Mendelian transmission in dentatorubral-pallidoluysian atrophy and Machado-Joseph disease: the mutant allele is preferentially transmitted in male meiosis. *American Journal of Human Genetics*, **58**, 730–3.

Imanishi, T., Azaka, T., Kimura, A., Tokunaga, K. and Gojobori, T. (1992) Allele and haplotype frequencies for HLA and complement loci in various ethnic groups. In *HLA 1991. Proceedings of the Eleventh International Histocompatibility Workshop and Conference*, vol. 5 (ed. K. Tsuji, M. Aizawa and T. Sasazuki), pp. 1065–1220. Oxford University Press, Oxford.

Imbert, G., Kretz, C., Johnson, K. and Mandel, J.-L. (1993) Origin of the expansion mutation in myotonic dystrophy. *Nature Genetics*, **4**, 72–6.

Imbert, G., Saudou, F., Yvert, G., Devys, D., Trottier, Y., Garnier, J.-M. *et al.* (1996) Cloning of the gene for spinocerebellar ataxia 2 reveals a locus with high sensitivity to expanded CAG/glutamine repeats. *Nature Genetics*, **14**, 285–91.

Ionov, Y., Peinado, M.A., Malkhosyan, S., Shibata, D. and Perucho, M. (1993) Ubiquitous somatic mutations in simple repeat sequences reveals a new mechanism for colorectal carcinogenesis. *Nature*, **363**, 558–61.

ISFH (1994) DNA Commission of the ISFH. Report concerning further recommendations of the DNA Commission of the ISFH regarding PCR-based polymorphisms in STR (short tandem repeats) *International Journal of Legal Medicine*, **107**, 159–60.

Ishii, S., Imamoto, F., Yamanashi, Y., Toyoshima, K. and Yamamoto, T. (1987) Characterization of the promoter region of the human c-erbB-2 protooncogene. *Proceedings of the National Academy of Sciences of the USA*, **84**, 4374–8.

Ito, K., Sato, K. and Endo, H. (1994) Cloning and characterization of a single-stranded DNA binding protein that specifically recognizes deoxycytidine stretch. *Nucleic Acids Research*, **22**, 53–8.

Jansen, G., Groenen, P.J., Bächner, D., Jap, P.H.K., Coerwinkel, M., Oerlemans, F. *et al.* (1996) Abnormal myotonic dystrophy protein kinase levels produce only mild myopathy in mice. *Nature Genetics*, **13**, 316–24.

Jarne, P. and Lagoda, J.L. (1996) Microsatellites, from molecules to populations and back. *Trends in Ecology and Evolution*, **11**, 424–9.

Jeeninga, R.E., Van Delft, Y., De Graaff-Vincent, M., Dirks-Mulder, A., Venema, J. and Raué, H.A. (1997) Variable regions V13 and V3 of Saccharomyces cerevisiae contain structural features essential for normal biogenesis and stability of 5.8S and 25S rRNA. *RNA*, **3**, 476–88.

Jeffreys, A.J. and Neumann, R. (1997) Somatic mutation processes at a human minisatellite. *Human Molecular Genetics*, **6**, 129–36.

Jeffreys, A.J., Wilson, V. and Thein, S.L. (1985a) Hypervariable 'minisatellite' regions in human DNA. *Nature*, **314**, 67–73.

Jeffreys, A.J., Wilson, V. and Thein, S.L. (1985b) Individual-specific 'fingerprints' of human DNA. *Nature*, **316**, 76–9.

Jeffreys, A.J., Royle, N.J., Wilson, V. and Wong, Z. (1988) Spontaneous mutation rates to new length alleles at tandem-repetitive hypervariable loci in human DNA. *Nature*, **332**, 278–81.

Jeffreys, A.J., Neumann, R. and Wilson, V. (1990) Repeat unit sequence variation in minisatellites: a novel source of DNA polymorphism for studying variation and mutation by single molecule analysis. *Cell*, **60**, 473–85.

Jeffreys, A.J., MacLeod, A., Tamaki, K., Neil, D.L. and Monckton, D.G. (1991a) Minisatellite repeat coding as a digital approach to DNA typing. *Nature*, **354**, 204–9.

Jeffreys, A.J., Turner, M. and Debenham, P. (1991b) The efficiency of multilocus DNA fingerprint probes for individualization and establishment of family relationships, determined from extensive casework. *American Journal of Human Genetics*, **48**, 824–40.

Jeffreys, A.J., Tamaki, K., MacLeod, A., Monckton, D.G., Neil, D.L. and Armour, J.A.L. (1994) Complex gene conversion events in germline mutation at human minisatellites. *Nature Genetics*, **6**, 136–45.

Jin, L. (1994) *Population genetics of VNT loci and their applications in evolutionary studies*. PhD thesis, University of Texas Graduate School of Biomedical Sciences, Houston, Texas, USA.

Jin, L., Macaubas, C., Hallmayer, J., Kimura, A. and Mignot, E. (1996) Mutation rate varies among alleles at a microsatellite locus: phylogenetic evidence. *Proceedings of the National Academy of Sciences of the USA*, **93**, 15285–8.

Jobling, M.A. and Tyler-Smith, C. (1995) Fathers and sons: the Y chromosome and human evolution. *Trends in Genetics*, **11**, 449–56.

Jobling, M.A., Pandya, A. and Tyler-Smith, C. (1997) The Y chromosome in forensic analysis and paternity testing. *International Journal of Legal Medicine*, **110**, 118–24.

Johnson, A.C., Jinno, Y. and Merlino, G.T. (1988) Modulation of epidermal growth factor receptor proto-oncogene transcription by a promoter site sensitive to S1 nuclease. *Molecular and Cellular Biology*, **8**, 4174–84.

Johnson, C.A., Densen, P., Hurford, R.K., Colten, H.R. and Wetsel, R.A. (1992) Type I human complement C2 deficiency. A 28-base pair gene deletion causes skipping of exon 6 during RNA splicing. *Journal of Biological Chemistry*, **13**, 9347–53.

Jones, M., Wagner, R. and Radman, M. (1987) Mismatch repair and recombination in *Escherichia coli. Cell*, **50**, 621–6.

Jorde, L.B. (1980) The genetic structure of subdivided human populations: a review. In *Current developments in anthropological genetics* (ed. J.H. Mielke and M.H. Crawford), pp. 133–206. Plenum Press, New York.

Jorde, L.B., Bamshad, M.J., Watkins, W.S., Zenger, R., Fraley, A.E., Krakowiak, P.A. *et al.* (1995) Origins and affinities of modern humans: a comparison of mitochondrial and nuclear genetic data. *American Journal of Human Genetics*, **57**, 523–38.

Jorde, L.B., Rogers, A.R., Bamshad, M., Watkins, W.S., Krakowiak, P., Sung, S. *et al.* (1997) Microsatellite diversity and the demographic history of modern humans. *Proceedings of the National Academy of Sciences of the USA*, **94**, 3100–3.

Joyce, P. (1994) Likelihood ratios for the infinite alleles model. *Journal of Applied Probability*, **31**, 595–605.

Kalchman, M.A., Graham, R.K., Xia, G., Koide, H.B., Hodgson, J.G., Graham, K.C. *et al.* (1996) Huntingtin is ubiquitinated and interacts with a specific ubiquitin conjugated enzyme. *Journal of Biological Chemistry*, **122**, 19385–94.

Kalchman, M.A., Koide, H.B., McCutcheon, K., Graham, R.K., Nichol, K., Nishiyama, K. *et al.* (1997) *HIP1*, a human homologue of *S. cerevisiae Sla2p*, interacts with membrane-associated huntingtin in the brain. *Nature Genetics*, **16**, 44–53.

Kandpal, R.P., Kandpal, G. and Weissman, S.M. (1994) Construction of libraries enriched for sequence repeats and jumping clones, and hybridization selection for region-specific markers. *Proceedings of the National Academy of Sciences of the USA*, **91**, 88–92.

Kang, S., Jaworski, A., Ohshima, K. and Wells, R.D. (1995) Expansion and deletion of CTG repeats from human disease genes are determined by the direction of replication in *E. coli. Nature Genetics*, **10**, 213–18.

Kang, S., Ohshima, K., Jaworski, A. and Wells, R.D. (1996) CTG triplet repeats from the myotonic dystrophy gene are expanded in Escherichia coli distal to the replication origin as a single large event. *Journal of Molecular Biology*, **258**, 543–47.

Kappe, A.L., Bijlsma, R., Osterhaus, A., Van Delden, W. and Van de Zande, L. (1997) Structure and amount of genetic variation at minisatellite loci within the subspecies complex of *Phoca vitulina* (the harbor seal). *Heredity*, **78**, 457–63.

Karafet, T.M. (1986) *Population and genetic structure of two Siberian indigenous groups: Forest Nentsi and Nganasans.* Ph.D. thesis. Vavilov Institute of General Genetics, Russian Academy of Sciences. Moscow (in Russian).

Karafet, T.M., Posukh, O.L. and Osipova, L.P. (1994) Population-genetic studies of North Siberian natives. *Siberian Journal of Ecology*, **2**, 105–118.

Karafet, T., de Knijff, P., Wood, E., Ragland, J., Clark, A. and Hammer, M.F. (1998) Different patterns of variation at the X- and Y-linked microsatellite loci *DXYS156X* and *DXYS156Y* in human populations. *Human Biology* **70**, 979–92.

Karlin, S. and Burge, C. (1996) Trinucleotide repeats and long homopeptides in genes and proteins associated with nervous system disease and development. *Proceedings of the National Academy of Sciences of the USA*, **93**, 1560–5.

Kashi, Y., King, D. and Soller, M. (1997) Simple sequence repeats as a source of quantitative genetic variation. *Trends in Genetics*, **13**, 74–8.

Kashi, Y., Tikochinsky, Y., Iraqi, F., Beckmann, J.S., Gruenbaum, Y. and Soller, M. (1990) Large DNA fragments containing poly(TG) are highly polymorphic in a variety of vertebrates. *Nucleic Acids Research*, **18**, 1129–32.

Kawaguchi, Y., Okamoto, T., Taniwaki, M., Aizawa, M., Inoue, M., Katayama, S. *et al.* (1994) CAG expansions in a novel gene for Machado-Joseph disease at chromosome 14q32.1. *Nature Genetics*, **8**, 221–7.

Kayser, M., Cagliá, A., Corach, D., Fretwell, N., Gehrig, C., Graziosi, C. *et al.* (1997) Evaluation of Y-chromosomal STRs: a multicenter study. *International Journal of Legal Medicine*, **110**, 125–33.

Kazemi-Esfarjani, P., Trifiro, M.A. and Pinsky, L. (1995) Evidence for a repressive function of the long polyglutamine tract in the human androgen receptor: Possible pathogenetic relevance for the $(CAG)_n$-expanded neuronopathies. *Human Molecular Genetics*, **4**, 523–7.

Keller, L.F., Arcese, P., Smith, J., Hochachka, W.M. and Stearns, S.C. (1994) Selection against inbred song sparrows during a natural population bottleneck. *Nature*, **372**, 356–7.

Kennedy, G.C., German, M.S. and Rutter, W.J. (1995) The minisatellite in the diabetes susceptibility locus *IDDM2* regulates insulin transcription. *Nature Genetics*, **9**, 293–8.

Kimmel, M. and Chakraborty, R. (1996) Measures of variation at DNA repeat loci under a general stepwise mutation model. *Theoretical Population Biology*, **50**, 345–67.

Kimmel, M. and Chakraborty, R. (1997) Dynamics of microsatellite loci under Markov-chain mutations and genetic drift, described by the Lyapunov differential equation. *American Journal of Human Genetics*, **61**, A203.

Kimmel, M., Chakraborty, R., Stivers, D.N. and Deka, R. (1996) Dynamics of repeat polymorphisms under a forward–backward mutation model: within- and between-population variability at microsatellite loci. *Genetics*, **143**, 549–55.

Kimmel, M., Chakraborty, R., King, J.P., Bamshad, M., Watkins, W.S. and Jorde, L.B. (1998) Signatures of population expansion in microsatellite repeat data, *Genetics*, **148**, 1921–30.

Kimpton, C., Fisher, D., Watson, S., Adams, M., Urqhart, A., Lygo, J. *et al.* (1994) Evaluation of an automated DNA profiling system employing multiplex amplification of four tetrameric STR loci. *International Journal of Legal Medicine*, **106**, 302–11.

Kimura, A. and Sasazuki, T. (1991) Eleventh International Histocompatibility Workshop reference protocol for the HLA DNA-typing technique. In *HLA 1991. Proceedings of the Eleventh International Histocompatibility Workshop and Conference* (vol 1) (ed. K. Tsuji, M. Aizawa and T. Sasazuki), p. 379. Oxford University Press, Oxford.

Kimura, M. (1983) *The neutral theory of molecular evolution*. Cambridge University Press, Cambridge.

Kimura, M. and Crow, J.F. (1964) The number of alleles that can be maintained in a finite population. *Genetics*, **49**, 725–38.

Kimura, M. and Ohta, T. (1978) Stepwise mutation model and distribution of allelic frequencies in a finite population. *Proceedings of the National Academy of Sciences of the USA*, **75**, 2868–72.

King, D.G. (1994) Triple repeat DNA as a highly mutable regulatory system. *Science*, **263**, 595–6.

King, D.G. and Soller, M. (1998) Variation and fidelity: the evolution of simple sequence repeats as functional elements in adjustable genes. In *Evolutionary theory and processes: modern perspectives*. (ed. S.P. Wasser) Oxford University Press, Oxford.

King, D.G., Soller, M. and Kashi, Y. (1997) Evolutionary tuning knobs. *Endeavour*, **21**, 36–40.

Kingman, J.F.C. (1982) Exchangeability and the evolution of large populations. In *Exchangeability in probability and statistics* (ed. G. Koch and F. Spizzichino), pp. 97–112. North-Holland, Amsterdam.

Kinzler, K.W. and Vogelstein, B. (1996) Lessons from hereditary colorectal cancer. *Cell*, **87**, 159–70.

Klesert, T.R., Otten, A.D., Bird, T.D. and Tapscott, S.J. (1997) Trinucleotide repeat expansion at the myotonic dystrophy locus reduces expression of *DMAHP*. *Nature Genetics*, **16**, 402–6.

Knight, S.J.L., Flannery, A.V., Hirst, M.C., Campbell, L., Christodoulou, Z., Phelps, S.R. *et al.* (1993) Trinucleotide repeat amplification and hypermethylation of a CpG island in *FRAXE* mental retardation. *Cell*, **74**, 127–34.

Koide, R., Ikeuchi, T., Onodera, O., Tanaka, H., Igarishi, S., Endo, K. *et al.* (1994) Unstable expansion of CAG repeat in hereditary detatorubral-pallidoluysian atrophy (DRPLA). *Nature Genetics*, **6**, 9–13.

Koide, R., Igarishi, S., Yamada, M., Takano, H., Date, H., Oyake, M. *et al.* (1997) Truncated DRPLA proteins harbouring an expanded polyglutamine tract produce aggregated bodies and induce apoptotic cell death. *American Journal of Human Genetics*, **61** (Supp), A52.

Koller, E., Hayman, A.R. and Trueb, B. (1991) The promoter of the chicken a2(VI) collagen gene has features characteristic of house-keeping genes and of proto-oncogenes. *Nucleic Acids Research*, **19**, 485–91.

Kolluri, R., Torrey, T.A., Kinniburgh, A.J. (1992) A CT promoter element binding protein: definition of a double-strand and a novel single-strand DNA binding motif. *Nucleic Acids Research*, **20**, 111–16.

Kolodner, R. (1996) Biochemistry and genetics of eukaryotic mismatch repair. *Genes and Development*, **10**, 1433–42.

Koretz, R.L. (1993) Malignant polyps: are they sheep in wolves' clothing? *Annals of Internal Medicine*, **118**, 63–8.

Koshy, B., Matilla, T., Burright, E.N., Merry, D.E., Fischbeck, K.H., Orr, H.T. *et al.* (1996) Spinocerebellar ataxia type-1 and spinobulbar muscular atrophy gene products interact with glyceraldehyde-3-phosphate dehydrogenase. *Human Molecular Genetics*, **5**, 1311–18.

Koutnikova, H., Campuzano, V., Foury, F., Dollé, P., Cazzalini, O. and Koenig, M. (1997) Studies of human, mouse and yeast homologues indicate mitochondrial function for frataxin. *Nature Genetics*, **16**, 345–51.

Krahe, R., Ashizawa, T., Abbruzzese, C., Roeder, E., Carango, P., Giacanelli, M. *et al.* (1995) Effect of myotonic dystrophy trinucleotide repeat expansion on *DMPK* transcription and processing. *Genomics*, **28**, 1–14.

Kramer, B., Kramer, W. and Fritz, H.J. (1984) Different base/base mismatches are corrected with different efficiencies by the methyl-directed DNA mismatch-repair system of E. coli. *Cell*, **38**, 879–87.

Kremer, B., Almqvist, E., Theilmann, J., Spence, N., Telenius, H., Goldberg, Y.P. *et al.* (1995) Sex-dependent mechanisms for expansions and contractions of the CAG repeat on affected Huntington disease chromosomes. *American Journal of Human Genetics*, **57**, 343–50.

Kremer, B., Squitieri, F., Telenius, H., Andrew, S.E., Theilmann, J., Spence, N. *et al.* (1993) Molecular analysis of late onset Huntington's disease. *Journal of Medical Genetics*, **30**, 991–3.

Krontiris, T.G., Devlin, B., Karp, D.D., Robert, N.J. and Risch, N. (1993) An association between the risk of cancer and mutations in the *HRAS1* minisatellite locus. *New England Journal of Medicine*, **329**, 517–23.

Kroutil, L.C., Register, K., Bebenek, K. and Kunkel, T.A. (1996) Exonucleolytic proofreading during replication of repetitive DNA. *Biochemistry*, **35**, 1046–53.

Kuhner, M.K., Yamato, J. and Felsenstein, J. (1995) Estimating effective population size and mutation rate from sequence data using Metropolis–Hastings sampling. *Genetics*, **140**, 1421–30.

Kunkel, T.A. (1986) Frameshift mutagenesis by eucaryotic DNA polymerases in vitro. *Journal of Biological Chemistry*, **261**, 13581–7.

Kunkel, T.A. (1990) Misalignment-mediated DNA synthesis errors. *Biochemistry*, **29**, 8004–11.

Kunkel, T.A. (1992) DNA replication fidelity. *Journal of Biological Chemistry*, **267**, 18251–4.

Kunkel, T.A., Patel, S.S. and Johnson, K.A. (1994) Error-prone replication of repeated DNA sequences by T7 DNA polymerase in the absence of its processivity subunit. *Proceedings of the National Academy of Sciences of the USA*, **91**, 6830–4.

Kunst, C.B. and Warren, S.T. (1994) Cryptic and polar variation of the fragile X repeat could result in predisposing normal alleles. *Cell*, **77**, 853–61.

Kunst, C.B., Leeflang, E.P., Iber, J.C., Arnheim, N. and Warren, S.T. (1997) The effect of FMR1 CGG repeat interruptions on mutation frequency as measured by sperm typing. *Journal of Medical Genetics*, **34**, 627–31.

Künzler, P., Matsuo, K. and Schaffner, W. (1995) Pathological, physiological, and evolutionary aspects of short unstable DNA repeats in the human genome. *Biological Chemistry Hoppe Seyler*, **376**, 201–11.

La Spada, A.R., Wilson, E.M., Lubahn, D.B., Harding, A.E. and Fischbeck, K.H. (1991) Androgen receptor gene mutations in X-linked spinal and bulbar muscular atrophy. *Nature*, **352**, 77–9.

La Spada, A.R., Roling, D.B., Harding, A.E., Warner, C.L., Spiegel, R., Hausmanowa-Petrusewicz, I. *et al.* (1992) Meiotic stability and genotype–phenotype correlation of the trinucleotide repeat in X-linked spinal and bulbar muscular atrophy. *Nature Genetics*, **2**, 301–4.

Laan, M. and Pääbo, S. (1997) Demographic history and linkage disequilibrium in human populations. *Nature Genetics*, **17**, 435–8.

Lafyatis, R., Denhez, F., Williams, T., Sporn, M. and Roberts, A. (1991) Sequence specific protein binding to and activation of the *TGF-beta3* promoter through a repeated TCCC motif. *Nucleic Acids Research*, **19**, 6419–25.

Lagercrantz, U., Ellegren, H. and Andersson, L. (1993) The abundance of various polymorphic microsatellite motifs differs between plants and vertebrates. *Nucleic Acids Research*, **21**, 1111–15

Lande, R. and Barrowclough, G.F. (1987) *Viable Populations for Conservation.* pp. 87–123. Cambridge University Press, Cambridge.

Lander, E.S. (1989) DNA fingerprinting on trial. *Nature*, **339**, 501–5.

Lanz, R.B., Wieland, S., Hug, M. and Rusconi, S. (1995) A transcriptional repressor obtained by alternative translation of a trinucleotide repeat. *Nucleic Acids Research*, **23**, 138–45.

Learn, B.A. and Grafstrom, R.H. (1989) Methyl-directed repair of frameshift heteroduplexes in cell extracts from *Escherichia coli*. *Journal of Bacteriology*, **171**, 6473–81.

LeClerc, J.E., Li, B., Payne, W.L. and Cebula, T.A. (1996) High mutation frequencies among *Escherichia coli* and *Salmonella* pathogens. *Science*, **274**, 1208–11.

Ledig, F.T. (1986) Heterozygosity, heterosis, and fitness in outbreeding plants. In *Conservation biology.* (ed. M.E. Soulé), pp. 77–104. Sinauer Press, Sunderland.

Levinson, G. and Gutman, G.A. (1987a) Slipped-strand mispairing: a major mechanism for DNA sequence evolution. *Molecular Biology and Evolution*, **4**, 203–21

Levinson, G. and Gutman, G.A. (1987b) High frequencies of short frameshifts in poly-CA/TG tandem repeats borne by bacteriophage M13 in *Escherichia coli* K-12. *Nucleic Acids Research*, **15**, 5323–38

Lewontin, R.C. (1972) The apportionment of human diversity. *Evolutionary Biology*, **6**, 381–98.

Li, X.-J., Li, S.-H., Sharp, A.H., Nucifora, F.C. Jr., Schilling, G., Lanahan, A. *et al.* (1995) A huntingtin-associated protein enriched in brain with implications for pathology. *Nature*, **378**, 398–402.

Libert, F., Cochaux, P., Beckman, G., Samson, M., Aksenova, M., Cao, A. *et al.* (1998) The deltaccr5 mutation conferring protection against HIV-1 in Caucasian populations has a single and recent origin in northeastern Europe. *Human Molecular Genetics*, **7**, 399–406.

Lin, A.A., Hebert, J.M., Mountain, J.L. and Cavalli-Sforza, L.L. (1994) Comparison of 79 DNA polymorphisms tested in Australians, Japanese and Papua New Guineans with those of five other human populations. *Gene Geography*, **8**, 191–214.

Lindsley, D.L. and Sandler, L. (1977) The genetic analysis of meiosis in female *Drosophila melanogaster*. *Philosophical Transactions of the Royal Society of London* B, **277**, 295–312.

Litt, M. and Luty, J.A. (1989) A hypervariable microsatellite revealed by *in vitro* amplification of a dinucleotide repeat within the cardiac muscle actin gene. *American Journal of Human Genetics*, **44**, 397–401.

Liu, R., Paxton, W.A., Choe, S., Ceradini, D., Martin, S.R., Horuk, R. *et al.* (1996) Homozygous defect in HIV-1 coreceptor accounts for resistance of some multiply-exposed individuals to HIV-1 infection. *Cell*, **86**, 367–77.

Loeb, L.A. (1991) Mutator phenotype may be required for multistage carcinogenesis. *Cancer Research*, **51**, 3075–9.

Long, J.C., Williams, R.C. and Urbanek, M. (1995) An E-M algorithm and testing strategy for multiple-locus haplotypes. *American Journal of Human Genetics*, **56**, 799–810.

Lue, N.L., Buchman, A.R. and Kornberg, R.D. (1989) Activation of yeast RNA polymerase II transcription by a thymidine-rich upstream element *in vitro*. *Proceedings of the National Academy of Sciences of the USA*, **86**, 486–90.

MacDonald, M.E., Barnes, G., Srinidhi, J., Duyao, M.P., Ambrose, C.M., Myers, R.H. *et al.* (1993) Gametic but not somatic instability of CAG repeat length in Huntington's disease. *Journal of Medical Genetics*, **30**, 982–6.

MacHugh, D.E., Shriver, M.D., Loftus, R.T., Cunningham, P. and Bradley, D.G. (1997) Microsatellite DNA variation and the evolution, domestication and phylogeography of taurine and Zebu cattle (*Bos taurus* and *Bos indicus*) *Genetics*, **146**, 1071–86.

McInnis, M.G. (1996) Anticipation: an old idea in new genes. *American Journal of Human Genetics*, **59**, 973–9.

Macintyre, G., Doiron, K.M. and Cupples, C.G. (1997) The Vsr endonuclease of *Escherichia coli*: an efficient DNA repair enzyme and a potent mutagen. *Journal of Bacteriology*, **179**, 6048–52.

Mackay, T. (1995) The genetic basis of quantitative variation: numbers of sensory bristles of *Drosophila melanogaster* as a model system. *Trends in Genetics*, **11**, 465–70.

McKeigue, P.M. (1997) Mapping genes underlying ethnic differences in disease risk by linkage disequilibrium in recently admixed populations. *American Journal of Human Genetics*, **60**, 188–96.

McKeon, C., Schmidt, A. and de Crombrugghe, B. (1984) A sequence conserved in both the chicken and mouse a2(I) collage promoter contains sites sensitive to S1 nuclease. *Journal of Biological Chemistry*, **259**, 6636–40.

Mahadevan, M., Tsilfidis, C., Sabourin, L., Shutler, G., Amemiya, C., Jansen, G. *et al.* (1992) Myotonic dystrophy mutation: an unstable CTG repeat in the 3′ untranslated region of the gene. *Science*, **255**, 1253–5.

Mahtani, M.M. and Willard, H.F. (1990) Pulsed-field gel analysis of alpha satellite DNA at the human X chromosome centromere: high frequency polymorphisms and array size estimate. *Genomics*, **7**, 607–13.

Mahtani, M.M., and Willard, H.F. (1993) A polymorphic X-linked tetranucleotide repeat locus displaying a high rate of new mutation: implications for mechanics of mutation at short tandem repeat loci. *Hum. Mol. Genet.*, **2**, 431–7.

Malter, H.E., Iber, J.C., Willemsen, R., de Graaff, E., Tarleton, J.C., Leisti, J. *et al.* (1997) Characterisation of the full fragile X syndrome mutation in fetal gametes. *Nature Genetics*, **15**, 165–9.

Mangiarini, L., Sathasivam, K., Seller, M., Cozens, B., Harper, A., Hetherington, C. *et al.* (1996) Exon 1 of the HD gene with an expanded CAG repeat is sufficient to cause a progressive neurological phenotype in transgenic mice. *Cell*, **87**, 493–506.

Mangiarini, L., Sathasivam, K., Mahal, A., Mott, R., Seller, M. and Bates, G.P. (1997) Instability of highly expanded CAG repeats in mice transgenic for the Huntington's disease mutation. *Nature Genetics*, **15**, 197–200.

Manivasakam, P., Rosenberg, S.M. and Hastings, P.J. (1996) Poorly repaired mismatches in heteroduplex DNA are hyper-recombinogenic in *Saccharomyces cerevisiae*. *Genetics*, **142**, 407–16.

Maroteaux, L., Heilig, R., Dupret, D. and Mandel, J.L. (1983) Repetitive satellite-like sequences are present within or upstream from 3 avian protein-coding genes. *Nucleic Acids Research*, **11**, 1227–43.

Marshall, T.C., Slate, J., Kruuk, L.E.B. and Pemberton, J.M. (1998) Statistical confidence for likelihood-based paternity inference in natural populations. *Molecular Ecology*, **7**, 639–55.

Marsischky, G.T., Filosi, N., Kane, M.F. and Kolodner, R. (1996) Redundancy of *Saccharomyces cerevisiae* MSH3 and MSH6 in MSH2-dependent mismatch repair. *Genes and Development*, **10**, 407–20.

Martin, M., Mann, D. and Carrington, M. (1995) Recombination rates across the HLA complex: use of microsatellites as a rapid screen for recombinant chromosomes. *Human Molecular Genetics*, **4**, 423–8.

Martin, M., Harding, A., Chadwick, R., Kronick, M., Cullen, M., Lin, L. *et al.* (1998) Characterization of 12 microsatellite loci of the human MHC in a panel of reference cell lines. *Immunogenetics*, **47**, 131–8.

Martinson, J.J., Chapman, N.H., Rees, D.C., Liu, Y.T. and Clegg, J.B. (1997) Global distribution of the CCR5 gene 32-basepair deletion. *Nature Genetics*, **16**, 100–3.

Matic, I., Radman, M., Taddei, F., Picard, B., Doit, C., Bingen, E. *et al.* (1997) Highly variable mutation rates in commensal and pathogenic *Escherichia coli*. *Science*, **277**, 1833–4.

Maurer, D.J., O'Callaghan, B.L. and Livingston, D.M. (1996) Orientation dependence of trinucleotide CAG repeat instability in *Saccharomyces cerevisiae*. *Molecular and Cellular Biology*, **16**, 6617–22.

May, C.A., Jeffreys, A.J. and Armour, J.A.L. (1996) Mutation rate heterogeneity and the generation of allele diversity at the human minisatellite MS205 (D16S309). *Human Molecular Genetics*, **5**, 1823–33.

Maynard Smith, J. and Haigh, J. (1974) The hitch-hiking effect of a favourable gene. *Genetical Research*, **23**, 23–35.

Meloni, R., Albanese, V., Ravassard, P., Treilhou, F. and Mallet, J. (1998) A tetranucleotide polymorphic microsatellite, located in the first intron of the tyrosine hydroxylase gene, acts as a transcription regulatory element *in vitro*. *Human Molecular Genetics*, **7**, 423–8.

Merila, J. (1997) Quantitative trait and allozyme divergence in the greenfinch (*Carduelis chloris*, Aves: Fringillidae). *Biological Journal of the Linnaean Society*, **61**, 243–66.

Messier, W., Li, S.-H. and Stewart, C.-B. (1996) The birth of microsatellites. *Nature*, **381**, 483.

Meyer, E., Wiegand, P., Rand, S.P., Kuhlman, E., Brack, M. and Brinkmann, B. (1995) Microsatellite polymorphisms reveal phylogenetic relationships in primates. *Journal of Molecular Evolution*, **41**, 10–14.

Michalakis, Y. and Excoffier, L. (1996) A genetic estimation of population subdivision using distances between alleles with special reference for microsatellite loci. *Genetics*, **142**, 1061–4.

Michalakis, Y. and Veuille, M. (1996) Length variation of CAG/CAA trinucleotide repeats in natural populations of *Drosophila melanogaster* and its relation to recombination rate. *Genetics*, **143**, 1713–25.

Mignot, E., Kimura, A., Lattermann, A., Lin, X., Yasunaga, S., Mueller-Eckhardt, G. *et al.* (1997) Extensive HLA class II studies in 58 non-DRB1*15 (DR2) narcoleptic patients with cataplexy. *Tissue Antigens*, **49**, 329–41.

Mitton, J.B. (1993) Theory and data pertinent to the relationship between heterozygosity and fitness. In *The natural history of inbreeding and outbreeding: theoretical and empirical perspectives*. (ed. N.W. Thornhill), pp. 17–41. University of Chicago Press, Chicago.

Modrich, P. (1991) Mechanisms and biological effects of mismatch repair. *Annual Review of Genetics*, **25**, 229–53.

Monckton, D. and Jeffreys, A.J. (1991) Minisatellite 'isoallele' discrimination in pseudohomozygotes by single molecule PCR and variant repeat mapping. *Genomics*, **11**, 465–467.

Monckton, D.G., Neumann, R., Guram, T., Fretwell, N., Tamaki, K., MacLeod, A. and Jeffreys A.J. (1994) Minisatellite mutation rate variation associated with a flanking DNA sequence polymorphism. *Nature Genetics*, **8**, 162–70.

Monckton, D.G., Coolbaugh, M.I., Ashizawa, K.T., Siciliano, M.J. and Caskey, C.T. (1997) Hypermutable myotonic dystrophy CTG repeats in transgenic mice. *Nature Genetics*, **15**, 193–6.

Montermini, L., Andermann, E., Labuda, M., Richter, A., Pandolfo, M. and Cavalcanti, F. (1997) The Friedreich ataxia GAA triplet repeat: premutation and normal alleles. *Human Molecular Genetics*, **6**, 1261–6.

Moore, S.S., Sargeant, L.L., King, T.J., Mattick, J.S., Georges, M. and Hetzel, D.J.S. (1991) The conservation of dinucleotide microsatellites among mammalian genomes allows the use of heterologous PCR primer pairs in closely related species. *Genomics*, **10**, 654–60.

Moran, P.A.P. (1975) Wandering distributions and the electrophoretic profile. *Theoretical Population Biology*, **8**, 318–30.

Morin, P.A. and Woodruff, D.S. (1996) Non-invasive genotyping for vertebrate conservation. In *Molecular Genetic Approaches in Conservation* (ed. T.B. Smith and R.K. Wayne), pp. 298–313. Oxford University Press, New York.

Morin, P.A., Wallis, J., Moore, J.J. and Woodruff, D.S. (1994*a*) Paternity exclusion in a community of wild chimpanzees using hypervariable simple sequence repeats. *Molecular Ecology*, **3**, 469–77.

Morin, P.A., Moore, J.J., Chakraborty, R., Jin, L., Goodall, J. and Woodruff, D.S. (1994*b*) Kin selection, social structure, gene flow and the evolution of chimpanzees. *Science*, **265**, 1193–201.

Morral, N., Nunes, V., Casals, T. and Estivill, X. (1991) CA/GT microsatellite alleles within the cystic fibrosis transmembrane conductance regulator (CFTR) gene are not generated by unequal crossingover. *Genomics*, **10**, 692–8.

Morral, N., Bertranpetit, J., Estivill, X., Nunes, V., Casals, T., Gimenez, J. *et al.* (1994) The origin of the major cystic fibrosis mutation (?F508) in European populations. *Nature Genetics*, **7**, 169–75.

Morrison, A., Johnson, A.L., Johnson, L.H. and Sugino, A. (1993) Pathway correcting DNA replication errors in Saccharomyces cerevisiae. *EMBO Journal*, **12**, 1467–73.

Morton, N.E. (1997) Genetic epidemiology. *Annals of Human Genetics*, **61**, 1–13.

Mountain, J.L. and Cavalli-Sforza, L.L. (1994) Inference of human evolution through cladistic analysis of nuclear DNA restriction polymorphisms. *Proceedings of the National Academy of Sciences of the USA*, **91**, 6515–19.

Mountain, J.L. and Cavalli-Sforza, L.L. (1997) Multilocus genotypes, a tree of individuals, and human evolutionary history. *American Journal of Human Genetics*, **61**, 705–18.

Moutou, C., Vincent, M.-C., Biancalana, V. and Mandel, J.-L. (1997) Transition from premutation to full mutation in fragile X syndrome is likely to be prezygotic. *Human Molecular Genetics*, **6**, 971–9.

Moxon, E.R., Rainey, P.B., Nowak, M.A. and Lenski, R.E. (1994) Adaptive evolution of highly mutable loci in pathogenic bacteria. *Current Biology*, **4**, 24–33.

Muelbert, M.M. and Bowen, W.D. (1993) Duration of lactation and postweaning changes in body composition of harbour seal, *Phoca vitulina*, pups. *Canadian Journal of Zoology*, **71**, 1405–14.

Mundy, N.I., Winchell, C.S., Burr, T. and Woodruff, D.S. (1997) Microsatellite variation and microevolution in the critically endangered San Clemente Island loggerhead shrike (*Lanius ludovicianus mearnsi*) *Proceedings of the Royal Society of London* B, **264**, 869–75.

Muraiso, T., Nomoto, S., Yamazaki, H., Mishima, Y. and Kominami, R. (1992) A single-stranded DNA binding protein from mouse tumor cells specifically recognizes the C-rich strand of the $(AGG:CCT)_n$ repeats that can alter DNA conformation. *Nucleic Acids Research*, **20**, 6631–5.

Murphy, G.L., Connell, T.D., Barritt, D.S., Koomey, M. and Cannon, J.G. (1989) Phase variation of gonococcal protein II—regulation of gene-expression by slipped-strand mispairing of a repetitive DNA sequence. *Cell*, **56**, 539–47.

Murray, A., Youings, S., Dennis, N., Latsky, L., Linehan, P., McKechnie, N. *et al.* (1996) Population screening at the FRAXA and FRAXE loci: molecular analyses of boys with learning difficulties. *Human Molecular Genetics*, **5**, 727–35.

Nadir, E., Margalit, H., Gallily, T. and Ben-Sasson, S.A. (1996) Microsatellite spreading in the human genome: Evolutionary mechanisms and structural implications. *Proceedings of the National Academy of Sciences of the USA*, **93**, 6470–5.

Nagafuchi, S., Yanagisawa, H., Sato, K., Shirayama, T., Ohsaki, E., Bundo, M., *et al.* (1994) Dentatorubral and pallidoluysian atrophy expansion of an unstable CAG trinucleotide on chromosome 12p. *Nature Genetics*, **6**, 14–8.

Nag, D.K., Scherthan, H., Rockmill, B., Bhargava, J. and Roeder, G.S. (1995) Heteroduplex DNA formation and homolog pairing in yeast meiotic mutants. *Genetics*, **141**, 75–86.

Nasmyth, K. (1985) A repetitive DNA sequence that confers cell-cycle START (CDC28)-dependent transcription of the *HO* gene in yeast. *Cell*, **42**, 225–35.

Nauta, M.J. and Weissing, F.J. (1996) Constraints on allele size at microsatellite loci: implications for genetic differentiation. *Genetics*, **143**, 1021–32.

Naylor, L.H. and Clark, E.M. (1990) $d(TG)_n \cdot d(CA)_n$ sequences upstream of the rat prolactin gene form Z-DNA and inhibit gene transcription. *Nucleic Acids Research*, **18**, 1595–601.

Nei, M. (1972) Genetic distances between populations. *American Naturalist*, **106**, 283–92.

Nei, M. (1973) Analysis of gene diversity in subdivided populations. *Proceedings of the National Academy of Sciences of the USA*, **70**, 3321–3.

Nei, M. (1978) Estimation of average heterozygosity and genetic distance from a small number of individuals. *Genetics*, **89**, 583–90.

Nei, M. (1987) *Molecular evolutionary genetics*. Columbia University Press, New York.

Nei, M. and Roychoudhury, A.K. (1974) Genetic variation within and between the three major races of man, Caucasoids, Negroids, and Mongoloids. *American Journal of Human Genetics*, **26**, 421–43.

Nei, M. and Roychoudhury, A.K. (1993) Evolutionary relationships of human populations on a global scale. *Molecular Biology and Evolution*, **10**, 927–43.

Neil, D.L. and Jeffreys, A.J. (1993) Digital DNA typing at a second hypervariable locus by minisatellite variant repeat mapping. *Human Molecular Genetics*, **2**, 1129–35.

Neri, C., Albanese, V., Lebre, A.-S., Holbert, S., Saada, C. *et al.* (1996) Survey of CAG/CTG repeats in human cDNAs representing new genes: candidates for inherited neurological disorders. *Human Molecular Genetics*, **7**, 1001–9.

Neuhauser, C. and Krone, S. (1997) The genealogy of samples in models with selection. *Genetics*, **145**, 519–34.

Neville, C.E., Mahadevan, M.S., Barceló, J.M. and Korneluk, R.G. (1994) High resolution genetic analysis suggests one ancestral predisposing haplotype for the origin of the myotonic dystrophy mutation. *Human Molecular Genetics*, **3**, 45–51.

Nielsen, J.E., Koefoed, P., Abell, K., Hasholt, L., Eiberg, H., Fenger, K. *et al.* (1997) CAG repeat expansion in autosomal dominant pure spastic paraplegia linked to chromosome 2p21-p24. *Human Molecular Genetics*, **6**, 1811–16.

Nielsen, R. (1997) A likelihood approach to population samples of microsatellite alleles. *Genetics*, **146**, 711–16.

Nielsen, R., Mountain, J.L., Huelsenbeck, J.P. and Slatkin, M. (1998) Maximum-likelihood estimation of population divergence times and population phylogeny in models without mutation. *Evolution*, **52**, 669–77.

Nowell, P.C. (1976) The clonal evolution of tumour cell populations. *Science*, **194**, 23–9.

NRC (1996) *The evaluation of forensic DNA evidence*. National Research Council, National Academy Press, Washington DC.

O'Donovan, M.C., Guy, C., Craddock, N., Murphy, K.C., Cardno, A.G., Jones, L.A. *et al.* (1995) Expanded CAG repeats in schizophrenia and bipolar disorder. *Nature Genetics*, **10**, 380–1.

O'Ryan, C., Harley, E.H., Bruford, M.W., Beaumont, M.A., Wayne, R.K. and Cherry, M.I. (1998) Microsatellite analysis of genetic diversity in fragmented South African buffalo (*Syncerus caffer*) populations. *Animal Conservation*, **1**, 85–95.

Oakey, R. and Tyler-Smith, C. (1990) Y Chromosome DNA haplotyping suggests that most European and Asian men are descended from one of two males. *Genomics*, **7**, 325–30.

Ohta, T. and Kimura, M. (1973) A model of mutation appropriate to estimate the number of electrophoretically detectable alleles in a finite population. *Genetical Research*, **22**, 201–4.

Oliver, S.G., van der Aart, Q.J., Agostoni-Carbone, M.L., Aigle, L., Alberghina, L., Alexandraki, D. *et al.* (1992) The complete DNA sequence of yeast chromosome III. *Nature*, **357**, 38–46.

Orita, M., Iwahana, H., Kanazawa, H., Hayashi, K. and Sekiya, T. (1989) Detection of polymorphism of human DNA by gel electrophoresis as single strand conformation polymorphisms. *Proceedings of the National Academy of Sciences of the USA*, **86**, 2766–70.

Orr, H.T., Chung, M.-Y., Banfi, S., Kwiatkowski, T.J., Jr., Servadio, A., Beaudet, A.L. *et al.* (1993) Expansion of an unstable trinucleotide (CAG) repeat in spinocerebellar ataxia type 1. *Nature Genetics*, **4**, 221–6.

Ostrander, E.A., Sprague, G.F. and Rine, J. (1993) Identification and characterization of dinucleotide repeat $(CA)_n$ markers for genetic mapping in dog. *Genomics*, **16**, 207–13.

Otchy, D.P., Ransohoff, D.F., Wolff, B.G., Waver, A., Ilstrup, D., Carlson, H. *et al.* (1996) Metachronous colon cancer in persons who have had a large adenomatous polyp. *American Journal of Gastroenterology*, **91**, 448–54.

Paetkau, D. and Strobeck, C. (1995) The molecular basis and evolutionary history of a microsatellite null allele in bears. *Molecular Ecology*, **4**, 519–20.

Paetkau, D., Calvert, W., Stirling, I. and Strobeck, C. (1995) Microsatellite analysis of population structure in Canadian polar bears. *Molecular Ecology*, **4**, 347–54.

Paetkau, D., Waits, L.P., Clarkson, P.L., Craighead, L. and Strobeck, C. (1997) An empirical evaluation of genetic distance statistics using microsatellite data from bear (Ursidae) populations. *Genetics*, **147**, 1943–57.

Pardue, M.L., Lowenhaupt, K., Rich, A. and Nordheim, A. (1987) $(dC-dA)_n$. $(dG-dT)_n$ sequences have evolutionarily conserved chromosomal locations in *Drosophila* with implications for roles in chromosome structure and function. *EMBO Journal*, **6**, 1781–9.

Parker, B.O. and Marinus, M.G. (1992) Repair of DNA heteroduplexes containing small heterologous sequences in *Escherichia coli*. *Proceedings of the National Academy of Sciences of the USA*, **89**, 1730–4.

Paulson, H.L. and Fischbeck, K.H. (1996) Trinucleotide repeats in neurogenetic disorders. *Annual Review of Neuroscience*, **19**, 79–107.

Paulson, H.L., Perez, M.K., Trottier, Y., Trojanowski, J.Q., Subramony, S.H., Das, S.S. *et al.* (1997) Intranuclear inclusions of expanded polyglutamine protein in spinocerebellar ataxia type 3. *Neuron*, **19**, 335–44.

Pemberton, J.M., Albon, S.D., Guinness, F.E., Clutton-Brock, T.H. and Dover, G.A. (1992) Behavioral estimates of male mating success tested by DNA fingerprinting in a polygynous mammal. *Behavioural Ecology*, **3**, 66–75.

Pemberton, J.M., Slate, J., Bancroft, D.R. and Barrett, J.A. (1995) Non amplifying alleles at microsatellite loci—a caution for parentage and population studies. *Molecular Ecology*, **4**, 249–52.

Pena, S.D., Santos, F.R., Bianchi, N.O., Bravi, C.M., Carnese, F.R. and Rothhammer, F. (1995) A major founder Y-chromosome haplotype in Amerindians. *Nature Genetics*, **11**, 15–16.

Pépin, L., Amigues, Y., Lépingle, A., Berthier, J.L., Bensaid, A. and Vaiman, D. (1995) Sequence conservation of microsatellites between cattle (*Bos taurus*), goat (*Capra hircus*) and related species. Examples of use in parentage testing and phylogeny analysis. *Heredity*, **74**, 53–61.

Pérez-Lezaun, A., Calafell, F., Mateu, E., Comas, D., Ruíz-Pachecho, R. and Bertranpetit, J. (1997) Microsatellite variation and the differentiation of modern humans. *Human Genetics*, **99**, 1–7.

Peters, J.M. (1997) Microsatellite loci for *Pseudomyrmex pallidus* (Hymenoptera: Formicidae). *Molecular Ecology*, **6**, 887–8.

Petes, T.D., Greenwell, P.W. and Dominska, M. (1997) Stabilization of microsatellite sequences by variant repeats in the yeast *Saccharomyces cerevisiae*. *Genetics*, **146**, 491–8.

Petrishchev, V.N., Kutueva, A.B. and Rychkov, Y.G. (1993) Deletion–insertion polymorphisms in ten Mongoloid populations of Siberia: frequency of deletion correlates with the geographical coordinates of the area. (in Russian). *Genetika*, **29**, 1196–204.

Phelan, C.M., Rebbeck, T.R., Weber, B.L., Devilee, P., Ruttledge, M.H., Lynch, H.T. *et al.* (1996) Ovarian risk in BRCA1 carriers is modified by the *HRAS1* variable number of tandem repeat (VNTR) locus. *Nature Genetics*, **12**, 309–11.

Phillips, A.V., Timchenko, L.T. and Cooper, T.A. (1998) Disruption of splicing regulated by a CUG-binding protein in a myotonic dystrophy. *Science*, **280**, 737–40.

Pieretti, M., Zhang, F.P., Fu, Y.-H., Warren, S.T., Oostra, B.A., Caskey, C.T. and Nelson, D.L. (1991) Absence of expression of the *FMR-1* gene in fragile X syndrome. *Cell*, **66**, 817–22.

Pollock, D.D., Bergman, A., Feldman, M.W. and Goldstein, D.B. (1998) Microsatellite behavior with range constraints: parameter estimation and improved distance estimation for use in phylogenetic reconstruction. *Theoretical Population Biology*, **53**, 256–71.

Poloni, E.S., Semino, O., Passarino, G., Santachiara-Benerecetti, A.S., Dupanloup, I., Langaney, I. *et al.* (1997) Human genetic affinities for Y chromosome p49a.f/TaqI haplotypes show strong correspondence with linguistics. *American Journal of Human Genetics*, **61**, 1015–35.

Potten, C.S. and Loeffler, M. (1990) Stem cells: attributes, cycles, spirals, pitfalls, and uncertainties. Lessons for and from the crypt. *Development*, **110**, 1001–20.

Pray, L.A., Schwartz, J.M., Goodknight, C.J. and Stevens, L. (1994) Environmental dependency of inbreeding depression: implications for conservation biology. *Conservation Biology*, **8**, 562–8.

Primmer, C.R. and Ellegren, H. (1998) Pattern of molecular evolution in avian microsatellites. *Molecular Biology and Evolution*, **15**, 997–1008.

Primmer, C.R., Ellegren, H., Saino, N. and Møller, A.P. (1996) Directional evolution in germline microsatellite mutations. *Nature Genetics*, **13**, 391–3.

Primmer, C.R., Raudsepp, T., Chowdhary, B.P., Møller, A.P. and Ellegren, H. (1997) Low frequency of microsatellites in the avian genome. *Genome Research*, **7**, 471–82.

Pritchard, J.K. and Feldman, M.W. (1996) Statistics for microsatellite variation based on coalescence. *Theoretical Population Biology*, **50**, 325–44.

Prokof'ev, G.N. (1940) Ethnology of the peoples of Ob–Yenisey basin. (in Russian). *Soviet Ethnography.*

Ptashne, M. (1988) How eukaryotic transcriptional activators work. *Nature*, **335**, 683–9.

Pulst, S.M., Nechiporuk, A., Nechiporuk, T., Gispert, S., Chen, X.-N., Lopes-Cendes, I. *et al.* (1996) Moderate expansion of a normally biallelic trinucleotide repeat in spinocerebellar ataxia type 2. *Nature Genetics*, **14**, 269–76.

Punt, P.J., Dingemanse, M.A., Kyuvenhoven, A., Soede, R.D.M., Pouwels, P.H. and van de Hondel, C.A.M.J.J. (1990) Functional elements in the promoter region of the *Aspergillus nidulans* gpd A gene encoding glyceraldehyde-3-phosphate dehydrogenase. *Gene*, **93**, 101–9.

Queller, D.C., Strassmann, J.E. and Hughes, C.R. (1993) Microsatellites and kinship. *Trends in Ecology Evolution*, **8**, 285.

Raha-Chowdhury, R., Bowen, D.J., Stone, C., Pointon, J.J., Terwillinger, J.D., Shearman, J.D. *et al.* (1995) New polymorphic microsatellite markers place the haemochromatosis gene telomeric to D6S105. *Human Molecular Genetics*, **4**, 1869–74.

Ralls, K., Harvey, P.H. and Lyles, A.M. (1986) Inbreeding in natural populations of birds and mammals. In *Conservation biology* (ed. M.E. Soulé), pp. 35–56. Sinauer Press, Sunderland.

Ramalingam, R., Blume, J.E., Ganguly, K. and Ennis, H.L. (1995) AT-rich upstream sequence elements regulate spore germination-specific expression of the Dictyostelium discoideum *cel A* gene. *Nucleic Acids Research*, **23**, 3018–25.

Rannala, B. and Hartigan, J.A. (1996) Estimating gene flow in island populations. *Genetical Research*, **67**, 147–58.

Rannala, B. and Mountain, J.L. (1997) Detecting immigration by using multilocus genotypes. *Proceedings of the National Academy of Sciences USA*, **94**, 9197–201.

Rannala, B. and Yang, Z. (1996) Probability distribution of molecular evolutionary trees: a new method of phylogenetic inference. *Journal of Molecular Evolution*, **43**, 304–11.

Ranum, L.P.W., Chung, M.-Y., Banfi, S., Bryer, A., Schut, L.J., Ramesar, R. *et al.* (1995) Molecular and clinical correlations in spinocerebellar ataxia type 1: evidence for familial effects on the age at onset. *American Journal of Human Genetics*, **55**, 244–52.

Reddy, P.H., Williams, M., Miller, G., Glass, M., Paylor, R. *et al.* (1997) Transgenic mouse models for Huntington's disease. *American Journal of Human Genetics*, **61**, A52.

Reddy, S., Smith, D.B.J., Rich, M.M., Leferovich, J.M., Reilly, P., Davis, B.M. *et al.* (1996) Mice lacking the myotonic dystrophy protein kinase develop a late onset progressive myopathy. *Nature Genetics*, **13**, 325–35.

Reich, D.E. and Goldstein, D.B. (1998) Genetic evidence for a paleolithic human population expansion in Africa. *Proceedings of the National Academy of Sciences of the USA*, **95**, 8119–25.

Relethford, J.H. (1995) Genetics and modern human origins. *Evolutionary Anthropology*, **4**, 53–63.

Reynolds, J., Weir, B.S. and Cockerham, C.C. (1983) Estimation of the coancestry coefficient, basis for a short-term genetic distance. *Genetics*, **105**, 767–79.

Richards, B., Zhang, H., Phear, G. and Meuth, M. (1997) Conditional mutator phenotypes in hMSH2-deficient tumour cell lines. *Science*, **277**, 1523–6.

Richards, R.I. and Sutherland, G.R. (1994) Simple repeat DNA is not replicated simply. *Nature Genetics*, **6**, 114–16.

Richards, R.I., Holman, K., Yu, S. and Sutherland, G.R. (1993) Fragile X syndrome unstable element, $p(CCG)_n$, and other simple tandem repeat sequences are binding sites for specific nuclear proteins. *Human Molecular Genetics*, **2**, 1429–35.

Rico, C., Rico, I. and Hewitt, G. (1996) 470 million years of conservation of microsatellite loci among fish species. *Proceedings of the Royal Society of London* B, **263**, 549–57.

Riess, O., Schöls, L., Böttger, H., Nolte, D., Vieira-Saecker, A.M.M., Schimming, C. *et al.* (1997) SCA6 is caused by moderate CAG expansion in the $a_{1A}$-voltage-dependent calcium channel gene. *Human Molecular Genetics*, **6**, 1289–93.

Risch, N. and Merikangas, K. (1996) The future of genetic studies of complex human diseases. *Science*, **273**, 1516–17.

Risch, N., de Leon, D., Ozelius, L., Kramer, P., Almasy, L., Fahn, S. *et al.* (1995) Genetic analysis of idiopathic torsion dystonia in Ashkenazi Jews and their recent descent from a small founder population. *Nature Genetics*, **9**, 153–9.

Robinson, W.P., Barbosa, J., Rich, S.S. and Thomson, G. (1993) Homozygous parent affected sib pair method for detecting disease predisposing variants: application to insulin dependent diabetes mellitus. *Genetic Epidemiology*, **10**, 273–88.

Roe, A. (1992) *Correlations and interactions in random walks and population genetics*. Ph.D. Thesis, University of London.

Roeder, K. (1994) DNA fingerprinting: a review of the controversy. *Statistical Science*, **9**, 222–78.

Roewer, L., Kayser, M., Dieltjes, P., Nagy, M., Bakker, E., Krawczak, M. *et al.* (1996) Analysis of molecular variance (AMOVA) of Y-chromosome-specific microsatellites in two closely related human populations. *Human Molecular Genetics*, **5**, 1029–33.

Rogers, A.R. and Harpending, H.C. (1992) Population growth makes waves in the distribution of pairwise genetic differences, *Molecular Biology and Evolution*, **9**, 552–69.

Rogers, A.R. and Jorde, L.B. (1996) Ascertainment bias in estimates of average heterozygosity. *American Journal of Human Genetics*, **58**, 1033–41.

Ross, C.A. (1995) When more is less: pathogenesis of glutamine repeat neurodegenerative diseases. *Neuron*, **15**, 493–6.

Rothuizen, J., Wolfswinkel, J., Lenstra, J.A. and Frants, R.R. (1994) The incidence of minisatellite and microsatellite repetitive DNA in the canine genome. *Theoretical and Applied Genetics*, **89**, 403–6.

Rousset, F. (1996) Equilibrium values of measures of population subdivision for stepwise mutation processes. *Genetics*, **142**, 1357–62.

Roy, M.S., Geffen, E., Smith, D., Ostrander, E.A. and Wayne, R.K. (1994) Patterns of differentiation in North American wolf-like canids, revealed by analysis of microsatellite loci. *Molecular Biology and Evolution*, **11**, 553–70.

Roy, M.S., Geffen, E., Smith, D. and Wayne, R.K. (1996) Molecular genetics of pre-1940 red wolves. *Conservation Biology*, **10**, 1413–24.

Rubinsztein, D.C. and Leggo, J. (1997) Non-Mendelian transmission at the Machado-Joseph disease locus in normal females: preferential transmission of alleles with smaller repeats. *Journal of Medical Genetics*, **34**, 234–6.

Rubinsztein, D.C., Amos, W., Leggo, J., Goodburn, S., Ramesar, R.S., Old, J. *et al.* (1994a) Mutational bias provides a model for the evolution of Huntington's disease and predicts a general increase in disease prevalence. *Nature Genetics*, **7**, 525–30.

Rubinsztein, D.C., Leggo, J., Amos, W., Barton, D.E. and Ferguson-Smith, M.A. (1994b) Myotonic dystrophy CTG repeats and the associated insertion/deletion polymorphism in human and primate populations. *Human Molecular Genetics*, **3**, 2031–5.

Rubinsztein, D.C., Amos, W., Leggo, J., Goodburn, S., Margolis, R.L., Ross, C.A. and Ferguson-Smith, M.A. (1995a) Microsatellite evolution—evidence for directionality and variation in rate between species. *Nature Genetics*, **10**, 337–43.

Rubinsztein, D.C., Leggo, J. and Amos, W. (1995b) Microsatellites evolve more rapidly in humans than in chimpanzees. *Genomics*, **30**, 610–12.

Rubinsztein, D.C., Leggo, J., Goodburn, S., Barton, D.E. and Ferguson-Smith, M.A. (1995c) Haplotype analysis of the $\Delta 2642$ and $(CAG)_n$ polymorphisms in the Huntington's disease (HD) gene provides and explanation for an apparent 'founder' HD haplotype. *Human Molecular Genetics*, **4**, 203–6.

Rubinsztein, D.C., Leggo, J., Coles, R., Almqvist, E., Biancalana, V., Cassiman, J.J. *et al.* (1996) Phenotypic characterisation of individuals with 30–40 CAG repeats in the Huntington's disease gene reveals HD cases with 36 repeats and apparently normal elderly individuals with 36–39 repeats. *American Journal of Human Genetics*, **56**, 16–22.

Rubinsztein, D.C., Leggo, J., Chiano, M., Dodge, A., Norbury, G., Rosser, E. and Craufurd, D. (1997) Genotypes at the GluR6 kainate receptor locus are associated with variation in the age of onset of Huntington disease. *Proceedings of the National Academy of Sciences of the USA*, **94**, 3872–6.

Ruiz Linares, A., Minch, E., Meyer, D. and Cavalli-Sforza, L.L. (1995) Analysis of classical and DNA markers for reconstructing human population history. In *The origin and past of* Homo sapiens sapiens *as viewed from DNA* (ed. S. Brenner and K. Hanihara), pp. 123–148. World Scientific Publishing Company, Singapore.

Ruiz Linares, A., Nayar, K., Goldstein, D.B., Hebert, J.M., Seielstad, M.T., Underhill, P.A. *et al.* (1996) Geographic clustering of human Y-chromosome haplotypes. *Annals of Human Genetics*, **60**, 401–8.

Russell, D.W., Smith, M., Cox, D., Williamson, V.M. and Young, E.T. (1983) DNA sequences of two yeast promoter-up mutants. *Nature*, **304**, 652–4.

Ruvolo, M. (1996) A new approach to studying modern humans origins, hypothesis testing with coalescence time distributions. *Molecular Phylogenetics and Evolution*, **5**, 202–19.

Rychkov, Y.G. (1969) Some approaches to Siberian anthropology based upon population genetics. (in Russian). *Voprosy Antropologii*, **33**, 16–33.

Rychkov, Y.G. and Sheremet'eva, V.A. (1980) The genetics of circumpolar populations of Eurasia related to the problem of human adaptation. In *The human biology of circumpolar populations* (ed. F. Milan), pp. 37–80. Cambridge University, London.

Saccheri, I.J. and Bruford, M.W. (1993) DNA fingerprinting in a butterfly *Bicyclus anynana* (Satyridae) *Journal of Heredity*, **84**, 195–200.

Saitou, N. and Nei, M. (1987) The neighbour-joining method: A new method for reconstructing phylogenetic trees. *Molecular Biology and Evolution*, **4**, 406–25.

Samadashwily, G.M., Raca, G. and Mirkin, S.M. (1997) Trinucleotide repeats affect DNA replication in vivo. *Nature Genetics*, **17**, 298–304.

Samadi, S., Erard, F., Estoup, A. and Jarne, P. (1998) The influence of mutation, selection and reproductive systems on microsatellite variability: a simulation approach. *Genetical Research*, **71**, 213–22.

Samson, M., Libert, F., Doranz, B.J., Rucker, J., Liesnard, C., Farber, C.M. *et al.* (1996) Resistance to HIV-1 infection in Caucasian individuals bearing mutant alleles of the CCR-5 chemokine receptor gene. *Nature*, **382**, 722–5.

Sandaltzopoulos, R., Mitchelmore, C., Bonte, E., Wall, G. and Becker P.B. (1995) Dual regulation of the *Drosophila hsp26* promoter *in vitro*. *Nucleic Acids Research*, **23**, 2479–87.

Sanpei, K., Takano, H., Igarashi, S., Sato, T., Oyake, M., Sasake, H. *et al.* (1996) Identification of the spinocerebellar ataxia type 2 gene using a direct identification of repeat expansion and cloning technique, DIRECT. *Nature Genetics*, **14**, 277–84.

Santos, F.R., Rodriguez-Delfin, L., Pena, S.D., Moore, J. and Weiss, K.M. (1996) North and South Amerindians may have the same major founder Y chromosome haplotype. *American Journal of Human Genetics*, **58**, 1369–70.

Sarkar, G., Paynton, C. and Sommer, S.S. (1991) Segments containing alternating purine and pyrimidine dinucleotides: patterns of polymorphism in humans and prevalence throughout phylogeny. *Nucleic Acids Research*, **19**, 631–6.

Satta, Y., O'huigin, C., Takahata, N. and Klein, J. (1995) Intensity of overdominant selection at the major histocompatibility complex loci. In *Current Topics in Molecular Evolution* (ed. M. Nei and N. Takahata), pp. 178–188. Institute for Molecular Evolutionary Genetics, Penn State University.

Sawyer, L.A., Hennessy, J.M., Peixoto, A.A., Rosato, E., Parkinson, H., Costa, R. *et al.* (1997) Natural variation in a *Drosophila* clock gene and temperature compensation. *Science*, **278**, 2117–20.

Scherzinger, E., Lurz, R., Turmaine, M., Mangiarini, L., Hollenbach, B., Hasenbank, R. *et al.* (1997) Huntingtin-encoded polyglutamine expansions form amyloid-like protein aggregates in vitro and in vivo. *Cell*, **90**, 549–58.

Schlötterer, C. and Pemberton, J. (1998) The use of microsatellites for genetic analysis of natural populations—a critical review. In *Molecular approaches to individuals, populations and species* (ed. R. DeSalle and B. Schierwater), pp.71–86. Birkhäuser, Basel.

Schlötterer, C. and Tautz, D. (1992) Slippage synthesis of simple sequence DNA. *Nucleic Acids Research*, **20**, 211–15.

Schlötterer, C., Amos, B. and Tautz, D. (1991) Conservation of polymorphic simple sequences in Cetacean species. *Nature*, **354**, 63–5.

Schlötterer, C., Vogl, C. and Tautz, D. (1997) Polymorphism and locus-specific effects on polymorphism at microsatellite loci in natural *Drosophila melanogaster* populations. *Genetics*, **146**, 309–20.

Schlötterer, C., Ritter, R., Harr, B. and Brem, G. (1998) High mutation rates of long microsatellite alleles in *Drosophila melanogaster* provide evidence for allele specific mutation rates. *Molecular Biology and Evolution*, **15**, 1269–74.

Schug, M.D., Mackay, T.F.C. and Aquadro, C.F. (1997*a*) Low mutation rates of microsatellite loci in *Drosophila melanogaster*. *Nature Genetics*, **15**, 99–102.

Schug, M.D., Hutter, C.M., Noor, M.A.F. and Aquadro, C.F. (1998) Mutation and evolution of microsatellites in *Drosophila melanogaster*. *Genetica*, **103**, 359–67.

Schug, M.D., Wetterstrand, K.A., Gaudette, M.S., Lim, R.H., Hutter, C.M. and Aquadro, C.F. (1998) The distribution and frequency of microsatellite loci in *Drosophila melanogaster*. *Molecular Ecology*, **7**, 57–69.

Scotland, R.W. (1992) Cladistic theory. In *Cladistics: a practical course in systematics* (ed. P.L. Forey, C.J. Humphries, I.J. Kitching, R.W. Scotland, D.J. Siebert and D.M. Williams), pp. 3–13. Clarendon Press, Oxford.

Scozzari, R., Cruciani, F., Malaspina, P., Santolamazza, P., Ciminelli, B.M., Torroni, A. *et al.* (1997) Different structuring of human populations for homologous X and Y microsatellite loci. *American Journal of Human Genetics*, **61**, 719–33.

Sheffield, V.C., Weber, J.L., Buetox, K.H., Murray, J.C., Even, D.A., Wiles, K. *et al.* (1995) A collection of tri- and tetranucleotide repeat markers used to generate high quality, high resolution human genome-wide linkage maps. *Human Molecular Genetics*, **4**, 1837–44.

Shen, L.-P. and Rutter, W.J. (1984) Sequence of the human somatostatin gene. *Science*, **224**, 168–70.

Sherman, S.L., Morton, N.E., Jacobs, P.A. and Turner, G. (1984) The marker (X) syndrome: a cytogenetic and genetic analysis. *Annals of Human Genetics*, **48**, 21–37.

Sherman, S.L., Jacobs, P.A., Morton, N.E., Froster-Iskenius, U., Howard-Peebles, P.N., Nielsen, K.B. *et al.* (1985) Further segregation analysis of the fragile X syndrome with special reference to transmitting males. *Human Genetics*, **69**, 289–99.

Shibata, D. (1997) Molecular tumour clocks and dynamic phenotype. *American Journal of Pathology*, **151**, 643–6.

Shibata, D., Peinado, M.A., Ionov, Y., Malkhosyan, S. and Perucho, M. (1994) Genomic instability in repeated sequences is an early somatic event in colorectal tumourigenesis that persists after transformation. *Nature Genetics*, **6**, 273–81.

Shibata, D., Navidi, W., Salovaara, R., Li, Z.H. and Aaltonen, L.A. (1996) Somatic microsatellite mutations as molecular tumour clocks. *Nature Medicine*, **2**, 676–81.

Shields, G.F., Schmiechen, A.M., Frazier, B.L., Redd, A., Voevoda, M.I., Reed, J.K. *et al.* (1993) mtDNA sequences suggest a recent evolutionary divergence for Beringian and northern North American populations. *American Journal of Human Genetics*, **53**, 549–62.

Shields, W.M. (1993) The natural and unnatural history of inbreeding and outbreeding. In *The natural history of inbreeding and outbreeding: theoretical and empirical perspectives* (ed. N.W. Thornhill), pp. 143–169. University of Chicago Press, Chicago.

Shriver, M.D. (1997) Ethnic variation as a key to the biology of human disease. *Annals of Internal Medicine*, **127**, 401–3.

Shriver, M.D., Jin, L., Chakraborty, R. and Boerwinkle, L.E. (1993) VNTR allele frequency distribution under the stepwise mutation model. *Genetics*, **134**, 983–93.

Shriver, M.D., Jin, L., Boerwinkle, L.E., Deka, R., Ferrell, R.E. and Chakraborty, R. (1995) A novel measure of genetic distance for highly polymorphic tandem repeat loci. *Molecular Biology and Evolution*, **12**, 914–20.

Shriver, M.D., Smith, M.W., Jin, L., Marcini, A., Akey, J.M., Deka, R. *et al.* (1997) Ethnic-affiliation estimation by use of population-specific DNA Markers. *American Journal of Human Genetics*, **60**, 957–64.

Sia, E.A., Jinks-Robertson, S. and Petes, T.D. (1997*a*) Genetic control of microsatellite stability. *Mutation Research*, **383**, 61–70.

Sia, E.A., Kokoska, R.J., Dominska, M., Greenwell, P. and Petes, T.D. (1997*b*) Microsatellite instability in yeast: dependence on repeat unit size and DNA mismatch repair genes. *Molecular and Cellular Biology*, **17**, 2851–8.

Sillero-Zubiri, C and Macdonald, D. (1997) *The Ethiopian wolf: status survey and conservation action plan.* IUCN/SSC Canid Specialist Group.

Simchenko, Y.B. (1968) Some data on ancient ethnic substrate in the composition of North Eurasia peoples. In *Problemi antropologii i istoricheskoy etnografii Azii.* Moscow (in Russian).

Sivolob, A.V. and Khrapunov, S.N. (1995) Translational positioning of nucleosomes on DNA: the role of sequence-dependent isotropic DNA bending stiffness. *Journal of Molecular Biology*, **247**, 918–31.

Skinner, P.J., Koshy, B.T., Cummings, C.J., Klement, I.A., Helin, K., Servadio, A. *et al.* (1997) Ataxin-1 with an expanded glutamine tract alters nuclear matrix-associated structures. *Nature*, **389**, 971–4.

Slatkin, M. (1995) A measure of population subdivision based on microsatellite allele frequencies. *Genetics*, **139**, 457–62.

Slatkin, M. and Rannala, B. (1997) Estimating the age of alleles by use of intrallelic variability. *American Journal of Human Genetics*, **60**, 447–58.

Slatkin, M. and Wiehe, T. (1998) Genetic hitchhiking in a subdivided population. *Genetical Research*, **71**, 155–60.

Smith, G.P. (1973) Unequal crossover and the evolution of multigene families. *Cold Spring Harbor Symposium on Quantitative Biology*, **38**, 507–13.

Smith, G.P. (1976) Evolution of repeated DNA sequences by unequal crossover. *Science*, **191**, 528–35.

Smith, M.W., Dean, M., Carrington, M., Winkler, C., Huttley, G.A., Lomb, D.A. *et al.* (1997) Contrasting genetic influence of CCR2 and CCR5 variants on HIV-1 infection and disease progression. *Science*, **277**, 959–65.

Smith, T.B. and Wayne, R.K. (1996) *Molecular genetic approaches in conservation*. Oxford University Press, New York.

Smith, T.B., Wayne, R.K., Girman, D.J. and Bruford, M.W. (1997) A role for ecotones in generating rainforest biodiversity. *Science*, **276**, 1855–7.

Smithers, R.H.E. (1983) *The mammals of the southern African subregion*. pp. 663–666. University of Pretoria, Pretoria.

Sokal, R.R. and Rohlf, F.J. (1981) *Biometry*. W.H. Freeman and Company, New York.

Sokal, R.R. and Rohlf, F.J. (1995) *Biometry*. W.H. Freeman and Company, New York.

Solomon, M.J., Straus, F. and Varshavsky, A. (1986) A mammalian high mobility group protein recognizes any stretch of six A-T base pairs in duplex DNA. *Proceedings of the National Academy of Sciences of the USA*, **83**, 1276–80.

Sparkes, R., Kimpton, C., Watson, S., Oldroyd, N., Clayton, T., Barnett, L. *et al.* (1996*a*) The validation of a 7-locus multiplex STR test for use in forensic casework (I) Mixtures, ageing, degradation and species studies. *International Journal of Legal Medicine*, **109**, 186–94.

Sparkes, R., Kimpton, C., Gilbard, S., Carne, P., Andersen, J., Oldroyd, N. *et al.* (1996*b*) The validation of a 7-locus multiplex STR test for use in forensic casework (II) Artefacts, casework studies and success rates. *International Journal of Legal Medicine*, **109**, 195–204.

Spencer, P.B.S., Odorico, D.M., Jones, S.J., Marsh, H.D. and Miller, D.J. (1995) Highly variable microsatellites in isolated colonies of the rock-wallaby (*Petrogale assimilis*) *Molecular Ecology*, **4**, 523–5.

Stallings, R.L. (1992) CpG suppression in vertebrate genomes does not account for the rarity of $(CpG)_n$ microsatellite repeats. *Genomics*, **17**, 890–1.

Stallings, R.L. (1994) Distribution of trinucleotide microsatellites in different categories of mammalian genomic sequence: implications for human genetic diseases. *Genomics*, **21**, 116–21.

Stallings, R.L. (1995) Conservation and evolution of $(CT)_n/(GA)_n$ microsatellite sequences at orthologous positions in diverse mammalian genomes. *Genomics*, **25**, 107–13.

Stallings, R.L., Ford, A.F., Nelson, D., Torney, D.C., Hildebrand, C.E. and Moyzis, R.K. (1991) Evolution and distribution of $(GT)_n$ repetitive sequences in mammalian genomes. *Genomics*, **10**, 807–15.

Steele, G.G. (1977) *Growth kinetics of tumours*, pp. 185–216. Clarendon Press, Oxford.

Stephens, J.C., Briscoe, D. and O'Brien, S.J. (1994) Mapping by admixture linkage disequilibrium in human populations: limits and guidelines. *American Journal of Human Genetics*, **55**, 809–24.

Stephens, J.C., Reich, D.E., Goldstein, D.B., Shin, H.D., Smith, M.W., Carrington, M. *et al.* (1998) Dating the origin of the CCR5-Δ32 AIDS resistance allele by the coalescence of haplotypes. *American Journal of Human Genetics*, **62**, 1507–15.

Stoneking, M. (1993) DNA and recent human evolution. *Evolutionary Anthropology*, **2**, 60–72.

Stott, K., Blackburn, J.M., Butler, P.J.G. and Perutz, M. (1995) Incorporation of glutamine repeats makes proteins oligomerize: implications for neurodegenerative diseases. *Proceedings of the National Academy of Sciences of the USA*, **92**, 6509–13.

Strand, M., Prolla, T.A., Liskay, R.M. and Petes, T.D. (1993) Destabilization of tracts of simple repetitive DNA in yeast by mutations affecting DNA mismatch repair. *Nature*, **365**, 274–6.

Strand, M., Earley, M., Crouse, G.F. and Petes, T.D. (1995) Mutations in the MSH3 gene preferentially lead to deletions within tracts of simple repetitive DNA in Saccharomyces cerevisiae. *Proceedings of the National Academy of Sciences of the USA*, **92**, 10418–21.

Strauss, B.S., Sagher, D. and Acharya, S. (1997) Role of proofreading and mismatch repair in maintaining the stability of nucleotide repeats in DNA. *Nucleic Acids Research*, **25**, 806–13.

Strauss, M., Lukas, J. and Bartek, J. (1995) Unrestricted cell cycling and cancer. *Nature Medicine*, **1**, 1245–6.

Streisinger, G. and Owen, J. (1985) Mechanisms of spontaneous and induced frameshift mutation in bacteriophage T4. *Genetics*, **109**, 633–59.

Streisinger, G., Okada, Y., Emrich, J., Newton, J., Tsugita, A., Terzaghi, E. *et al.* (1966) Frameshift mutations and the genetic code. *Cold Spring Harbor Symposium on Quantitative Biology*, **31**, 77–84.

Stringer, C.B. and Andrews P. (1988) Genetic and fossil evidence for the origin of modern humans. *Science*, **158**, 1263–8.

Struhl, K. (1985) Naturally occurring poly(dA-dT) sequences are upstream promoter elements for constitutive transcription in yeast. *Proceedings of the National Academy of Sciences of the USA*, **82**, 8419–23.

Suen, T.-C. and Hung, M.-C. (1990) Multiple cis- and trans-acting elements involved in regulation of the neu gene. *Molecular and Cellular Biology*, **10**, 6306–15.

Sukernik, R.I., Karafet, T.M. and Osipova, L.P. (1978) Distribution of blood groups serum markers and red cell enzymes in two human populations from Northern Siberia. *Human Heredity*, **28**, 321–7.

Sukernik, R.I., Osipova, L.P., Karafet, T.M. and Abanina, T.A. (1980) Studies on blood groups and other genetic markers in Forest Nentsi: variation among the subpopulations. *Human Genetics*, **55**, 397–404.

Sukernik, R.I., Schurr, T.G., Starikovskaya, E.B. and Wallace, D.C. (1996) Mitochondrial DNA variation in native Siberians, with special reference to the evolutionary history of American Indians. Studies on restriction polymorphism. (in Russian). *Genetika*, **32**, 432–9.

Sunnucks, P., England, P.R., Taylor, A.C. and Hales, D.F. (1996) Microsatellite and chromosome evolution of parthenogenetic *Sitobion* aphids in Australia. *Genetics*, **144**, 747–56.

Sutcliffe, J.S., Nelson, D.L., Zhang, F., Pieretti, M., Caskey, C.T., Saxe, D. *et al.* (1992) DNA methylation represses transcription in fragile X syndrome. *Human Molecular Genetics*, **1**, 397–400.

Sutherland, G.R. and Richards, R.I. (1995) Simple tandem DNA repeats and human genetic disease. *Proceedings of the National Academy of Sciences of the USA*, **92**, 3636–41.

Swarbrick, P.A. and Crawford, A.M. (1992) Ovine dinucleotide repeat polymorphism at the MAF109 locus. *Animal Genetics*, **23**, 84.

Swarbrick, P.A., Buchanon, F.C. and Crawford, A.M. (1991) Ovine dinucleotide repeat polymorphism at the MAF35 locus. *Animal Genetics*, **22**, 369–70.

Szathmary, E.J.E. (1981) Genetic markers in Siberian and Northern North American Populations. *Yearbook of Physical Anthropology*, **24**, 37–74.

Szostak, J.W., Orr-Weaver, T.L. and Rothstein, R.J. (1983) The double strand break repair model for recombination. *Cell*, **33**, 25–35.

Taberlet, P., Griffin, S., Goossens, B., Questiau, S., Manceau, V., Escaravage, N., Waits, L.P. *et al.* (1996) Reliable genotyping of samples with very low DNA quantities using PCR. *Nucleic Acids Research*, **24**, 3189–94.

Taberlet, P., Camarra, J.J., Griffin, S., Uhres, E., Hanotte, O., Waits, L.P. *et al.* (1997) Noninvasive genetic tracking of the endangered Pyrenean brown bear population. *Molecular Ecology*, **6**, 869–76.

Tachida, H. and Iizuka, M. (1992) Persistence of repeated sequences that evolve by replication slippage. *Genetics*, **131**, 471–8.

Taddei, F., Matic, I., Godelle, B. and Radman, M. (1997a) To be a mutator, or how pathogenic and commensal bacteria can evolve rapidly. *Trends in Microbiology*, **5**, 427–8.

Taddei, F., Radman, M., Maynard-Smith, J., Toupance, B., Gouyon, P.H. and Godelle, B. (1997b) Role of mutator alleles in adaptive evolution. *Nature*, **387**, 700–2.

Tajima, F. (1989) Statistical method for testing the neutral mutation hypothesis by DNA polymorphism. *Genetics*, **123**, 585–95.

Takano, H., Onodera, O., Takahashi, H., Igarashi, S., Yamada, M., Oyake, M. *et al.* (1996) Somatic mosaicism of expanded CAG repeats in brains of patients with dentatorubral-pallidoluysian atrophy: cellular population-dependent dynamics of mitotic instability. *American Journal of Human Genetics*, **58**, 1212–22.

Takezaki, N. and Nei, M. (1996) Genetic distances and reconstruction of phylogenetic trees from microsatellite DNA. *Genetics*, **144**, 389–99.

Takiyama, Y., Sakoe, K., Soutome, M., Namekawa, M., Ogawa, T., Nakano, I. *et al.* (1997) Single sperm analysis of the CAG repeats in the gene for Machado-Joseph disease (*MJD1*): evidence for non-Mendelian transmission of the *MJD1* gene and for the effect of the intragenic CGG/GGG polymorphism in the intergenerational instability. *Human Molecular Genetics*, **6**, 1063–8.

Tal, M., King, R., Kraus, M.H., Ullrich, A., Schlessinger, J. and Givol, D. (1987) Human HER2 (neu) promoter: Evidence for multiple mechanisms for transcriptional initiation. *Molecular and Cellular Biology*, **7**, 2597–601.

Talbot, C.C., Avramopoulos, D., Gerken, S., Chakravarti, A., Armour, J.A., Matsunami, N. *et al.* (1995) The tetranucleotide repeat polymorphism D21S1245 demonstrates hypermutability in germline and somatic cells. *Human Molecular Genetics*, **4**, 1193–9.

Tanner, M.A. (1993) *Tools for statistical inference*. Springer-Verlag, New York.

Tautz, D. (1989) Hypervariability of simple sequences as a general source for polymorphic DNA markers. *Nucleic Acids Research*, **17**, 6463–71.

Tautz, D. and Schlötterer, C. (1994) Simple sequences. *Current Opinion in Genetics and Development*, **4**, 832–7.

Tautz, D., Trick, M. and Dover, G.A. (1986) Cryptic simplicity in DNA is a major source of genetic variation. *Nature*, **322**, 652–6.

Tautz, D., Lehmann, R., Schnürch, H., Schuh, R., Seifert, E., Kienlin, A. *et al.* (1987) Finger protein of novel structure encoded by hunchback, a second member of the gap class of Drosophila segmentation genes. *Nature*, **327**, 383–9.

Tavaré, S. (1984) Line-of-descent and genealogical processes, and their applications in population genetics models. *Theoretical Population Biology*, **46**, 119–64.

Taylor, A.C., Sherwin, W.B. and Wayne, R.K. (1994) Genetic variation of microsatellite loci in a bottlenecked species: the northern hairy-nosed wombat *Lasiorhinus kreffti*. *Molecular Ecology*, **3**, 277–90.

Taylor, M.F.J., Shen, Y. and Kreitman, M.E. (1995) A population genetic test of selection at the molecular level. *Science*, **270**, 1497–9.

Telenius, H., Kremer, B., Goldberg, Y.P., Theilmann, J., Andrew, S.E., Zeisler, J. *et al.* (1994) Somatic and gonadal mosaicism of the Huntington disease gene CAG repeat in brain and sperm. *Nature Genetics*, **6**, 409–14.

Templeton, A.R. (1986) Coadaptation and outbreeding depression. In *Conservation biology* (ed. M.E. Soulé), pp. 105–116. Sinauer Press, Sunderland.

Templeton, A.R. (1993) The 'Eve' hypothesis, a genetic critique and reanalysis. *American Anthropologist*, **95**, 51–72.

Thibodeau, S.N., Bren, G. and Schaid, D. (1993) Microsatellite instability in cancer of the proximal colon. *Science*, **260**, 816–19.

Thompson, E.A. and Neel, J.V. (1997) Allelic disequilibrium and allele frequency distribution as a function of social and demographic history. *American Journal of Human Genetics*, **60**, 197–204.

Thompson, P.M., Kovacs, K.M. and McConnell, B.J. (1994) Natal dispersal of harbour seals (*Phoca vitulina*) from breeding sites in Orkney, Scotland. *Journal of Zoology*, **234**, 668–73.

Thornton, C.A., Wymer, J.P., Simmons, Z., McClain, C. and Moxley, R.T., III. (1997) Expansion of the myotonic dystrophy CTG repeat reduces expression of the flanking DMAHP gene. *Nature Genetics*, **16**, 407–9.

Timchenko, L.T., Timchenko, N.A., Caskey, C.T. and Roberts, R. (1996) Novel proteins with binding specificity for DNA CTG repeats and RNA CUG repeats: Implications for myotonic dystrophy. *Human Molecular Genetics*, **5**, 115–21.

Tisch, R. and McDevitt, H. (1996) Insulin-dependent diabetes mellitus. *Cell*, **85**, 291–7.

Tishkoff, S.A., Goldman, A., Speed, W.C., Kidd, J.R., Jenkins, T. and Kidd, K.K. (1995) A global haplotype analysis of myotonic dystrophy CTG repeats in humans and other primates. *American Journal of Human Genetics*, **57**, (supp) A42.

Tiwari, J. and Terasaki, P. (1985) *HLA and disease associations*. Springer-Verlag, New York.

Torkelson, J., Jarris, R.S., Lombardo, M.-J., Nagendran, J., Thulin, C. and Rosenberg, S.M. (1997) Genome-wide hypermutation in a subpopulation of stationary-phase cells underlies recombination-dependent adaptive mutation. *EMBO Journal*, **16**, 3303–11.

Torroni, A., Sukernik, R.I., Schurr, T.G., Starikovskaya, Y.B., Cabell, M.F., Crawford, M.H. *et al.* (1993) mt-DNA variation of aboriginal Siberians reveals distinct genetic affinities with Native Americans. *American Journal of Human Genetics*, **53**, 591–608.

Tran, H.T., Keen, J.D., Kricker, M., Resnick, M.A. and Gordenin, D.A. (1997) Hypermutability of homonucleotide runs in mismatch repair and DNA polymerase proofreading yeast mutants. *Molecular and Cellular Biology*, **17**, 2859–65.

Treier, M., Pfeifle, C. and Tautz, D. (1989) Comparison of the gap segmentation gene hunchback between *Drosophila melanogaster* and *Drosophila virilis* reveals novel modes of evolutionary change. *EMBO Journal*, **8**, 1517–25.

Trepicchio, W.L. and Krontiris, T.G. (1992) Members of the rel/NF-cB family of transcriptional regulatory proteins bind the *HRAS1* minisatellite DNA sequence. *Nucleic Acids Research*, **20**, 2427–34.

Trepicchio, W.T. and Krontiris, T.G. (1993) IGH minisatellite suppression of USF-binding-site and Em-mediated transcriptional activation of the adenovirus major late promoter. *Nucleic Acids Research*, **21**, 977–85.

Tsao, J.L., Davis, S.D., Baker, S.M., Liskay, R.M. and Shibata, D. (1997) Intestinal stem cell divisions and genetic diversity: a computer and experimental analysis. *American Journal of Pathology*, **151**, 573–9.

Tsuji, K., Aizawa, M. and Sasazuki, T. (1992) *HLA 1991. Proceedings of the eleventh international histocompatibility workshop and conference.* Oxford University Press, Oxford.

Tut, T.G., Ghadessy, F.J., Trifiro, M.A., Pinsky, L. and Yong, E.L. (1997) Long polyglutamine tracts in the androgen receptor are associated with reduced trans-activation, impaired sperm production, and male infertility. *Journal of Clinical Endocrinology and Metabolism*, **82**, 3777–82.

Underhill, P.A., Jin, L., Zemans, R., Oefner, P.J. and Cavalli-Sforza, L.L. (1996) A pre-Colombian Y chromosome-specific transition and its implications for human evolutionary history. *Proceedings of the National Academy of Sciences of the USA*, **93**, 196–200.

Underhill, P.A., Jin, L., Lin, A.A., Mehdi, S.Q., Jenkins, T., Vollrath, D. *et al.* (1997) Detection of numerous Y chromosome biallelic polymorphisms by denaturing high-performance liquid chromatography. *Genome Research*, **7**, 996–1005.

Valdes, A.M., Slatkin, M. and Freiner, N.B. (1993) Allele frequencies at microsatellite loci: the stepwise mutation model revised. *Genetics*, **133**, 737–49.

Valle, G. (1993) TA-repeat microsatellites are closely associated with ARS consensus sequences in yeast chromosome III. *Yeast*, **9**, 753–9.

Valsecchi, E, Palsbøll, P., Hale, P. *et al.* (1997) Microsatellite genetic distances between oceanic populations of the humpback whale (*Megaptera novaeangliae*). *Molecular Biology Evolution*, **14**, 355–62.

Van Nues, R.W., Venema, J., Planta, R.J. and Raué, H.A. (1997) Variable region V1 of Saccharomyces cerevisiae 18S rRNA participates in biogenesis and function of the small ribosomal subunit. *Chromosoma*, **105**, 523–31.

Van Tol, H.H.M., Wu, C.M., Guan, H.-C., Ohara, K., Bunzow, J.R., Civelli, O. *et al.* (1992) Multiple dopamine D4 receptor variants in the human population. *Nature*, **358**, 149–52.

Van Treuren, R., Kuittinen, H., Kärkkäinen, K., Baena-Gonzalez, E. and Savolainen, O. (1997) Evolution of microsatellites in *Arabis petra* and *Arabis lyrata*, outcrossing relatives of *Arabidopsis thaliana*. *Molecular Biology and Evolution*, **14**, 220–9.

Vashakidze, R.P., Chelidze, M.G., Mamulashvili, N.A., Kalandarishvili, K.G. and Tsalkalamanidze, N.V. (1988) Nuclear proteins from *Drosophila* melanogaster embryos which specifically bind to simple homopolymeric sequences poly(dT-dG)-(dC-dA) *Nucleic Acids Research*, **16**, 4989–94.

Vasilyev, V.I. (1985) Main problems of ethnic history of North Samoyeds. (In Russian). In *Uralo-Altaistika. Archeology, ethnography, language.* pp. 119–123. Nauka, Novosibirsk.

Vergnaud, G., Mariat, D., Apiou, F., Aurias, A., Lathrop, M. and Lauthier, V. (1991) The use of synthetic tandem repeats to isolate new VNTR loci: cloning of a human hypermutable sequence. *Genomics*, **11**, 135–44.

Viard, F., Brémond, P., Labbo, R., Justy, F., Delay, B. and Jarne, P. (1996) Microsatellites and the genetics of highly selfing populations in the freshwater snail *Bulinus truncatus*. *Genetics*, **142**, 1237–47.

Viard, F., Franck, P., Dubois, M.-P., Estoup, A. and Jarne, P. (1998) Variation of microsatellite size homoplasy across electromorphs, loci and populations in three invertebrate species. *Journal of Molecular Evolution*, **47**, 42–51.

Virtaneva, K., D'Amato, E., Miao, J., Kiskiniemi, M., Norio, R. *et al.* (1997) Unstable minisatellite expansion causing recessively inherited myoclonus epilepsy, EPM1. *Nature Genetics*, **15**, 393–6.

Vogel, F. and Motulsky, A.G. (1997) *Human genetics: problems and approaches*. 3rd edn., pp. 385–429. Springer-Verlag, Berlin.

Vogler, A.P., Welsh, A. and Hancock, J.M. (1997) Phylogenetic analysis of slippage-like sequence variation in the V4 rRNA expansion segment in Tiger Beetles (Cicindelidae). *Molecular Biology and Evolution*, **14**, 6–19.

Wahls, W.P., Wallace, L.J. and Moore, P.D. (1990) The Z-DNA motif d(TG)$_{30}$ promotes reception of information during gene conversion events while stimulating homologous recombination in human cells in culture. *Molecular and Cellular Biology*, **10**, 785–93.

Wang, G., Seidman, M.M. and Glazer, P.M. (1996) Mutagenesis in mammalian cells induced by triple helix formation and transcription-coupled repair. *Science*, **271**, 802–5.

Wang, J., Pegoraro, E., Menegazzo, E., Gennarelli, M., Hoop, R.C., Angelini, C. *et al.* (1995) Myotonic dystrophy: evidence for a possible dominant-negative RNA mutation. *Human Molecular Genetics*, **4**, 599–606.

Wanker, E.E., Rovira, C., Scherzinger, E., Hasenbank, R., Walter, S., Tait, D., Colicelli, J. and Lehrach, H. (1997) HIP-I: a huntingtin interacting protein isolated by the yeast two-hybrid system. *Human Molecular Genetics*, **6**, 487–95.

Warburton, P.E. and Willard, H.F. (1996) Evolution of centromeric alpha satellite DNA: molecular organization within and between human and primate chromosomes. In *Human Genome Evolution* (ed. M. Jackson, T. Strachan and G. Dover) BIOS Scientific Publishers, Oxford.

Waser, N.M. (1993) Sex, mating systems, inbreeding and outbreeding. In *The natural history of inbreeding and outbreeding: theoretical and empirical perspectives*. (ed. N.W. Thornhill), pp. 1–13. University of Chicago Press, Chicago.

Wayne, R.K. and Jenks, S.M. (1991) Mitochondrial DNA analysis implying extensive hybridization of the endangered red wolf *Canis rufus*. *Nature*, **351**, 565–8.

Weber, J.L. (1990) Informativeness of human $(dC-dA)_n \cdot (dG-dT)_n$ polymorphisms. *Genomics* **7**, 524–30.

Weber, J.L. and May, P.E. (1989) Abundant class of human DNA polymorphisms which can be typed using the polymerase chain reaction. *American Journal of Human Genetics*, **44**, 388–96.

Weber, J.L. and Wong, C. (1993) Mutation of human short tandem repeats. *Human Molecular Genetics*, **2**, 1123–8.

Weir, B.S. (1990) *Genetic Data Analysis*. Sinauer Associates, Sunderland, MA.

Weir, B.S. (1996) *Genetic Data Analysis II*. Sinauer Associates, Sunderland, MA.

Weir, B.S., Triggs, C.M., Starling, L., Stowell, L.I. and Walsh, K.A.J. (1997) Interpreting DNA mixtures. *Journal of Forensic Sci.*, **42**, 213–221.

Weiss, K. (1973) Demographic models for anthropology. *American Antiq.*, **38**, 1–186.

Wells, R.D., Jacobs, T.M., Narang, S.A. and Khorana, H.G. (1967) Studies on polynucleotides LXIX. Synthetic deoxyribopolynucleotides as templates for the DNA polymerase of Escherichia coli: DNA-like polymers containing repeating trinucleotide sequences. *Journal of Molecular Biology*, **27**, 237–63.

Wexler, N.S., Young, A.B., Tanzi, R.E., Travers, H., Starosta-Rubistein, S., Penney, J.B. *et al.* (1987) Homozygotes for Huntington disease. *Nature*, **326**, 194–7.

Wharton, K.A., Yedvobnick, B., Finnerty, V.G. and Artavanis-Tsakonas, S. (1985) opa: a novel family of transcribed repeats shared by the Notch locus and other developmentally regulated loci in D. melanogaster. *Cell*, **40**, 55–62.

White, J.H.M., Johnson, A.L., Lowndes, N.F. and Johnston L.H. (1991) The yeast DNA ligase gene CDC9 is controlled by six orientation specific upstream activating sequences that respond to cellular proliferation but which alone cannot mediate cell cycle regulation. *Nucleic Acids Research*, **19**, 359–64.

Whitfield, L.S., Sulston, J.E. and Goodfellow, P.N. (1995) Sequence variation of the human Y chromosome. *Nature*, **378**, 379–80.

Wiehe, T. (1998) The effect of selective sweeps on the variance of the allele distribution of a linked multi-allele locus-hitchhiking of microsatellites. *Theoretical Population Biology*, **53**, 272–83.

Wierdl, M., Dominska, M. and Petes, T.D. (1997) Microsatellite instability in yeast: dependence on the length of the microsatellite. *Genetics*, **146**, 769–79.

Wierdl, M., Greene, C.N., Datta, A., Jinks-Robertson, S. and Petes, T.D. (1996) Destabilization of simple repetitive DNA sequences by transcription in yeast. *Genetics*, **143**, 713–21.

Wilkie, A.O.M. and Higgs, D.R. (1992) An unusually large $(CA)_n$ repeat in the region of divergence between subtelomeric alleles of human chromosome 16p. *Genomics*, **13**, 81–8.

Wilson, A.C., Cann, R.L., Carr, S.M. *et al.* (1985) Mitochondrial DNA and two perspectives on evolutionary genetics. *Biological Journal of the Linnaean Society*, **26**, 375–400.

Wilson, I.J. and Balding, D.J. (1998) Genealogical inference from microsatellite data. *Genetics*, **150**, 499–510.

Wilson, R.B. and Roof, D.M. (1997) Respiratory deficiency due to loss of mitochondrial DNA in yeast lacking the frataxin homologue. *Nature Genetics*, **16**, 352–7.

Winter, E. and Varshavsky, A. (1989) A DNA binding protein that recognizes oligo(dA)-oligo(dT) tracts. *EMBO Journal*, **8**, 1867–77.

Wise, C.A., Sraml, M., Rubinsztein, D.C. and Easteal, S. (1997) Comparative nuclear and mitochondrial diversity in humans and chimpanzees. *Molecular Biology and Evolution*, **14**, 707–16.

Wolfe, K., Li, W.H. and Sharp, P. (1987) Rates of nucleotide substitution vary greatly among plant mitochondrial, chloroplast and nuclear DNAs. *Proceedings of the National Academy of Sciences of the USA*, **84**, 9054–8.

Wolfe, K.H., Sharp, P.M. and Li, W.H. (1989) Mutation rates differ among regions of the mammalian genome. *Nature*, **337**, 283–5.

Wolpoff, M.H., Wu, X.Z. and Thorne, A.G. (1984) Modern *Homo sapiens* origins, a general theory of hominid evolution involving the fossil evidence from East Asia. In *Origins of modern humans, a world survey of the fossil evidence* (ed. F.H. Smith and F. Spencer), pp. 411–484. Alan R. Liss, New York.

Wong, L.-J.C., Ashizawa, T., Monckton, D.G., Caskey, C.T., and Richards, C.S. (1995) Somatic heterogeneity of the CTG repeat in myotonic dystrophy is age and size dependent. *American Journal of Human Genetics*, **56**, 114–22.

Wong, Z., Wilson, V., Jeffreys, A.J. and Thein, S.L. (1986) Cloning a selected fragment from a human DNA 'fingerprint': isolation of an extremely polymorphic minisatellite. *Nucleic Acids Research*, **14**, 4605–16.

Wong, Z., Wilson, V., Patel, I., Povey, S. and Jeffreys, A.J. (1987) Characterization of a panel of highly variable minisatellites cloned from human DNA. *Annals of Human Genetics*, **51**, 269–88.

Woodruff, R.C. (1989) Genetic anomalies associated with *Cerion* hybrid zones: the origin and maintenance of new electrophoretic variants called hybrizymes. *Biological Journal of the Linnaean Society*, **36**, 281–94.

Wright, S. (1943) Isolation by distance. *Genetics*, **28**, 114–38.

Wright, S. (1951) The genetical structure of populations. *Annals of Eugenics*, **15**, 323–54.

Xu, W., Liu, L., Emson, P.C., Harrington, C.R. and Charles, I.G. (1997) Evolution of a homopurine–homopyrimidine pentanucleotide repeat sequence upstream of the human inducible nitric oxide synthase gene. *Gene*, **204**, 165–70.

Yandava, C., Gastier, J.M., Pulido, J.C., Brody, T., Sheffield, V., Murray, J. *et al.* (1997) Characterization of Alu repeats that are associated with trinucleotide and tetranucleotide repeat microsatellites. *Genome Research*, **7**, 716–24.

Yang, S.Y., Milford, E., Hammerling, V. and Dupont, B. (1989) Description of the reference panel of B-lymphoblastoid cell lines for factors of the HLA system: the B-cell line panel designed for the Tenth International Histocompatibility Workshop. *Immunobiology of HLA*, vol. 1 (ed. B. Dupont), p. 11. Springer-Verlag, New York, 1989.

Yang, Z. and Rannala, B. (1997) Bayesian phylogenetic inference using DNA sequences: a Markov chain Monte Carlo method. *Molecular Biology and Evolution*, **14**, 717–24.

Yano-Yanagisawa, H., Li, Y., Wang, H. and Kohwi, Y. (1995) Single-stranded DNA binding proteins isolated form mouse brain recognize specific trinucleotide repeat sequences *in vitro*. *Nucleic Acids Research*, **23**, 2654–60.

Yee, H.A., Wong, A.K.C., van de Sande, J.H. and Rattner, J.B. (1991) Identification of novel single-stranded d(TC)$_n$ binding proteins in several mammalian species. *Nucleic Acids Research*, **19**, 949–53.

Zahn, L.M. and Kwiatkowski, D.J. (1995) A 37-marker PCR-based genetic linkage map of human chromosome 9: observation on mutations and positive interference. *Genomics*, **28**, 140–6.

Zar, J.H. (1984) *Biostatistical Analysis*. Prentice-Hall, Englewood Cliffs, NJ.

Zeitlin, S., Liu, J.-P., Chapman, D.L., Papaioannou, V.E. and Efstratiadis, A. (1995) Increased apoptosis and early embryonic lethality in mice nullizygous for the Huntington's disease gene homologue. *Nature Genetics*, **11**, 155–63.

Zerjal, T., Dashnyam, B., Pandya, A., Kayser, M., Roewer, L., Santos, F.R. *et al.* (1997) Genetic relationships of Asians and northern Europeans, revealed by Y-chromosomal DNA analysis. *American Journal of Human Genetics*, **60**, 1174–83.

Zerylnick, C., Torroni, A., Sherman, S.L. and Warren, S.T. (1995) Normal variation at the myotonic dystrophy locus in global human populations. *American Journal of Human Genetics*, **56**, 123–30.

Zhao, Y., Cheng, W., Gibb, C.L.D., Gupta, G. and Kallenbach, N.R. (1996) HMG box proteins interact with multiple tandemly repeated (GCC)$_n \cdot$(GGC)$_m$ DNA sequences. *Journal of Biomolecular Structure and Dynamics*, **14**, 235–8.

Zheng, L., Collins, F.H., Kumar, V. and Kafatos, F.C. (1993) A detailed genetic map for the X chromosome of the malaria vector, *Anopheles gambiae*. *Science*, **261**, 605–8.

Zhivotovsky, L.A. and Feldman, M.W. (1995) Microsatellite variability and genetic distances. *Proceedings of the National Academy of Sciences of the USA*, **92**, 11549–52.

Zhivotovsky, L.A., Feldman, M.W. and Grishechkin, S.A. (1997) Biased mutation and microsatellite variation. *Molecular Biology and Evolution*, **14**, 926–33.

Zhuchenko, O., Bailey, J., Bonnen, P., Ashizawa, T., Stockton, D.W., Amos, C. *et al.* (1997) Autosomal dominant cerebellar ataxia (SCA6) associated with small polyglutamine expansions in the $\alpha$1A-voltage dependent calcium channel. *Nature Genetics*, **15**, 62–9.

Zischler, H., Kammerbauer, C., Studer, R., Grzeschik, K.-H. and Epplen, J.T. (1992) Dissecting (CAC)$_5$/(GTG)$_5$ multilocus fingerprints from man into individual locus-specific, hypervariable components. *Genomics*, **13**, 983–90.

# Glossary

**Allele size**

Indicates the size of a PCR fragment containing a microsatellite locus when the corresponding number of repeats is not known. When the number of repeats can be inferred *repeat count* should be used instead of allele size.

**Anticipation**

A form of inheritance in which penetrance, age of onset, or severity of disease increases from generation to generation through an affected pedigree. The genetic basis of many forms of anticipation is now known to be trinucleotide microsatellites in which the probability, magnitude, and directional bias of mutations increase with the repeat count.

**Ascertainment bias**

The hypothesis that a microsatellite selected in a focal species will differ systematically from its orthologues in related species due to the criteria used to isolate it in the focal species.

**Bottleneck**

A dramatic reduction in population size. If the bottleneck has occurred sufficiently recently in the past a population will not be in mutation–drift equilibrium, in which case the effect of the bottleneck may be detectable in the amount and other characteristics of the genetic variation in the population.

**Coalescence**

The point of time (in the past) at which two or more alleles were derived from a single ancestor.

**Coalescent theory**

A modelling framework for describing and making inferences based upon the ancestral relationships among a set of extant alleles. Coalescent theory begins with a set of extant alleles and steps backwards through time, describing the history of which pairs of alleles coalesce when. For this reason, it is often referred to as a backward approach, in comparison with classical forward approaches which follow sampling and inheritance forward through time. The coalescent very closely represents populations in which sampling follows the Wright–Fisher process, but many other sampling processes also result in configurations of ancestral relationships that approximate the coalescent.

**Codominant marker**

A genetic marker in which heterozygotes can be distinguished from either homozygote.

**Cross species amplification**

Primer binding sites sufficiently conserved in a related species to allow amplification using primers designed in a different species. As the microsatellite sequence is often present and polymorphic in closely related species, orthologous microsatellites are an increasingly important source of markers. The exact repeat count can be inferred only if the flanking region has been sequenced.

**Cryptic simplicity**

A sequence stretch containing more structure in terms of repetitions of sequence motifs than expected under a model of random ordering of the four bases given their local frequencies.

**Delta mu square**

The genetic distance $(\delta\mu)^2$ is defined as the square of the difference between the average repeat size in two populations. Under the assumptions of reproductive isolation between populations, mutation–drift equilibrium within the ancestral and derivative populations, and stepwise mutations, $(\delta\mu)^2$ is a linear function of the separation time between the populations with the slope given by the mutation rate. While the theoretical properties of $(\delta\mu)^2$ are attractive, the utility of it and other stepwise distances remains unclear due to its high variances and departures from stepwise mutations.

**Directional evolution**

Used in the microsatellite literature to refer to either biases in the distribution of mutation sizes, or to differences between species in the average length of microsatellites thought to be due, at least in part, to such biases. In reference to the mutation process, the more precise description is biased mutations or asymmetry in the distribution of mutation sizes.

**Effective population size**

Variable population size, variance in reproductive output among individuals, unequal sex ratio and other factors create a situation in which populations of large census size have certain evolutionary properties similar to idealized populations of a smaller size. The size of the idealized population that would behave similarly to the real population is called the effective population size. Depending on the properties of interest, various formulae are available for relating census size to effective size. It must be kept in mind, however, that the smaller, idealized population cannot possibly have the same evolutionary properties as the larger population in any general sense. The equivalence applies only in the very narrow sense that a specific summary statistic, such as the heterozygosity, has been made to match in the two cases.

**Fingerprinting**

The identification of individuals based upon their multilocus genotypes. Earlier approaches to fingerprinting relied on minisatellite loci, but most approaches now make use of a standardized set of PCR format microsatellite loci in which case the preferred term is DNA profiling.

**Flanking region**

The single-copy DNA sequence immediately upstream and downstream of a microsatellite locus that allows the design of specific primers that preferentially amplify the target microsatellite.

**Frameshift mutation**

An insertion or deletion mutation of a size that is not a multiple of three and therefore alters the reading frame for DNA translation. The tendency of microsatellite motif sizes other than three to produce frameshift mutations by and large restricts non-trinucleotide microsatellites to non-coding DNA regions.

$F_{st}$

A genetic distance measure usually used to estimate migration or separation time in geographically structured populations that focuses on changes in gene frequency caused by genetic drift.

**Gene conversion**

A process involving the correction of heteroduplex DNA following a recombination event that results in sequence from one homologous gene replacing the sequence of the other homologue. Gene conversion was initially described for Ascomycetes, in which allele frequencies in spores were found to deviate from the expectation.

**Gene flow**

Migration of individuals among subpopulations with successful reproduction in the new population.

**Genetic distance**

An estimator of the separation time among populations based upon measurable genetic differences within and between the populations. Some genetic distances focus on changes in allele frequencies (e.g. $F_{st}$) while others incorporate the mutation process and in the case of microsatellites are called stepwise distances. Matrices of genetic distances among populations can be converted into evolutionary trees using clustering algorithms such as UPGMA or Neighbour Joining.

**Haplotype**

Term describing a particular compound haploid genotype involving a specific set of alleles across two or more loci.

**Heterozygosity (gene diversity)**

The observed or expected proportion of individuals whose homologous chromosomes carry distinguishable alleles. Expected heterozygosity is calculated as one minus the sum of squared allele frequencies, and is not restricted to diploid organisms.

**Hitchhiking**

The phenomenon by which a selectively favoured mutation at a linked site sweeps through a population carrying with it a specific haplotype in the surrounding region. For a period of time following the sweep variation in the surrounding region will be reduced, and the pattern of the variation may also be sufficiently affected to allow detection using tests of selective neutrality.

**Homoplasy**

At a microsatellite locus, homplasy occurs when two allelic lineages converge on the same size but have different histories of mutations increasing and decreasing size. Size identity, therefore, is due to the mutation process as opposed to ancestry. Thus, identity by state does not always entail identity by decent.

**Inbreeding**

Reproduction involving genetically related individuals.

**Inbreeding depression**

A reduction in fitness due to mating between genetically related individuals. The reduction may be due to the bringing together of deleterious recessive alleles at the same loci, or due to loss of heterozygosity at loci showing heterozygous advantage.

**Indel**

Term used for an insertion or deletion if the ancestral state is unknown.

**Infinite alleles model (IAM)**

Mutation model in which all mutations are to novel alleles not previously represented in the population.

**$K$ allele model (KAM)**

Mutation model in which $K$ different alleles are allowed, and new mutations are to any of the $K$ alleles regardless of the state of the parental allele.

**Library (genomic)**

A set of clones containing all or most of the genomic DNA of an individual(s) from a species. Genomic libraries (or repeat enriched genome libraries) are used to identify microsatellite loci and to determine the flanking sequences required for primer design.

**Likelihood ratio**

The ratio of probabilities of obtaining the observed data under different hypotheses concerning the assumed model used to generate the data. Likelihood ratios can be used to estimate confidence intervals around a maximum likelihood estimator of parameters such as the migration rate, separation time, or $\theta$, or to compare alternative hypotheses such as biased or unbiased mutation distributions.

**Linkage**

The tendency of loci in the genome to have a better than even chance of being co-inherited due to their physical proximity. Genetic linkage is only approximately correlated, however, with the physical distance separating loci.

**Linkage disequilibrium**

The statistical association between alleles at different (linked or unlinked) loci. That is, haplotype frequencies in the population differ from what would be predicted based on the frequencies of the constituent alleles under the assumption that they are independently drawn in constructing the haplotype. Linkage disequilibrium may be caused by recent mutation, selection, or population structure.

**MALD**

Mapping by Admixture Linkage Disequilibrium. A method for mapping genes that makes use of the linkage disequilibrium created when differentiated populations hybridize.

**Maximum likelihood**

A method of statistical inference in which the probability of obtaining the observed data is calculated (or estimated) under specific assumptions about the process (model) giving rise to the data. The assumptions resulting in the highest probability of obtaining the observed data give rise to the maximum likelihood estimate.

**Microsatellite**

Tandemly repeated DNA sequences of 1–6 bases.

**Microsatellite terminology**

*Perfect microsatellite* microsatellite that consists of a single repeat motif and is not interrupted anywhere by a base that does not match the repeat pattern, e.g. ctctctctctctctctctctctctct

*Imperfect microsatellite* microsatellite in which one or more repeats carry a base pair that does not fit the repeat structure, e.g.: ctctctctctgtctct

*Interrupted microsatellite* microsatellite with the insertion of a small number of basepairs that do not fit the repeat structure, e.g.: ctctctctct**ggg**ctctct

*Compound microsatellite* Microsatellite consisting of two or more adjacent microsatellites with different repeat types, e.g.: ctctctctctct**gatgatgatgatgatgat**

**Migration**

Movement of individuals between subpopulations which may or may not lead to gene flow.

**Minisatellite**
Tandemly repeated sequences in which the iterated motif ranges in size from 14–100 bp, often iterated hundreds of times. The predominant mutation mechanisms appear to be gene conversion and unequal crossing over. DNA slippage is apparently of minor importance. The distinction between microsatellites and minisatellites is often arbitrary for repeat units between 8 and 15 bp. Given the different mutation mechanisms of the repeat types, it has been suggested that sequences that form loops that are not repaired by the mismatch repair system be called minisatellites. In yeast, experiments show that tandemly repeated sequences with repeat units longer than 13 bp should be classified as minisatellites.

**Mismatch repair system**
Summary term for the proteins involved in the recognition and removal of errors occurring during DNA synthesis that lead to mismatches between the parent and daughter strands.

**MRCA**
Acronym for 'most recent common ancestor', a term used in coalescence theory to describe the coalescence event uniting all lineages under consideration.

**Mutation**
Microsatellite mutations most frequently are the gain or loss of one or more repeat units, leading to an increase or decrease of the repeat count (or allele size).

**Mutation–drift equilibrium**
The long-term balance between the generation of novel alleles by mutation and their elimination by genetic drift.

**Mutation model**
Description of the probabilities of change among the assumed, finite or infinite range of possible allelic states.

**Mutation rate**
Mutations per replication (cell division), or more generally, per organismal generation. Microsatellite mutation rates in mammals range from $10^{-2}$ to $10^{-5}$. Microsatellite mutation rates may vary among taxa, loci, and among alleles at a locus.

**Mutational bias**

A tendency to mutate toward larger or smaller size. For microsatellites a mutational bias towards longer alleles has often been reported, indicating that additions of repeat units are more frequent than the loss of repeat units.

**Mutator phenotype**

Genetic deficiency resulting in an increased mutation rate.

**Neutrality**

Mutations that do not influence the fitness of the organism are called neutral mutations. The proportion of mutations that are neutral has been the subject of extensive debate since the proposal of the neutral theory. Currently, neutrality is often used as a null hypothesis, and a variety of statistical test of neutrality based on observed variation are available. With a few notable exceptions (e.g. tri-nucleotide expansions), most microsatellites are regarded as neutral markers.

**Orthologous microsatellite loci**

Microsatellite loci in the same genome position in two or more species due to their common derivation from an ancestor carrying the microsatellite.

**Outbreeding depression**

When mating occurs between individuals of different subpopulations sufficiently diverged that their genomes are no longer fully compatible. Genetically the effect may result from the breaking up of coadapted gene complexes.

**PCR**

Polymerase Chain Reaction, in vitro amplification of a specific DNA region. PCR requires the design of unique sequence stretches of about 20 bases pairs, called primers, flanking the region to be amplified. Due to cryptic simplicity that often occurs near microsatellites, primers to amplify microsatellite loci are sometimes difficult to design.

**Polymorphism**

Condition under which more than a single allele at a locus is present. Traditionally, a polymorphism has been defined as a locus in which at least 5% of the alleles in a population are different from the most common one.

**Premutation**

Certain allele size range at loci for which tri-nucleotide expansions have been described. Premutations are not associated with a disease phenotype of the carrier, but have a higher probability of leading to affected offspring with an expanded tri-nucleotide repeat.

**Probe**

Labelled DNA used to detect complementary DNA (e.g. on a membrane).

**Range constraint**

Describes a microsatellite locus in which only a specific set of sizes are attainable. The lower boundary is well defined with a minimum of one or two repeats. An upper boundary may result from the mutation process or from selection. At present, relatively little is known about the ranges of microsatellites and even less about the mechanisms imposing them.

**Recombination**

Exchange of genetic material between chromosomes.

**Repeat count**

Repeat count, or repeat number, is the actual number of repeats in a microsatellite array.

**Repeat number**

*See* repeat count.

**Repeat unit**

The smallest entity of a microsatellite, e.g. mono-, di-, trinucleotide.

**Secondary structure**

Structure of DNA and RNA molecules that is generated and maintained through intramolecular Watson–Crick pairings. Paired regions form stems and unpaired regions loops or bulges.

**Selection**

The tendency of the bearers of particular genotypes to reproduce more or less than others in the population, thereby systematically altering allele frequencies.

**Simple sequence**

Synonym for microsatellite, but often used to include cryptically simple sequences.

**Sister chromatid exchange**

Exchange of material between the two identical strands (chromatids) of one chromosome. Sister chromatid exchange does not result in an exchange of flanking markers, because both chromatids are identical.

**Size range**

Spectrum of all possible allelic states of a microsatellite.

**Slippage**

(synonyms: slipped-strand mispairing, replication slippage) An error during DNA replication whereby the nascent DNA strand becomes attached out of register.

**SSCP**

Single Strand Confirmation Polymorphism, a widely used technique for the detection of polymorphisms without recourse to DNA sequencing. Double stranded DNA is heat denatured, chilled on ice, where the single-stranded DNA forms a sequence-specific secondary structure. The structure leads to a characteristic migration behaviour in a non-denaturing gel which usually allows different sequences to be distinguished.

**Stepwise mutation model**

Transitions to new alleles are by single steps, which correspond to the addition or deletion of single repeat units. If the model permits changes by single repeat units only, the model is called a strict stepwise mutation model. More general models allow for changes by multiple repeat units (generalized model). A special case of the generalized model is the two phase model, which consists of a strict stepwise mutation process with occasional changes by multiple repeat units.

**STR**

Short Tandem Repeat, a synonym for microsatellite.

**Stutter (shadow) bands**

PCR artefacts, which are especially pronounced at long dinucleotide repeats. DNA slippage during the PCR amplification produces DNA fragments that are one or several repeats shorter than the actual allele. Stutter bands can be a severe problem in analysing heterozygous individuals with similar alleles. Very rarely, stutter bands are also observed longer than the actual allele. Dinucleotide microsatellites are highly prone to stuttering, trinucleotides less so, and the larger motif sizes stutter very little if at all.

**Transposable element**

Transposable elements are also called mobile DNA, because they can move within and between genomes. Several classes of transposable elements, including DNA and RNA transposons, have been described. It has been suggested that microsatellite genesis may be associated with the integration of retrotransposons, which move through an RNA intermediate (see Chapter 1).

**Tri-nucleotide expansion**

Some tri-nucleotide repeats have been described that have the capacity to expand to very large repeat counts leading to various clinical conditions.

**Unequal crossing over (UCO)**

Recombination between non-homologous sites. UCO is likely to occur at tandemly repeated genomic regions, resulting in one chromosome with more repeats and one chromosome with fewer repeats.

**VNTR**

Variable Number of Tandem Repeats, a terminology introduced by Nakamura *et al.* in 1987 for minisatellite loci. Since then, the term has been applied to both minisatellites and microsatellites. We prefer the more specific terms minisatellites and microsatellites rather than VNTR, and do not see the need for a term combing these very different classes of marker.

**Wright–Fisher model**

A widely used model of genetic drift introduced by R. Fisher and S. Wright. The model assumes $N$ individuals ($2N$ gametes) with discrete, non-overlapping generations. The gametes in the current generation are determined by drawing gametes with replacement from the previous generation until $2N$ gametes are selected.

# Index